Microprobe Analysis of Biological Systems

ACADEMIC PRESS RAPID MANUSCRIPT REPRODUCTION

Proceedings of the Conference on Techniques and Applications
of Microprobe Analysis of Cells and Tissues
Held in Seattle, Washington, 30 July – 1 August 1980

Microprobe Analysis of Biological Systems

Edited by

THOMAS E. HUTCHINSON

Center for Bioengineering
University of Washington
Seattle, Washington

ANDREW P. SOMLYO

Muscle Institute
University of Pennsylvania Medical School
Philadelphia, Pennsylvania

ACADEMIC PRESS
A Subsidiary of Harcourt Brace Jovanovich, Publishers
New York London Toronto Sydney San Francisco 1981

ACADEMIC PRESS, INC.
111 Fifth Avenue, New York, New York 10003

United Kingdom Edition published by
ACADEMIC PRESS, INC. (LONDON) LTD.
24/28 Oval Road, London NW1 7DX

Library of Congress Catalog Card Number: 81-15015
ISBN 0-12-362880-6

PRINTED IN THE UNITED STATES OF AMERICA

81 82 83 84   9 8 7 6 5 4 3 2 1

Contents

PART III

PART IV

Contributors and Participants

Numbers in parentheses indicate the pages on which the authors' contributions begin.

K. Adamson-Sharp (309), Ontario Cancer Institute, Toronto, Ontario, Canada M4X 1K9

S. Brian Andrews (21), Department of Physiology, School of Medicine, Yale University, New Haven, Connecticut 06510

D. Bazett-Jones (309), Ontario Cancer Institute, Toronto, Ontario, Canada M4X 1K9

L. Buja (127), Department of Pathology, Southwestern Medical School, The University of Texas Health Science Center, Dallas, Texas 75235

K. Burton (127), Department of Pathology, Southwestern Medical School, The University of Texas Health Science Center, Dallas, Texas 75235

Marie Cantino (65), Center for Bioengineering WD-12, University of Washington, Seattle, Washington 98195

C. Colliex (251), Laboratoire de Physique des Solides Centre D'Orsay, Universite de Paris-Sud, Orsay, France 91405

L. Daniels,* Battelle, Pacific Northwest Laboratories, Richland, Washington 99352

A. Dörge (47), Physiologisches Institut der Universität München, 8000 München 2, Pettenkoferstrasse 12, West Germany

R. P. Ferrier (231), Department of Natural Philosophy, University of Glasgow, Glasgow, Scotland G12 8QQ

B. Gupta (3), Department of Zoology, University of Cambridge, Cambridge, England CB2 3EJ

H. Hagler (127), Department of Pathology, Southwestern Medical School, The University of Texas Health Science Center, Dallas, Texas 75235

T. A. Hall (3, 423), Department of Zoology, University of Cambridge, Cambridge, England CB2 3EJ

James G. Hecker (83), Center for Bioengineering WD-12, University of Washington, Seattle, Washington 98195

Thomas E. Hutchinson (83, 157), Center for Bioengineering WD-12, University of Washington, Seattle, Washington 98195

M. Isaacson (289), School of Applied and Engineering Physics, Cornell University, Ithaca, New York 14853

C. Jeanquillaume (251), Laboratoire de Physique des Solides Centre D'Orsay, Universite de Paris-Sud, Orsay, France 91405

Dale E. Johnson (351), Center for Bioengineering WD-12, University of Washington, Seattle, Washington 98195

D. Joy (325), Bell Laboratories, Murray Hill, New Jersey 07974

R. Gary Kirk (21, 367), Department of Physiology, School of Medicine, Yale University, New Haven, Connecticut 06510

C. Lechene,* Biotechnology Research Center, Harvard Medical School, Boston, Massachusetts 02115

Ping Lee (367), Department of Physiology, West Virginia University, Morgantown, Virginia 26505

Alan P. Mackenzie (397), Center for Bioengineering WD-12, University of Washington, Seattle, Washington 98195

D. Maher (325), Bell Laboratories, Murray Hill, New Jersey 07974

Joseph E. Mazurkiewicz (21), Department of Anatomy, Albany Medical College of Union University, Albany, New York 12208

Keith L. Monson (157), Center for Bioengineering WD-12, University of Washington, Seattle, Washington 98195

R. Ornberg (213), Laboratory of Neuroanatomical and Neuropathological Sciences, National Institute of Neurological Communicative Disorders and Stroke, National Institutes of Health, Bethesda, Maryland 20205

F. Ottensmeyer (309), Ontario Cancer Institute, Toronto, Ontario, Canada M4X 1K9

P. Peters,* Boeing Commercial Airlines, Seattle, Washington 98124

T. Reese (213), Laboratory of Neuroanatomical and Neuropathological Sciences, National Institute of Neurological Communicative Disorders and Stroke, National Institutes of Health, Bethesda, Maryland 20205

R. Rick (47), Physiologisches Institut der Universität München, 8000 München 2, Pettenkoferstrasse 12, West Germany

C. Roloff (47), Physiologisches Institut der Universität München, 8000 München 2, Pettenkoferstrasse 12, West Germany

A. Saubermann (377), Department of Anaësthesia, Harvard Medical School at Beth Israel Hospital, Boston, Massachusetts 02215

H. Shuman (103, 273), Pennsylvania Muscle Institute, School of Medicine, University of Pennsylvania, Philadelphia, Pennsylvania 19104

Andrew P. Somlyo (103, 273), Pennsylvania Muscle Institute, School of Medicine, University of Pennsylvania, Philadelphia, Pennsylvania 19104

Avril V. Somlyo (103, 273), Pennsylvania Muscle Institute, School of Medicine, University of Pennsylvania, Philadelphia, Pennsylvania 19104

D. A. Taylor (197), The Proctor & Gamble Company, Miami Valley Laboratories, Cincinnati, Ohio 45247

M. Thompson,* Philips Electronic Instruments, Mahwah, New Jersey 07430

K. Thurau (47), Physiologisches Institut der Universität München, 8000 München 2, Pettenkoferstrasse 12, West Germany

John McD. Tormey (177), Department of Physiology, School of Medicine, University of California, Los Angeles, California 90024

P. Trebbia (251), Laboratoire de Physique des Solides Centre D'Orsay, Universite de Paris-Sud, Orsay, France 91405

R. R. Warner (197), The Proctor & Gamble Company, Miami Valley Laboratories, Cincinnati, Ohio 45247

*Participant.

Participants

Top row (left to right): Dr. S. Csillag; Dr. P. Peters; Dr. R. Ferrier; Dr. F. Ottensmeyer; Dr. P. Trebbia; Dr. D. Maher; Dr. M. Thompson; Dr. K. Monson; Dr. A. V. Somlyo; Dr. A. Saubermann; and Dr. A. P. MacKenzie.

Middle row: Dr. M. Isaacson; Dr. R. Rick; Mr. L. Daniels; Dr. R. Warner; Dr. T. Hall; Dr. H. Hagler; Dr. B. Andrews; Dr. G. Kirk; and Dr. D. E. Johnson.

Bottom row: Ms. D. Scearce; Dr. T. E. Hutchinson; Dr. A. P. Somlyo; Dr. M. Cantino; Dr. L. Buja; Dr. J. Hecker; Dr. C. Lechene; Dr. J. Tormey; and Dr. R. Ornberg.

Preface

This volume is the outgrowth of a conference held at Battelle Conference Center in Seattle, Washington, in the summer of 1980. The meeting was limited to thirty participants in order to maximize interactions and thereby provide an atmosphere conducive to in-depth discussion. Most of the major laboratories in the field of biological microanalysis in the United States, England, Scotland, France, and Germany were represented. Each contributor was asked to present the latest findings, theories, techniques, and procedures of the laboratory represented, no matter how tentative and exploratory.

Thus, this volume constitutes a rather complete picture of the state of the art in the early months of 1981. Further, since the conference was held without parallel sessions and with generous time for discussion, numerous areas of endeavor heretofore unexposed to an open forum were introduced and debated. These discussions were recorded, transcribed, and sent to the authors for comment. They are presented here as nearly as possible in verbatim form, and comprise material of equal importance to that presented in the papers. They often contain unique insights into details of current research and pinpoint areas of controversy.

The contrast between this volume and the widely used and referenced proceedings of the Battelle Conference of 1973 chaired by Ted Hall reflects the progress that has occurred in the past eight years. Numerous advances in techniques and instrumentation are presented with particular emphasis on quantification. Many more of the presentations describe the application of the technique to important biological systems. Also apparent is the greater sophistication of methods and procedures for specimen preparation, greater sensitivity to elements in low concentration through higher efficiency of x-ray detector systems, and computer data treatment largely brought about by extensive software development employed in background removal and spectral peak "stripping." Moreover, much higher spatial resolution using advanced scanning transmission electron microscopy is reported. Of particular significance is the advent of electron energy loss spectroscopy (EELS) applied to biological systems. This technique extends the range of elements detectable by microprobe analysis into the low atomic number elements while offering the possibility of obtaining characterizations of bonding states. A substantial portion of this volume is devoted to the theory and application of EELS.

In total, we hope this volume will be of value to scientists interested in elemental (and ion) transport within cells and between cells and extracellular compartments. In addition, although its focus of application is primarily biological, it is hoped that the book will also serve as a reference for the wider range of fields to which analytical microscopy is currently being applied as it was our aim to bring together in one volume the latest developments in the techniques, instrumentation, applicability and limitations, and fundamental theory of microprobe analysis.

Acknowledgments

The editors are indebted to the participants for their extreme cooperation in submitting both papers and edited versions of the discussion. We are also most grateful to Jolene Kitzerow and her staff at the Battelle Center in Seattle and for their assistance in organizing the conference, and to Julie Eulenberg for her help in word processing. Winnie Notske and Ken Requa, at Battelle, Seattle, are also particularly commended for their production, editing, layout, and word processing abilities. Each has contributed beyond compensated expectation. We are especially pleased to acknowledge the Battelle Institute and the Center for Advanced Studies in Biological Sciences, University of Washington, funded by the Reynolds Foundation, for the bulk of financial support. A very special debt of gratitude is due Robert Rushmer, Professor of Bioengineering and Director of the Center for Advanced Studies without whose enthusiasm and guidance neither the conference nor the resulting proceedings would have been possible. We are additionally indebted to JEOL, KEVEX, Phillips, and the Electron Microscope Society of America for contributions to the conference. We are indebted to Debby Scearce, formerly of the Center for Bioengineering, University of Washington, for her help in laying out the manuscript, typing, and coordination with Battelle, Seattle, staff in work processing. We are also deeply indebted to Sandi Klein of the Center for Bioengineering for her invaluable aid in all aspects of manuscript preparation during the final highly important stages of proofing, alteration, and assembly. Her thoroughness and patience during this period is much to be commended.

Part I

SOME RESULTS OF MICROPROBE ANALYSIS IN THE STUDY OF EPITHELIAL TRANSPORT

T. Hall
B. Gupta

Department of Zoology
University of Cambridge
Cambridge, U. K.

INTRODUCTION

The standard procedure in our laboratory, described in detail elsewhere (Gupta, Hall, and Moreton, 1977), is to analyze 1- μm sections of quench-frozen tissue mounted on a cold-stage at -170° C in a JEOL JXA-50A scanning microanalyzer. The sections may be studied either hydrated or dehydrated, and are often analyzed first in the hydrated state and again after dehydration within the specimen chamber of the microanalyzer. We use the transmission scanning mode for imaging, with beam voltages near 50 kV and probe currents in the range 1-5 nA. Our quantitative analyses are generally based on the ratios of x-ray characteristic line to x-ray continuum intensities (Hall and Gupta, 1979), with the characteristic intensities for sodium and calcium usually obtained from wave-length spectrometers while intensities for other elements and for the continuum are obtained from an energy-dispersive spectrometer (at present a Kevex detector feeding into a Link Systems spectrometer). In this paper we shall exemplify what can be done with such a system in the investigation of several aspects of epithelial transport.

ISBN 0-12-362880-6

OSMOLALITY OF INTERCELLULAR SPACES

A major objective of our microprobe laboratory from its inception has been the direct investigation of the hypothesis of the "hypertonic interspace." This hypothesis, proposed in early forms by Curran and MacIntosh (1962) and Diamond and Bossert (1967), is that the flow of water and ions across many epithelia is driven by the pumping ions by the epithelial cells into intercellular spaces; water is then supposed to be drawn into the spaces by osmosis and to flow from there into the lumen or serosa under hydrostatic pressure. Microprobe analysis offers one of the very few ways to check directly the central postulate that the interspaces are hyperosmotic. However, the analysis of interspace fluids should be performed in specimens which have not been dehydrated, a fact which has led us to concentrate on the analysis of frozen hydrated tissue sections.

Table 1 recapitulates data obtained from microprobe analyses of intercellular spaces in a variety of epithelial tissues during a period of several years. The epithelia are quite diverse: the first two secrete into a lumen isotonic with the serosal side; the ileum tissue absorbs when the lumen is isotonic or hypertonic (Curran and MacIntosh, 1962); and *Calliphora* rectal papillae, under conditions of water deprivation, absorb from a very hypertonic lumen. As indicated in the table, in all of the epithelia where we have analyzed the relevant intercellular fluid, we have found the total concentration of the measured electrolytes to be higher than in the fluid space (serosal or lumenal) from which water is drawn. In fact, this intercellular total is also higher than the totals in the epithelial cells themselves and in the fluid space into which the water is expelled.

Table 1 strongly supports the hypothesis that hypertonic interspaces exist in transporting epithelia. However, we are mainly concerned here with an exposition of the uncertainties, approximations, and limitations of technique inherent in the tabulations. Several explanatory and qualifying remarks are necessary:

1. We have tabulated "mM/l H_2O." To estimate osmolarity, one must multiply this quantity by an osmotic activity coefficient, and to estimate the magnitude of the osmotic driving force, one should multiply the "excess" column by this activity coefficient, which we must (reasonably) assume to be the same in the intercellular and source fluid spaces.

TABLE 1. Hypertonicity of some cell interspaces

Tissue	Space	Estimated mM/l H_2O of electrolytes	Excell mM/l H_2O	Ref.
Rhodnius Malpighian tubule (stimulated)	apical brush border	440	120	1
Calliphora salivary gland	canaliculi	400	80	2
Calliphora rectal papillae	"intercellular" "extra stack"	300 380	200 280	3,5
Calliphora rectal water-deprived	"intercellular" "extra stack"	1270 1720	(500) (1000)	3,5
Rabbit ileum	lateral intercellular	360	40	4,5

*Excess above the fluid space, lumenal or serosal, from which water is drawn.

1. Gupta, Hall, Maddress and Moreton (1976)
2. Gupta, Berridge, Hall and Moreton (1978a)
3. Gupta, Wall, Oschman and Hall (1980)
4. Gupta, Hall and Naftalin (1978b)
5. Gupta and Hall (in press)

2. The quantities we actually measure are mM/kg of the electrolyte elements Na, K, and Cl (also Ca and Mg, but these do not contribute significantly to the total). To arrive at a total of mM/l H_2O, two important steps are involved: (a) As the tissues are run in known media *in vitro* prior to quench-freezing, we generally know *a priori* the value of total mM/l H_2O in the lumenal and serosal spaces. For the intercellular spaces we must divide the total measured mM/kg by (1-f) where *f* is the measured dry-weight fraction (Gupta and Hall, 1979; Hall and Gupta, 1979). The quantity *f* is generally less than 10% and the associated uncertainty in conversion to mM/l is still smaller. (b) We have no measurements for osmotically active substances like small organic molecules or bicarbonate ion. In fluid spaces where

[Na + K] considerably exceeds [Cl] we have estimated the total mM/l as 2 [Na = K[, assuming that charge neutrality is preserved through the presence of unmeasured negative ions.

3. Lack of contrast in the image and/or the narrowness of intercellular channels may prevent the full exclusion of cytoplasmic contributions during the analysis of an intercellular space. In this situation, the relative size of the cytoplasmic contribution can be estimated from the measured concentration of an element like phosphorus, which is predominantly intracellular. The problem is well illustrated in Calliphora salivary gland (Figures 1A and 1B). Analysis can be quite well localized within the main channel of a canaliculus, and the tabulated value is based on measurements in this region. But when one attempts to analyze

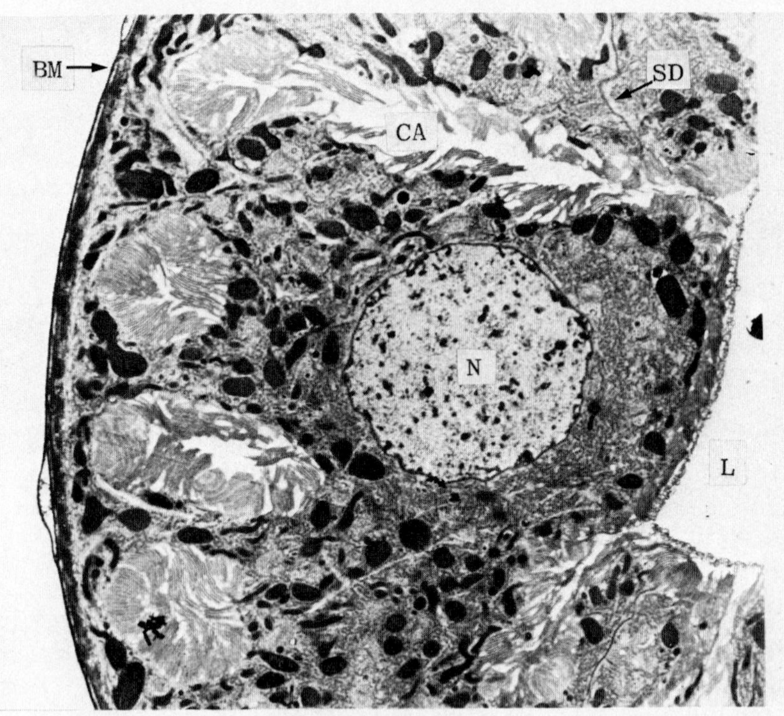

FIGURE 1. Electron micrographs of transverse sections of Calliphora salivary gland.

1a. Ultrathin section fixed in glutaraldehyde/osmium, embedded in Araldite and stained in uranyl acetate and lead citrate. BM, basement membrane; N, nucleus; CA, canaliculus; SD, septate desmosomal junctions; L, lumen. (Micrograph by Dr. J. L. Oschman).

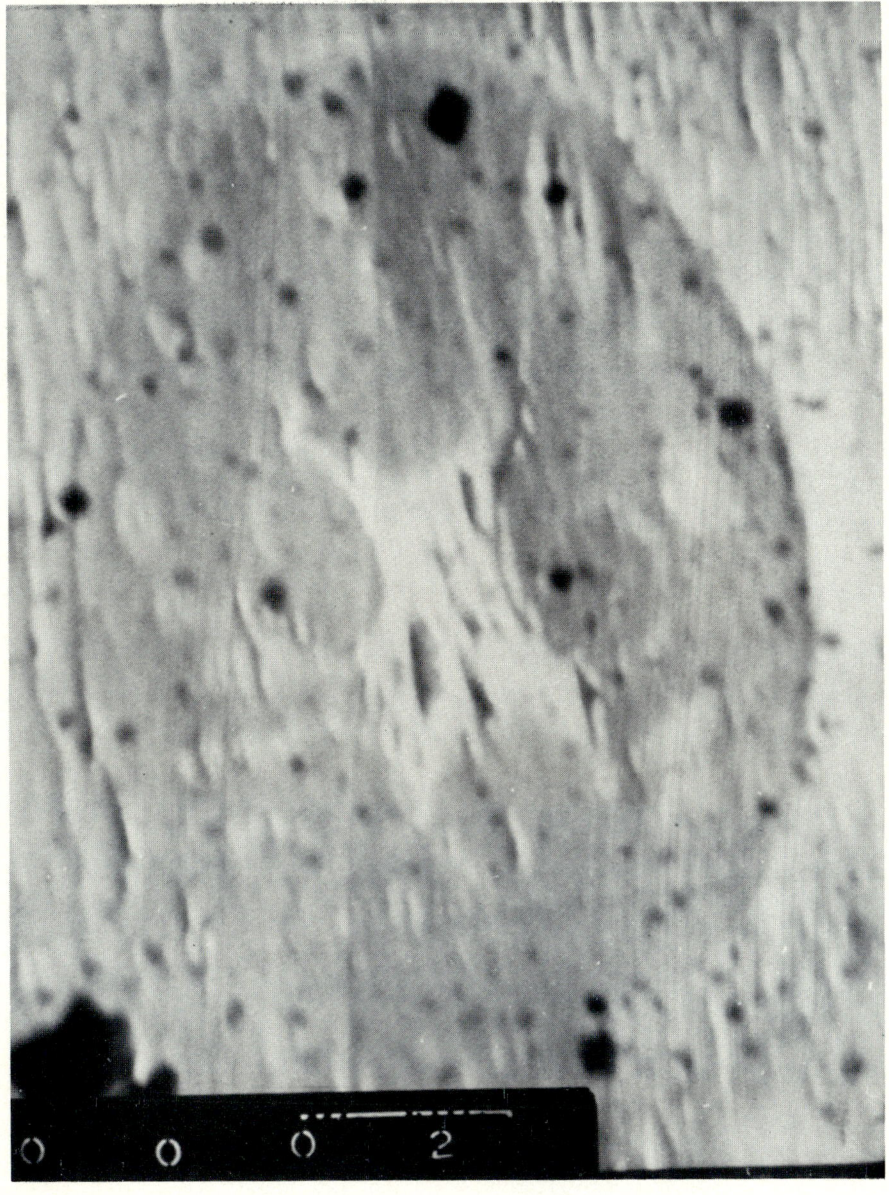

1b. Scanning transmission image of hydrated, 1- μm section. Gland diameter approx. 80 μm. (Contrast not maximal but set for sufficient range to distinguish medium, cells, and lumen.)

fluid spaces within the apical folds, it is clear that the localization is imperfect. When the measurements within the folds are corrected for a cytoplasmic contribution, it appears that the interfold space has a total electrolyte concentration even higher than the main channel (although some of the K and Na may be osmotically inactive due to binding to fixed negative charges at the cell surface).

Thus, our technique is adequate to demonstrate that a hypertonic interspace exists in this tissue, but the mapping of electrolyte concentration in correlation with the fine structure is very difficult. Indeed, since the fold channels are only of the order of 10 nm in width, we could not expect to confine x-ray excitation entirely within a channel with a 50-kV beam impinging on a 1- μm hydrated section. It remains to be seen whether these channels can be very selectively analyzed even with finer probes within thinner sections; image contrast is certainly inadequate in our sections (Figure 1B), and may still be inadequate even in thinner hydrated sections.

4. The most convincing evidence for local osmosis in Table 1 comes from the observations in Calliphora rectal papillae. In this tissue ions are believed to be pumped through finely folded and stacked cell membranes into extracellular channels ("extra-stack" in the table). From there the fluid goes into larger channels ("intercellular" in the table) and then into still larger sinuses and onwards toward the serosa. See Gupta et al. (1980) for gross morphology; Gupta and Berridge (1966) for fine structure. Ionic concentrations could be cleanly measured in the intercellular spaces and sinuses and were virtually the same in both. It was impossible to resolve the extracellular stack space; but after correction for the cytoplasmic contribution, it appeared that this space is the site of maximal osmolarity. In tissue from water-deprived flies we did not know or measure the osmolarity of the lumen, as it rises to at lease 700 mOsmol due to the presence of organic excreta (Phillips, 1969). The tabulated excess is put in parentheses because we do not know the lumenal osmolarity accurately. But under water deprivation, the intercellular concentrations are seen to rise drastically so that water can still be absorbed by local osmosis from a lumen assumed to be 700 mOsmolar.

We have discussed elsewhere the uncertainties in the actual measurement of elemental mM/kg (Gupta and Hall, 1979; Hall and Gupta, 1979). With respect to the hypothesis of local osmosis, the greater uncertainties are associated with the points raised above in the interpretation of the measured values. It should be noted, however, that under all plausible assumptions, whether one refers to [Na + K + Cl] or 2[Na + K] and whether or not one corrects for cytoplasmic contributions, the intercellular spaces are found to be hypertonic; the particular assumptions affect only the magnitude of the excess.

PARACELLULAR TRANSPORT

In most (though not all) models of epithelial transport, it has been assumed that all the water which crosses the epithelium goes through the cells. However, our microprobe data in two studies have incidentally suggested substantial transport through paracellular routes, i.e, routes which bypass the cells, going through intercellular spaces and the apical "tight" junction.

In the rabbit ileum, the relevant observation (Gupta et al., 1978b) was that the difference in sodium concentration between lumen and the junctional end of the lateral intercellular space (L.I.S.) is much less than the difference between the latter and the maximal sodium concentration which is reached halfway along the L.I.S. The likely explanation is a substantial flow of water coming through the junction and diluting the fluid at the narrow end of the L.I.S.

In Calliphora salivary gland the relevant observation (Gupta et al., 1978a; Gupta and Hall, in press) was that a canaliculus is typically highly hyperosmolar from its blind basal end almost to the lumen, where the osmolarity drops almost to the lumenal value in the neighborhood of the septate junction. This effect obviously could be caused by a flow of water through the junction into the neck of the canaliculus. In this tissue, the microprobe observations as a whole indicate that transcellular and paracellular routes are both important: the water which must be drawn into the hyperosmotic blind end of the canaliculus can come only from the cell.

More evidence of the water transport routes in Calliphora salivary gland is coming from a study now in progress (Gupta, Berridge, and Hall) combining microprobe and microelectrode observations. Here we recapitulate data from a preliminary report (Gupta and Hall, in press). The data compare glands working *in vitro* in a conventional Ringer and in a hypotonic Ringer in which the NaCl concentration has been reduced to 50 mM/l. Some results to date are summarized in Table 2.

Table 2 shows that the hypotonic Ringer induces a drop in cytoplasmic osmolarity in both unstimulated and 5-HT stimulated glands, and an increase in cell H_2O/dry mass in unstimulated glands. These effects are as one would expect if water from the medium goes through the cells. (In 5-HT stimulated glands the rapid secretion dilates the canaliculi and lumen, compressing the cells; we postulate that the hypotonic medium does not affect the value of cell H_2O/dry mass further because the cells are kept maximally compressed.) Another strong indication of water transport through the cells is the relationship between canalicular and cytoplasmic osmolarity: while the osmolarity at the base of the canaliculus is much reduced in the hypotonic medium, it seems to remain high enough to draw water from the cell.

While these data give evidence of the transcellular movement of water, evidence for paracellular transport has been obtained during the first minute of the immersion of the glands in the hypotonic medium. The response to the medium is rapid, and the values listed in Table 2 (for 5-minute

TABLE 2. Comparison of *Calliphora* salivary glands working in conventional and in hypotonic Ringer.

	Known	Estimates from Microprobe*				
	Ringer	mOsmol				Cell H_2O
	mOsmol	Canaliculus†		Lumen	Cell†	kg H_2O/kg dry
		Basal	Neck			
Unstim.	310	400	320	310	290	4.6
	150	240	160	150	200	5.7
5-HT	310	350	320	310	330	3.6
	150			150	250	3.6

Experimental glands run for 5 min in hypotonic Ringer

*Measured mM/l x activity coefficient assumed to be 0.85

†Osmolarity estimated as 2 x .85 x [K + Na] mM/l.

For the method of measurement of cell H_2O, see Gupta and Hall (1979) or Hall and Gupta, (1979).

immersion) remain the same over a 30-minute period. But microelectrode recordings by Dr. Berridge indicate that during the first minute, lumen-K^+ equilibrates with the hypotonic Ringer faster than the cell contents do. One would expect the reverse if the water flux were all transcellular. We hope to obtain microprobe data as well characterizing this early transitional stage.

Technically, our assessment of paracellular transport has depended largely on the variations in concentration appearing along single intercellular channels. We believe that these variations are real rather than artifacts of quench-freezing, although modern diffusion theory does show that concentration gradients may be generated in homogenous fluids under thermal shock (Huppert and Linden, 1979).

In the study of transepithelial routes, the microprobe localization of transported tracer elements is a technique remaining to be exploited.

RECYCLING OF IONS

Recent studies show that ionic recycling is a basic feature of the transport process in two epithelia.

In Calliphora rectal papillae, as noted above, the cells pump ions through membrane "stacks" into extracellular channels which are the sites of maximal osmolarity. From there the fluid follows a tortuous extracellular path, progressively decreasing in osmolarity until it enters the serosal space. Table 3, abstracted from Table 8 in Gupta et al. (1980) and Figure 4 in Gupta and Hall (in press), shows that the osmolarity is reduced by means of a selective reabsorption of potassium into the cells. The particular sites are listed in the "order of appearance" along the path from the stack channels to the serosa.

The recycling of potassium might well be anticipated in rectal papillae where the extracellular fluid runs along a lengthy path lined with cells. The geometry is quite different in a second epithelium which we recently studied (Civan, Hall, and Gupta, 1980), toad urinary bladder, where the cells form a simple thin sheet separating lumen and serosa. The study was intended to elucidate a peculiar phenomenon noted when the tissue functions in vitro in a potassium-free serosal medium: most of the intracellular

TABLE 3. K concentrations (mM/kg wet) and K/Na molar ratios in extracellular fluid in Calliphora rectal papillae.

	From Water-fed Animal		From Water-deprived Animal	
	[K]	K/Na	[K]	K/Na
Extracellular stack channels	131	1.8	640	3.2
"Intercellular space"	88	1.3	430	2.4
Intercellular sinus	63	1.5	350	1.7
Infundibular space	51	0.7	45	0.7

potassium seems to leave the cells very slowly, with a half-time of about 9 hours, as if it is compartmentalized or immobilized within the cells (Robinson and Macknight, 1976). We compared intracellular potassium levels in bladder tissue functioning for 83 to 133 minutes in an ordinary medium and in K-free serosal media with or without 10^{-2}M ouabain. The results are shown in Table 4, which is abstracted from Table 4 of Civan et al. (1980).

Removal of K from the serosal medium reduced intracellular K by about 20%; but with the addition of ouabain to the K-free medium, the granular epithelial cells lost about 70% or more of their potassium. (The potassium concentration in the basal epithelial cells was much less affected by the addition of ouabain to the medium.) Since ouabain inhibits the normal pumping of potassium into the cells, the straightforward interpretation is that potassium is not mainly compartmentalized and can readily leave the cells, but in the zero-K medium in the absence of ouabain, it is likely to be pumped back before it escapes the extracellular vicinity. Hence recycling is an important feature of the dynamics of the transport process. The implications for the basal lamina should be considered in conjunction with the earlier microprobe observations of elevated concentrations of K in the basement membranes of several insect epithelia (Gupta et al., 1977, p. 135). It appears that the

TABLE 4. Effect of Ouabain on intracellular K in K-free Serosal Medium.

Medium	[K] Cyto	Nuc	IC
Baseline	491 ±13	594 ±57	543 ±38
O Serosal K	399 ±37	477 ±38	444 ±16
O Serosal K + Ouabain	168 ±36	159 ±34	210 ±42

[K] is potassium concentration, mM per kg dry tissue. "Cyto" = cytoplasmic site. "Nuc" = nuclear site. "IC" = intracellular but not further identified. Dry-mass fraction (average for all measurements) was 22.2% ± 1.6 (S.E.M.), dry mass/total mass.

extracellular structure on the basal side of the cells acts as more than a mere skeleton.

Technically, it is noteworthy that fully hydrated sections were not required to obtain the data on K/Na ratios in rectal papillae (Table 3); the K/Na ratio in a fluid space is presumably not affected by a partial dehydration. For the study bearing on recycling in toad urinary bladder, fully dehydrated sections could be used. We analyzed both hydrated and dehydrated sections, with consistent results, but localization and stability were better in the dehydrated material. We are informed that the Munich group has obtained similar data on potassium movements in toad bladder, using dehydrated sections (Dörge and Rick, personal communication).

CELL CALCIUM

Changes in the level of cytosolic calcium activity from about 10^{-7}M to about 10^{-6}M are often associated with the triggering of cell secretion. While these levels are too low for microprobe analysis, one can use the microprobe to analyze presumed sites of calcium storage and regulation. In collaboration with Dr. Tudor Barnard, we have recently been measuring calcium in the mitochondria of Calliphora salivary gland cells, especially to see what the mitochondrial level is when secretion finally stops with the gland working in a calcium-free medium. As the investigation is not completed and the experimental conditions have not yet been optimized, we do not want to present biological results here, but we do want to remark on the suitability of the analytical system for the intended studies.

We measure calcium by means of both a Si-Li spectrometer and a wave-length spectrometer with a P.E.T. diffracting crystal. In our system (with the 30-mm^2 Si-Li detector presently 50 mm from the specimen) the diffracting crystal gives the better limit detectability in 1-μm tissue sections.

For tissues functioning in normal Ringer containing 2mMCa^{2+}, both cell cytoplasm and mitochondria contained about 10 mM of calcium per kg dry mass. When the glands were depleted by incubation in Ringer containing no free calcium, the few measurements made thus far did not reveal any calcium in the cytoplasm (-0.1 ± 1 mM/Kg dry mass) but mitochondrial calcium was about 3.3 ± 1 mM/kg dry mass. If confirmed, these data suggest that the calcium in mitochondria cannot all be mobilized to maintain Ca^{2-} activity.

Three mM Ca/kg dry mass can in practice indeed be measured in dehydrated 1- μm sections by means of the diffracting spectrometer, but only with care in lengthy analyses. A special procedure is needed to correct for background under the calcium peak: the customary separate runs at each analyzed site with the diffractor offset from the peak are not practicable for reasons of time, stability, and counting statistics. Instead the calcium count in a run is obtained from an equation of the form

$$S = T - C - k(B + F). \tag{1}$$

Here T is the total count from the diffracting spectrometer set on the calcium peak; C is the instrumental background (cosmic rays, etc.) which occurs even with the beam switched off; B and F are the continuum counts generated, respectively, in the specimen itself and in the specimen support film, recorded in an energy-band in the simultaneously operated Si-Li spectrometer; k is a fixed proportionality constant; and S is the background-corrected signal. C and k are instrumental factors established by extensive separate runs; F is determined from separate runs on the support film; and it is necessary to determine the contribution of the bulk surround to the Si-Li continuum and to leave this term out of the sum (B + F) since the continuum from the bulk surround is rejected by the fully focusing diffracting crystal. If we analyze a field containing 3 mM Ca/kg in a 1-μm dried section mounted on a Formvar-coated nickel grid, running for 1000 seconds with a probe current of 5 nA, the magnitudes for the terms in equation (1) are $T \sim 57$, $C \sim 27$, $kB \sim 64$, and $kF \sim 15$; hence $S \sim 51$ (and signal/background is approximately 1/2).

Under these conditions measurements are possible. The main problem is that one cannot be sure of localizing an analysis to a mitochondrion in a 1-μm section since the apparent mitochondrial fields must often include some cytoplasm above or below the organelle. For mitochondrial analysis, the sections should really be much thinner. But the x-ray intensity would then be too low for the diffracting spectrometer, and energy-dispersive spectrometry with the detector close to the specimen may prove more effective.

REFERENCES

Civan, M. M., Hall, T. A., and Gupta, B. L. (1980). J. Membrane Biol. 55:187-202.

Curran, P. F., and MacIntosh, J. R. (1962). Nature 193:347-48.

Diamond, J. M., and Bossert, W. H. (1967). J. Gen. Physiol. 50:2061-83.

Gupta, B. L., and Berridge, M. J. (1968). J. Morph. 120:23-82.

Gupta, B. L., and Hall, T. A. (1979). Fedn Proc. Fedn Am. Socs exp. Biol. 38:144-53.

Gupta, B. L., and Hall, T. A. (In press). "Microprobe Analysis of Fluid Transporting Epithelia; Evidence for Local Osmosis and Solute Recycling." 15th Alfred Benzon Symposium (Copenhagen 1980). Munksgaard, Copenhagen.

Gupta, B. L., Hall, T. A., Maddrell, S. H. P., and Moreton, R. B. (1976). Nature, Lond. 264:284-87.

Gupta, B. L., Hall, T. A., and Moreton, R. B. (1977). In "Transport of Ions and Water in Animals" (B. L. Gupta, R. B. Moreton, J. L. Oschman, and B. J. Wall, eds.), Chapter 4, pp. 83-145. Academic Press, London.

Gupta, B. L., Berridge, M. J., Hall, T. A., and Moreton, R. B. (1978a). J. Exp. Biol. 72:261-84.

Gupta, B. L., Hall, T. A., and Naftalin, R. J. (1978b). Nature, Lond. 272:70-73.

Gupta, B. L., Wall, B. J., Oschman, J. L., and Hall, T. A. (1980). J. Exp. Biol. 88:21-47.

Hall, T. A., and Gupta, B. L. (1979). In "Introduction to Analytical Electron Microscopy" (J. J. Hren, J. I. Goldstein, and D. C. Joy, eds.), pp. 169-97. Plenum Press, New York.

Huppert, H. E., and Linden, P. F. (1979). J. Fluid Mech. 95: 431-64.

Phillips, J. E. (1969). Can. J. Zool. 47:851-63.

Robinson, B. A., and Macknight, A. D. C. (1976). J. Membrane Biol. 26:269-86.

DISCUSSION

SPEAKER: T. A. Hall.

HUTCHINSON: What do you feel is appropriate statistical analysis of your data?

HALL: We have put down standard errors of the mean.

HUTCHINSON: Standard errors of the mean is what you feel is appropriate? I think this will come up in discussions with Professor Somlyo also.

HALL: I think the standard error of the mean is appropriate when you want to establish that a quantity has really changed in response to a change in experimental conditions. (Of course absolute accuracy of the measurements is another matter; they might all be wrong by a fixed factor due to a fault in standardization without affecting the relative standard errors at all). I agree that a tabulation like Table 4 should show numbers of measurements as well, so that one can estimate standard deviations and the reliability of apparent differences. I did not want to put too much material into a slide, but numbers of measurements are shown in the tables in the original publication (Civan et al., 1980). In fact there is no doubt at all that in Table 4, the very large effects apparently caused by ouabain are statistically significant.

LECHENE: How many measurements are there in each entry?

HALL: Table 4 is based on a total of about 300 measurements for all of the entries (hence an average of about 30 measurements per entry). (Explanatory note added later: in the particular case of Table 4, each entry was not obtained by averaging the concentrations given by the individual measurements, but rather by summing separately the characteristic and the continuum x-ray counts and taking the ratio of the sums. The tabulated "S. E. M.," however, was estimated from the run-to-run variations in the x-ray counts. As discussed in Civan et al. (1980), while this procedure gives the best estimate of mean concentrations, it grossly overestimates the S. E. M. since the count variations are due chiefly to thickness differences which are mainly compensated in the ratio method.)

RICK: There seems to be one important factor which you have not discussed in your presentation, which is that the loss of potassium also depends critically on whether the epithelium is intact or not. I think when it is shortcircuited, you allow sodium to get into epithelial cells from the outside at a much faster rate, so that, on the other hand, potassium is pushed out of the cell. It is not only a leak (tightening) a question of what forces exist to actually push potassium out of the cell. Under open-circuited conditions there is practically no driving force for potassium to get out of the cell, or at least it is much smaller. It cannot be replaced by another cation.

WARNER: I have a couple of points. One was that you made a point that in a hypotonic medium, the cell concentration fell. Was that an argument for water transport through the cell?

HALL: The observation is that the cellular ionic concentration decreases in the hypotonic medium. The natural interpretation is that water enters the cell.

WARNER: But what I could not see is two processes. One, the cell just wants to regulate its osmotic strength to the outside, which has nothing to do with transport. The transport process is entirely different from the osmotic equilibration of the cell.

HALL: Yes. But the data indicate (despite the technical difficulties in comparing cell with canaliculus) that the canaliculus remains hyperosmotic to the cell, so that in the hypotonic medium, water must be drawn both from medium to cell and from cell to canaliculus.

WARNER: The other question I have has to do with the use of cryoprotectants from these intracellular channels that you measured. Can you make some comment on what problems you had?

HALL: Since the cryoprotectant is in the medium and not in the cells, lumen or intercellular spaces, I agree that one must consider its possible effect on the osmolarity of the medium relative to these other spaces. Dr. Tudor Barnard, during his recent stay in our laboratory, evaluated this effect by means of several criteria, including the effect of the added cryoprotectant on the tonicity of the secretion, effects on other ions as manifested through ionic activity, and freeze-point depression (manuscript in preparation). Under our conditions (15% W/W Dextran, molecular weight 250,000) the data show that the agent adds 10 mOsmol or less to the medium.

WARNER: What about the beam stability and beam interactions with your medium, which is the cryoprotectant free region: Have you noticed anything unusual when you do bombard the intercellular space?

HALL: The intercellular spaces are not more beam-sensitive than the others. Our method of mounting does not insure that the section is everywhere in good contact with the support film, and every compartment occasionally shows damage (manifested most readily by a decreasing continuum count) when local contact is poor. But we have not noticed that the fluid spaces, either "cryoprotected" or not, are especially sensitive.

WARNER: Have you tried carbon measurements in your sections?

HALL: No, we have not gotten around to that yet.

RICK: I have one biological question. I guess that in the water-supplied state you have seen considerable increase in the intracellular osmolarity as estimated from the microprobe data and about twice as high an increase in the extracellular space and in the canaliculus. From looking at the numbers I have a question that the osmolarity of the cell is just in between the osmolarity of the canaliculus and the outer bathing solution. So one would roughly calculate that the water permeability of the outer facing membranes or the paratubular membranes is about the same. The overall water permeability is about the same as that of the membranes lining the canalicular space. That seems a little surprising to me, because just from the images I had the impression that the membrane area of those intracellular spaces is much larger than the membrane area that is facing the paratubular compartment. Would you expect that per membrane area there is a significant difference in the water permeability of these two different barriers or what is your estimation?

HALL: It seems possible that there is a difference in permeability per unit area.

RICK: I thought that when you see such an extraordinary enlargement of the membrane area then one would think, okay, that is good for equilibration of the osmolarity between the two spaces and obviously the equilibration is as poor as the equilibration across the paratubular membrane. So to me it does not make sense.

HALL: But this is going to be affected by the fact that the membrane is pumping.

RICK: That might be another possible meaning of the increase in the membrane area --that it simply adds new pump sites.

HALL: Well, I think it very likely is. I do not see what this very elegant membrane structure is doing, unless it is therefore providing a tremendous pumping surface.

ELECTRON MICROPROBE ANALYSIS OF SECRETORY EPITHELIA: AVIAN SALT GLAND

S. Brian Andrews
R. Gary Kirk

Department of Physiology
School of Medicine
Yale University
New Haven, Connecticut

Joseph E. Mazurkiewicz

Department of Anatomy
Albany Medical College
of Union University
Albany, New York

INTRODUCTION

Electron microprobe analysis (EMA) is the only current technology that is capable of quantitative, in situ elemental analysis at the subcellular level; consequently, it has been evident for some time that this method could in principle provide unique and important information to cell physiologists. However, EMA experiments of such significance have become feasible only within the last few years, as a result of technical and methodological advances from a number of laboratories (Shuman et al., 1976; Gupta et al., 1977; Dörge et al., 1978). These advances have facilitated the development of a new system for the quantitative energy-dispersive (EDS) microprobe analysis of freeze-dried cryosections of tissues, and this system is described in the first part of this contribution.

ISBN 0-12-362880-6

The remainder of this account describes the application of this EMA system to the determination of intracellular ion concentrations in the erythrocytes and the secretory epithelium of the duckling salt gland. This epithelium, which is the major component of the compound tubular salt gland, is capable of secreting a sodium chloride solution that can approach six times the osmolarity of the plasma when the animal is salt-loaded (Schmidt-Nielson, 1960). This secretion contributes to the ability of marine birds to maintain salt and water balance even while imbibing sea water. The transport mechanism(s) of the principal cells of this gland is of considerable interest (Ernst and Mills, 1977; Ellis et al., 1977) and a knowledge of the cellular electrolyte composition is crucial to an understanding of this process. The results of the microprobe and chemical analysis reported here support a mechanism for salt secretion based on sodium-coupled chloride transport (Frizzell et al., 1979; Ernst and Mills, 1977) and allow us to formulate a model which predicts some of the details of the regulation of ion transport in this tissue.

METHODS

Animals and Tissues

The blood and salt glands of the domestic duckling, *Anas platyrhynchos*, were studied in two distinct developmental and physiological states. In the case of normal (unstressed) ducklings, the animals were maintained for fifteen days after hatching on an *ad libitum* supply of duck mash and fresh water, prior to sacrifice by decapitation. The salt-stressed ducklings were subjected to the same regimen, except that at day nine they were switched to a 1% saline solution for drinking. Compared to the unstressed ducklings, who were robust and healthy and gained weight rapidly, the stressed ducklings appeared weak and sickly and gained little, if any, weight during the six days of salt stress. Immediately following sacrifice, aliquots of whole blood were collected into heparinized vessels, and the salt glands were rapidly excised and quench-frozen by plunging into supercooled Freon 22 (-169 C) (Somlyo et al., 1977). The glands of the normal animals were crescent-shaped tissues approximately 2 mm in diameter by 5 mm long and could be frozen in one piece; in contrast, the glands of the stressed ducklings were plump, blood-engorged tissues approximately three times larger than the unstressed gland, and therefore were cut into 2-3 pieces before freezing. In either case, the entire procedure from excision to freezing required less than fifteen seconds. Samples of packed pellets of red blood cells in plasma were quench-frozen on small copper pins as described above. The remainder of the blood was used for flame photometric measurement of the sodium and potassium concentrations of erythrocytes and plasma, and for gravimetric determination of the dry mass fraction.

Cryoultramicrotomy

Cryosections of approximately 130 nm thickness (as determined by continuum x-ray production and STEM contrast) were cut from the natural face of the salt gland at -105° C using a modified Sorvall MT-2B microtome equipped with an FTS cryokit. The frozen pellets of erythrocytes were sectioned similarly, but at -90° C. The cryosections were picked up from the back of the dry glass knife and mounted on 100-mesh copper grids covered with a carbon film (ca. 45nm thick). The remainder of the procedure--sandwiching with a second coated grid, pressing with a chilled brass rod, freeze-drying, separating the grids, and coating the dry sections with a thin carbon film--was carried out essentially as described in detail by Somlyo et al. (1977).

Electron Microprobe Analysis

Energy-dispersive x-ray spectra were obtained using an ETEC Autoscan electron microscope equipped with a 30 mm^2 Kevex Si(Li) detector and a Kevex 7000 series x-ray spectrometer interfaced to a DEC PDP 11V03-L computer. X-ray spectra were acquired at an accelerating voltage of 30 kV and a beam current of 1.0 nA for 100 seconds (livetime), using small scanning rasters of 0.2-1.0 μm^2; the specimen stage was at ambient temperature. The extraction of quantitative results from raw EDS spectra was carried out by the multiple least-squares fitting method (Schamber, 1977); our computer implementation of this method is essentially that described by Shuman et al. (1976), particularly with respect to the correction of the continuum for grid-generated and other extraneous contributions. In addition, the continuum was also corrected for: (1) the contribution of the support film, by determining the weighted average of the carbon film continuum and subtracting this from the continuum for the support plus specimen; and (2) contamination, as discussed in the next paragraph. For both of these corrections, the variance of the specimen continuum was recalculated appropriately. To obtain elemental concentrations from x-ray intensities, both the continuum normalization method (Hall, 1971) and the internal standard method (Dörge et al., 1978) were employed. In favorable cases, both methods can be shown to give self-consistent results, as described below.

RESULTS AND DISCUSSION

Analytical Considerations

It has been the experience of many workers that microprobe analysis is methodologically difficult. Thus, maximizing experimental convenience was a major consideration in choosing the characteristics of a new EMA system. However, it was also clear that many of the biological problems of interest to this laboratory would require the elemental analysis of small and specific subcellular domains on the order of 0.1 μm^2. This implied the need for relatively high spatial resolution, both morphologically and analytically, and therefore a requirement for thin sections, no more than 100-200 nm thick. The need to maintain the distribution of diffusible ions in the tissues dictated that the sections be cryosections of unfixed, quench-frozen material. In addition, it was also decided to utilize freeze-dried sections, partly for reasons of experimental convenience and partly for the improved morphological detail accompanying the increased contrast due to water loss. Some sacrifice in applicability is inherent in the decision to use freeze-dried tissues, since it is becoming increasingly apparent that frozen-hydrated materials will be necessary to obtain satisfactory data from tissue domains of high water content (Gupta, 1979); this restriction appears to apply to virtually all extracellular compartments. Further, artifacts arising from the freeze-drying process itself have been the source of some concern, particularly regarding ion dislocation and lateral shrinkage (Lechene et al., 1979; Gupta, 1979). These problems notwithstanding, it has been amply demonstrated that with the appropriate choice of techniques and tissues, good EMA analyses of cellular and subcellular compartments can be obtained using freeze-dried specimens (Somlyo et al., 1977; Rick et al., 1978a,b; Somlyo et al., 1979).

As a further consequence of choosing thin sections, the analyzed microvolumes are very small and the x-ray yields low. Thus, energy-dispersive spectroscopy is the clear choice for analyzing most of the biologically relevant elements. This position has become even more compelling in recent years with the increasing availability of fast and reliable software for the deconvolution of EDS spectra. Finally, and purely as a matter of convenience, it was found that this system could be operated with an ambient temperature specimen stage. From the outset, it was evident that a microprobe that irradiated thin sections of biological material at room temperature with beam current densities of 1-10 nA/ μm^2 would be particularly sensitive to artifacts arising from electron beam/specimen interactions. Therefore, the influence of mass loss and contamination on the performance of this instrument was examined in some detail. Cryosections of quench-frozen 20% albumin solutions containing known amounts of electrolytes (Dörge et al., 1978) were suitable specimens for determining the effects of mass loss on a typical biological matrix. Characteristic, continuum and total x-ray counts were acquired from such sections as a function of time during irradiation with current densities ranging from 6 x 10^{-5} -1 nA/ μm^2. Regardless of the current density, the total and continuum x-rays decreased by approximately 25% with the loss being complete after a dose of approximately 0.2 nC/ μm^2. In contrast, none of the characteristic x-ray peaks, with the sole exception of the sulfur K peak, were affected by doses as high as 1000 nC/ μm^2; it was specifically noted that the chlorine K lines were stable under these conditions. Essentially similar results were obtained on sections of other biological materials, including human and duckling erythrocytes, duckling salt gland, and rat kidney. These observations are consistent with previous reports on the nature of beam-induced mass loss from organic materials (Bahr et al., 1965) and, in particular, are very similar to the results of Dörge et al. (1978) on sections of similar composition. The implications of these observations for quantitation of data from experimental tissues can be appreciated by considering how the calibration constants for the quantitation routine were derived. After determining the relative sensitivity of our spectrometer to various elements using binary crystals as suggested by Shuman et al. (1976), the absolute calibration constants for converting peak/continuum ratios to mmols/kg dry weight were derived from sections of electrolyte-doped albumin of the kind described above. Since these calibration spectra were obtained using current densities of 1 nA/μm^2, it is evident that full mass loss must have occurred within the first second of these analyses; this will be equally true of all experimental spectra, as these were always acquired with similar or higher current densities. Therefore, both the calibration spectra and the experimental spectra reflect full mass loss conditions and quantitation errors will be introduced only to the extent that the mass loss from the experimental tissue differs from the 25% built into the calibration constants via the albumin standards. Our data on a variety of tissues suggest that this error is not likely to exceed 5%, since all specimens so far examined exhibited losses of 20%-30%.

The dose dependence of specimen contamination was found to be in marked contrast to that described for mass loss. Whereas mass loss was complete at doses less than 1 nC/μm^2, the onset on contamination could only be detected as an increase in the total and continuum count rate at doses on the order of 10 nC/ μm^2; further, this increase was linearly dependent on the dose (and therefore the analysis time) at least up to 1000 nC/ μm^2. In order to avoid overestimating the true tissue continuum due to this artifact, the following precautions and/or corrections were employed. A spectrum of the total or (uncorrected) continuum x-ray count rate vs. time was acquired simultaneously with each experimental EDS spectrum. Figure 1 illustrates an example of such a curve. An estimate of the contamination contribution to the x-ray spectrum can be obtained from this data by using the linear regression method to determine the slope of the count rate increase and calculating the area of the right triangle which has this line as its hypotenuse.

This area, divided by the total area, is the fraction of the continuum due to contamination and, therefore, the continuum can be corrected for this contribution. This approach has been found to be accurate for contamination fractions as large as 30% using albumin cryosections as test specimens. However, as a matter of prudence, experimental spectra with contamination fraction greater than 10% are rejected. As a practical limit of this EMA system, we have found that as long as beam current densities $< 5nA/ \mu m^2$ are used, the contamination can routinely be maintained at $<10\%$ using only the cold traps and anticontaminators provided by the microscope manufacturer.

The precautions and corrections just described are intended to ensure an accurate estimate of the mass of the analyzed microvolume. A major justification for the use of these methods is their efficacy in improving the precision of concentration per unit dry weight calculations based on the

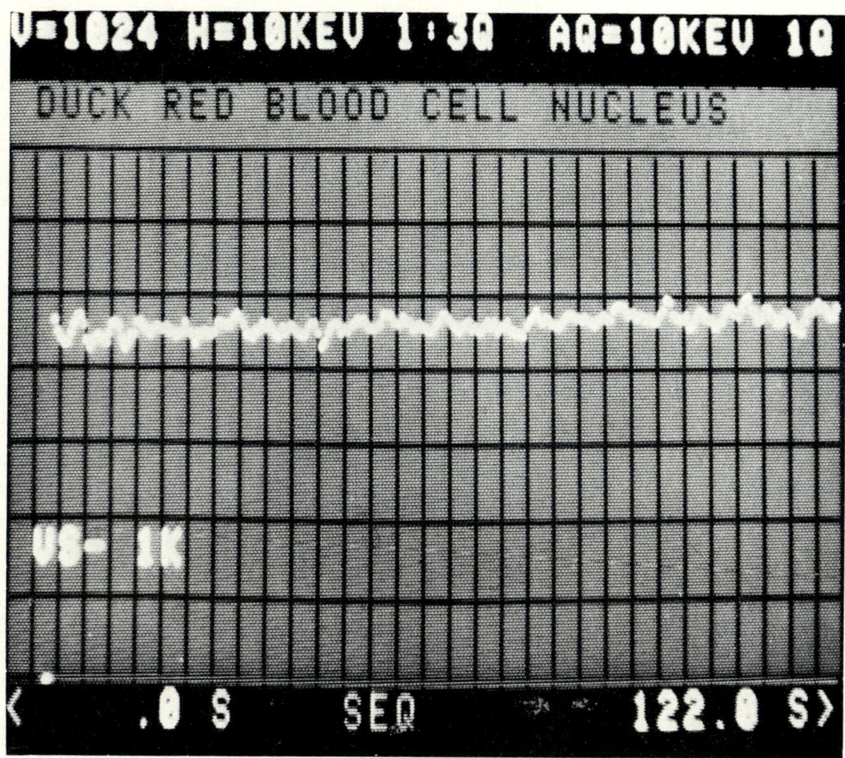

FIGURE 1. A representative graph of the intensity of total detected x-rays (0-10 keV) from the nucleus of a cryosectioned duckling erythrocyte as a function of time. Beam current was approximately 2nA/ μm^2. The fraction of x-rays arising from contamination was calculated to be 4% after 120 seconds of irradiation.

Hall formalism (Hall, 1971). However, such improvements often cannot be exploited in practice because concentrations normalized to tissue dry weight are frequently inadequate. For many tissues, and in particular for transporting epithelia, a measure of the local water content is indispensable for inferring the physiological relevance of the results. There is presently only one documented approach to obtaining concentrations per unit volume (approximately equivalent to concentrations per unit wet weight) and tissue dry mass fractions (Appleton and Newell, 1977; Dörge et al., 1978). This method, which depends on the use of peripheral standards, is subject to reservations regarding its sensitivity to variations in section thickness and lateral shrinkage artifacts (Hall, 1979). In the case of the present studies, we have found it possible and desirable to express the results as wet weight concentrations. This was achieved by using either a variant of the internal standard method, an independent determination of the tissue water content, or a complete compartmental analysis. In favorable cases, these methods can be shown to give results which are consistent with the dry weight concentrations directly calculated by the computer program. The details of these calculations are illustrated in the following paragraphs.

Duckling Erythrocytes

Although the tissue of ultimate interest in this study was the salt gland epithelium, it was desirable for a number of reasons to first obtain electron microprobe results on freeze-dried cryosections of packed red blood cell pellets. The nucleated erythrocytes of the duck are interesting in their own right because of the general and unresolved issue of ion compartmentalization in the nuclei of such cells, and also because of the possibility that the composition of the red cells may reflect the salt load of the blood in a stressed bird. In this study the red cells also served a methodological purpose, in that they were a logical next choice as specimens for testing the suitability of our freezing, sectioning, analysis, and quantitation methods. Using both morphological (Figure 2) and spectroscopic (Figure 3) criteria, the preparations were found to be satisfactory.

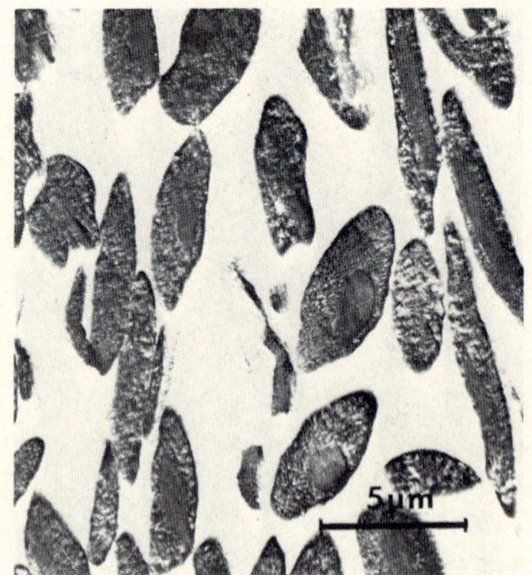

FIGURE 2. Transmission electron micrograph of a typical freeze-dried cryosection of unstressed duckling erythrocytes. Bar, 5 μm.

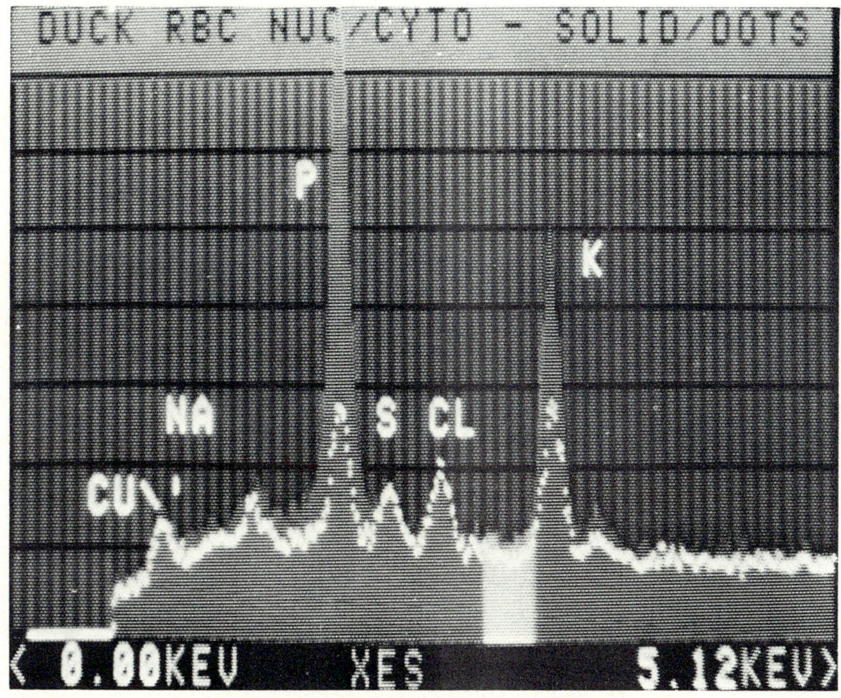

FIGURE 3. EDS x-ray obtained from the nucleus (bars) and the cytoplasm (dots) of unstressed duckling erythrocyte; 100 seconds analysis time.

However, a more interesting and rigorous test was available to assess the accuracy of our quantitation protocol, particularly with respect to tissue mass estimates based on the corrected continuum. As a two-compartment system, the wet weight concentrations of a given element in these erythrocytes are uniquely determined by the ratio of characteristic x-ray intensities from the compartments, the total elemental concentration per liter of packed cells (by chemical analysis), and the volume fraction of the nuclear compartment (by morphometric estimate). The following equations illustrate the generalized relationships:

$$C_{X,T} = V_N C_{X,N} + V_C C_{X,C} \quad (1)$$

$$C_{X,N} = (I_{X,N}/I_{X,C}) C_{X,C} \quad (2)$$

$$V_N + V_C = 1 \quad (3)$$

where V = volume fraction, C = concentration in mmols/kg wet weight, and I = x-ray counts. The element (or continuum) and cell compartment are indicated in subscripts. An example of this calculation for potassium concentrations is given in Table 1; the chemical data in this table are in reasonable agreement with literature values (Schmidt and McManus, 1977).

TABLE 1. Calculation of potassium concentrations and dry mass fractions of normal duckling erythrocytes[a].

	X-ray counts		K Concentration			
	I_K	I_C	C_K'(dry)	C_K(dry)	C_K(wet)	dry wt. fract.
				mmol/kg		g/100g
Red Blood Cells						
Microprobe Analysis						
Nucleus	5875	2494	436±15	439±14	190±5	43±1
Cytoplasm	2559	2273	211±7	211±7	83±2	39±1
Total				247	100	40
Chemical Analysis				252	100±4	39.7±0.4

[a]Data are the weighted average ± SEM for the paired analysis of nucleus/cytoplasm for 10 red cells. Symbols are defined in the text. The weighted average of the ratio $I_{K,N}/I_{K,C}$ was 2.29±0.10 (SEM); for the continuum, the ratio $I_{C,N}/I_{C,C}$ was 1.10±0.03 (SEM). The volume-fraction of the nucleus was 0.158±0.014.

The compartmental concentrations C_K (wet) given in column 5 are specified by the ratio of characteristic potassium x-rays, $I_{K,N}/I_{K,C}$ (column 1) according to the above equations. Similarly, the compartmental dry mass fractions (column 6) follow from the ratio of corrected continuum intensities (column 2). This manipulation allows the dry weight concentrations to be calculated in two independent ways: (1) C_K (dry) (column 4), calculated as column 5 divided by column 6; or (2) C_K' (dry) (column 3), calculated from peak/continuum ratios, that is, from column 1 divided by column 2, multiplied by a calibration constant. These two calculations of dry weight concentrations will be in agreement only if there is no significant absolute error in the estimation of the true tissue continuum. The good agreement found experimentally indicates that our conditions are adequate for approximating the local tissue dry mass from continuum intensities with a useful degree of accuracy. As a corollary, this calculation illustrates a method for deriving wet weight concentrations without recourse to peripheral standards.

This strategy was used to determine the wet weight concentrations of all analyzed elements for both stressed and unstressed duckling erythrocytes (Table 2). No major differences between stressed and unstressed red cells were found, even though plasma sodium was found to be 170 mM in the stressed bird as compared with 155 mM in the normal animal; high plasma sodium has also been reported in the salt-loading herring gull (Schmidt-Nielsen, 1976). For both populations of erythrocytes, the EMA results indicated high concentrations of potassium and phosphorus in the nuclei of these cells, while chloride was slightly elevated in the cytoplasm, and iron was detected only in this compartment. Intracellular sodium levels were too low to quantitate by EMA. This observation is consistent with the chemical determination of $C_{Na,T}$(wet) = 4.8 mmols/L of packed cells. We do not detect iron in the nucleus of the duckling red cells, but otherwise the present results are consistent with the elemental distributions reported by the Cambridge group for chick embryo erythrocytes (Jones et al., 1979). The similarity in the data includes the dramatic localization of potassium and phosphorus in the nuclei of these cells; this finding is interesting if only because such large concentration differences are seldom encountered when cellular compartments are analyzed by microprobe methods. The biological significance of this ion compartmentalization remains to be determined, but the atypical elemental

TABLE 2 - Elemental concentrations in duckling erythrocytes and salt gland principal cells[a].

	No. of Analyses	Na	P	Cl	K	Dry wt fraction
			mmol/kg wet wt.			g/100g
Red Blood Cells						
Nucleus	10	–[b]	385±13	64±3	190±5	43±1
Cytoplasm	10	–[b]	67±3	72±3	83±2	39±1
			mmol/kg dry wt.			
Salt Gland Epithelium						
Stressed						
Apical Cytoplasm	9	54±22	493±16	125±7	447±14	24±1
Nucleus	14	49±18	559±15	173±7	595±16	19±1
Basal Cytoplasm[c]	16	114±16	512±12	149±5	451±10	27±1
Unstressed						
Apical Cytoplasm	9	42±10	542±23	163±8	444±19	25[d]
Nucleus	15	53±10	551±22	137±7	448±17	25[d]
Basal Cytoplasm[c]	7	110±15	514±22	179±13	447±29	25[d]

[a]Data are the weighted average ± SEM; [b]Values not significantly different from zero; [c]Refer to the text for a discussion of the components of this compartment; and [d]the origin of this estimate is discussed in the text.

composition of these nuclei may well be related to the unusually dense, heterochromatic nature of nuclei of avian erythrocytes.

Salt Gland Epithelium

During the past few years, it has become increasingly apparent that epithelial electrolyte secretion is a more universal and physiologically important process than had been previously appreciated (Binder, 1979). This realization has intensified interest in the mechanism by which the avian salt gland, a favored model tissue for over two decades (Schmidt-Nielsen et al., 1957), secretes a concentrated sodium chloride solution. Although the duck salt gland consists of at least four cell types, the principal cell has been the focus of most investigations since this is the most abundant cell type and has been reasonably presumed to dominate the exocrine function. Moreover, it is the principal cell which is primarily responsible for the dramatic developmental response of the salt gland to dietary salt stress. In normal ducklings the principal cells of the non-secreting gland are relatively cuboidal with only a moderate degree of basolateral infolding and a moderate number of ouabain binding sites (Ernst and Ellis, 1969; Ernst and Mills, 1977; Hossler et al., 1978). The morphology of unstressed tissue is illustrated in Figure 4 using both conventional plastic sections and freeze-dried cryosections; these electron micrographs may also be compared to demonstrate the satisfactory preservation of characteristic morphology and the reasonable level of ice crystal damage in the cryosections. In ducklings subjected to salt loading the epithelial cells undergo a dramatic amplification of basolateral membrane area, ouabain binding sites, and NaK-ATPase activity, as well as a substantial increase in the complexity of the interdigitations between basolateral cell processes (Ernst and Ellis, 1969; Ernst and Mills, 1977; Hossler et al., 1978). Furthermore, these changes correlate with the onset of saline secretion. The morphological consequences of salt stress are illustrated in Figure 5, again comparing conventional sections with cryosections.

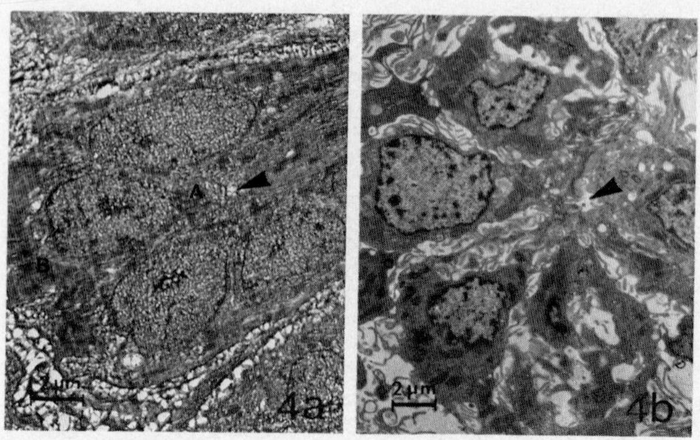

FIGURE 4. Transmission electron micrographs of secretory portions of salt gland tubules from an unstressed duckling. A = unfixed, unstained, freeze-dried cryosection. Bar, 2 µm. B = conventional Epon section. Bar, 2 µm. Both micrographs illustrate the compressed lumen (arrowhead) and moderate lateral interdigitations that are characteristic of the nonsecreting state. The areas designated apical and basal cytoplasm are indicated on Figure 4a as A and B, respectively.

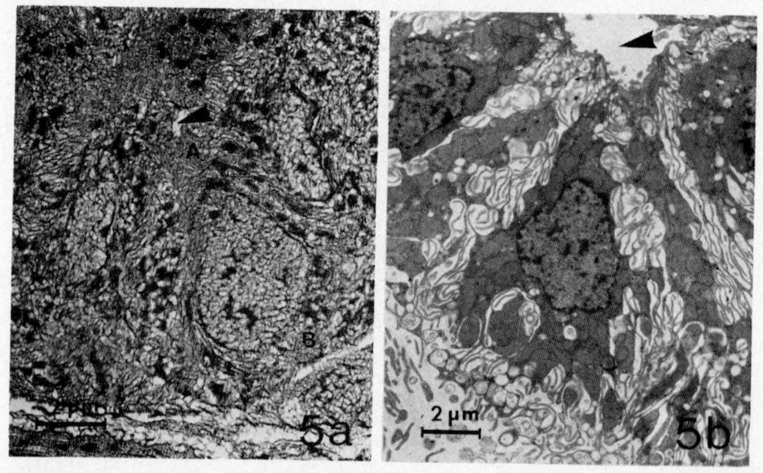

FIGURE 5. Transmission electron micrographs of secretory portions of salt gland tubules from a salt-stressed duckling. A = unfixed, unstained, freeze-dried cryosection. Bar, 2 µm. B = conventional Epon section. Bar, 2 µm. In comparison to Figure 4, the lateral and especially the basal membrane infoldings are highly amplified and very tortuous; the lumen is more evident (arrowhead). Apical and basal areas of analysis are indicated on Figure 5a as A and B, respectively.

Many models have been proposed to explain salt secretion in this tissue (Peaker and Linzell, 1975; Ernst and Mills, 1977; Ellis et al., 1977), but few of these satisfactorily reconcile the results which show that ouabain-sensitive NaK-ATPase is exclusively located in the basolateral membrane (Ernst and Mills, 1977) and that the tight junctions of this tissue are "leaky" (Ernst and Riddle, 1979). In addition, most of these models require assumptions regarding the electrical and permeability properties of the cell membranes which are difficult to verify in a tissue with the complex macroscopic and microscopic organization of the salt gland. These considerations suggested that it might be profitable to examine the salt gland by microprobe analysis, not only because the intracellular electrolyte concentrations should be valuable for deducing the mechanism of secretion, but also because the EMA results--although they are complementary to, rather than substitutes for, activity and potential measurements--might be of some use in predicting or delimiting the electrical behavior of these cells.

The microprobe studies reported here were carried out on the principal cells of salt glands directly excised from normal and stressed ducklings. Measurements were made in three subcellular regions: the nucleus, the apical cytoplasm, and the basal compartment. The morphological definition of these regions is indicated in Figures 4 and 5, an example of a typical EDS x-ray spectra is shown in Figure 6 and the results, as calculated in mmols/kg dry weight directly from peak/continuum ratios, are given in Table 2. For reasons previously discussed, it was necessary to convert these values to wet weight concentrations. This was accomplished in two different ways. In the case of the stressed gland, it was generally possible to find several well-preserved erythrocytes in the intertubular microvasculature. Since the elemental concentrations and dry mass fractions of these red cells were well- known from EMA and chemical analyses of the blood of the same animal, the _in situ_ red cells could be used as natural internal standards. This approach was used to calculate the results given in Figure 8 for the stressed principal cells. Unfortunately, this method could not be applied to the unstressed gland since this tissue is poorly perfused _in vivo_ and thus erythrocytes were rarely found in the cryosections. However, the cell water content of unstressed herring gull salt gland has been reported as 74% (Schmidt-Nielsen, 1976) and this value should mainly reflect the hydration state of the principal cells. Further, the x-ray continuum data indicate only minor intracellular variations in local water content in the unstressed cells. Therefore, we have estimated the wet weight concentration shown in Figure 7 by assuming a dry mass fraction of 0.26 for all regions of the unstressed cells. It is emphasized that this manipulation is only intended to facilitate the comparison of stressed and unstressed cells in physiologically useful units and that no conclusions rely on this estimate of the unstressed dry mass.

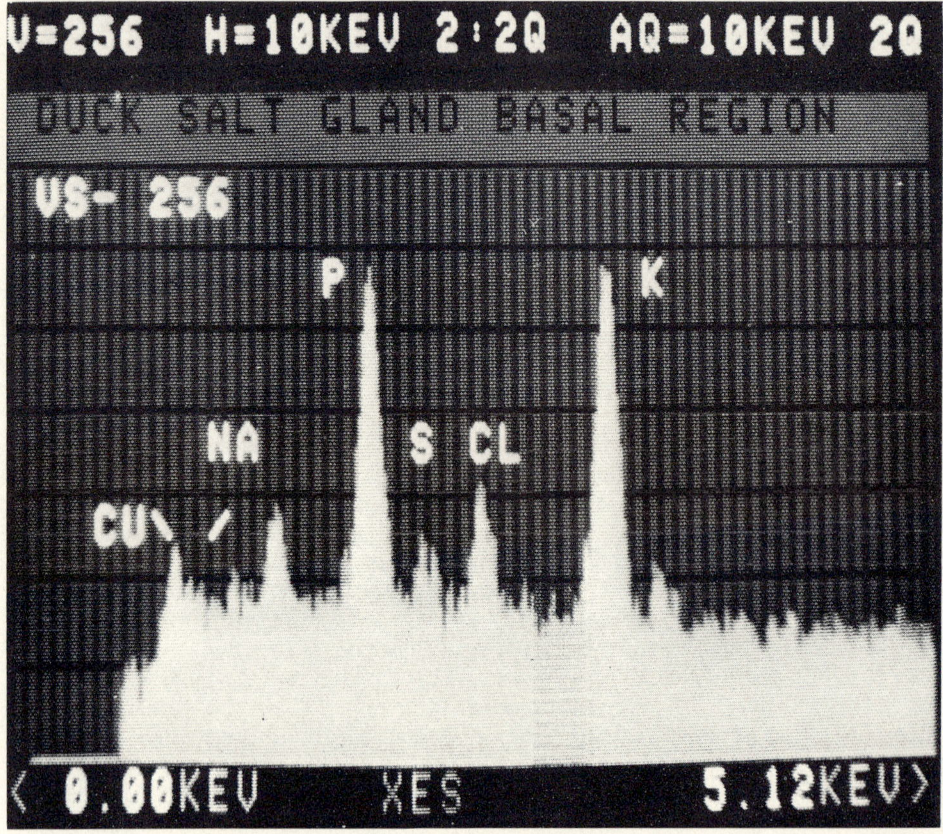

FIGURE 6. EDS x-ray spectrum obtained from the basal region of a stressed salt gland principal cell; 100 seconds analysis time.

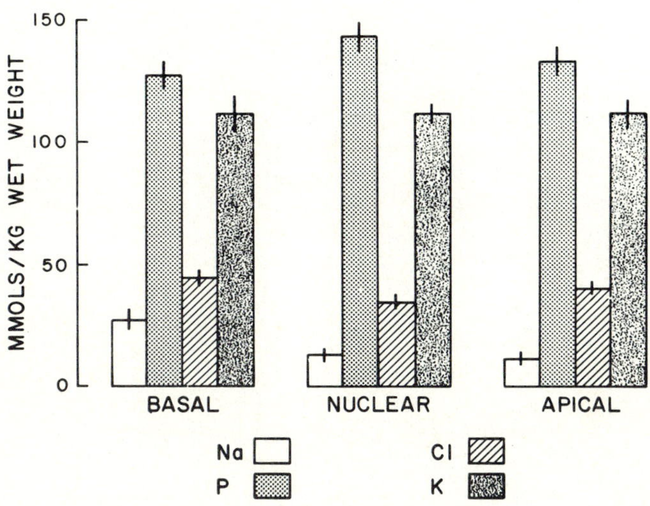

FIGURE 7. Wet weight concentrations of electrolytes in different subcellular regions of the principal cells from an unstressed duckling salt gland. These results were calculated from the dry weight concentrations and dry mass fractions given in Table 2.

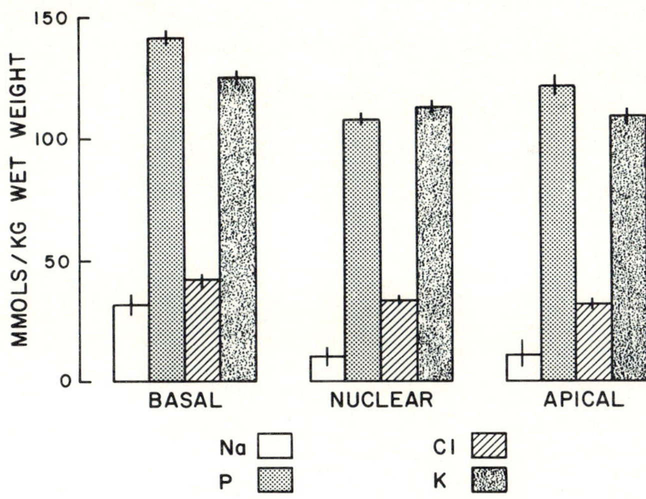

FIGURE 8. Wet weight concentrations of electrolytes in different subcellular regions of the principal cells from a salt-stressed duckling salt gland. These results were obtained using in situ erythrocytes as internal standards. The corresponding dry mass fractions are given in Table 2.

With reference to Figures 7 and 8, the results indicate a relatively uniform distribution of electrolytes between the nucleus and apical cytoplasm of both stressed and unstressed cells. When the data are expressed as mmols/L cell water, the concentrations of sodium (10-15 mM), potassium (140 mM), and chloride (40-50 mM) are suggestive of a prototypical epithelial cell. In comparison to chemical determinations of intracellular ion concentrations in avian salt glands (Hokin, 1967; Peaker, 1971; Schmidt-Nielsen, 1976), microprobe results indicate a significantly lower cellular sodium and high potassium. This is a common observation when comparing quantitative EMA results with chemical analyses and has been attributed to difficulties in the estimation of extracellular space associated with the chemical method (Rick et al., 1978a). These conclusions regarding elemental concentrations in the apical cytoplasm are relatively clear, since the electron probe could be unequivocally located on the apical region of a single cell. It would be desirable to have the same certitude regarding probe localization for analyses in the basolateral regions, especially because this is the cellular domain that is enzymatically and ultrastructurally specialized for salt transport and that responds developmentally and physiologically to salt loading. However, the complex geometry of the stressed basolateral regions and resolution limitations of our microprobe make it unlikely that the data from basal cell regions are not influenced by contributions from multiple cells, intercellular spaces, and abundant organelles (e.g., mitochondria). Nevertheless, the results are of some interest, in that they indicate elevated concentrations of all ions (and increased dry mass fractions) in this region, and they cannot be accounted for by trivial explanations. In particular, the high concentrations of potassium and phosphorus found here are not consistent with explanations based on cell damage or extracellular space contributions. However, alternative models can be proposed. The high cation concentration in the "basal cytoplasm" may in fact reflect increased intracellular sodium concentration; this in turn may be related to the increased activity of the sodium pump and/or its rate-limiting role in sodium recycling. In this case, elevated levels of phosphorus and chloride might be expected, considering the increased ATP demands of the stimulated sodium pumps, the possibility of Na-coupled chloride transport, and the need to maintain electrical neutrality. Finally, the accumulation of ions and organic mass in one region of a cell relative to another may pose some interesting problems regarding osmotic water flow in this cell. Microprobe experiments at higher spatial resolution along with results from correlative methods, as discussed below, will be necessary to address these possibilities. However, certain conclusions regarding the mechanism and regulation of ion secretion can be inferred without reliance on the data from the basal cell regions. These points are integrated into a working model of the stressed salt gland principal cell, as depicted in Figure 9; this model is based on the sodium-coupled chloride transport hypothesis, as detailed for absorptive epithelia (Frizzell et al., 1979) and proposed for the salt gland (Ernst and Mills, 1977) in analogy with the shark rectal gland (Silva et al., 1977).

The rationale for presuming active chloride transport in the salt gland depends on the microprobe determination of the intracellular Cl concentration, which indicates that the ratio of extra-/intracellular chloride is approximately 3; therefore, chloride would be in electrochemical equilibrium across the basolateral membrane if the membrane potential were about -30 mV (inside negative). Although this potential has not been measured experimentally, it can be estimated by considering that if this cell is in fact involved in active Cl uptake, then the basolateral membrane should behave roughly as a potassium electrode. In this case a potential of ca. -80 mV is predicted from the ratio of potassium concentrations by the Nernst equation; such a potential would be generally consistent with other secretory epithelia. Since the predicted membrane potential is 50 mV more negative than the equilibrium potential, it follows that the measured intracellular Cl concentration can reasonably be explained by active uptake. If Cl transport is energized directly by linkage to sodium diffusion and indirectly by ATP via the sodium pump, then local accumulations of diffusible ions in the vicinity of the transport sites become a plausible consequence of this process. The situation is somewhat different at the apical membrane since it is an open question whether the electrolyte composition of the primary secretion is the same as measured in the effluent of the main duct. We note, however, that if the primary secretion is ca. 650 mM saline, chloride would be in electrochemical equilibrium at an apical membrane potential of approximately -70 mV. This is a reasonable value in relation to the basolateral potential considering that the epithelium is "leaky" and that the postulated secretion mechanism requires paracellular transport of sodium by electrical forces. Thus, these data do not rule out the possibility that the primary secretion may be very hypertonic and, if this is the case, they suggest that the apical membrane presents no effective barrier to

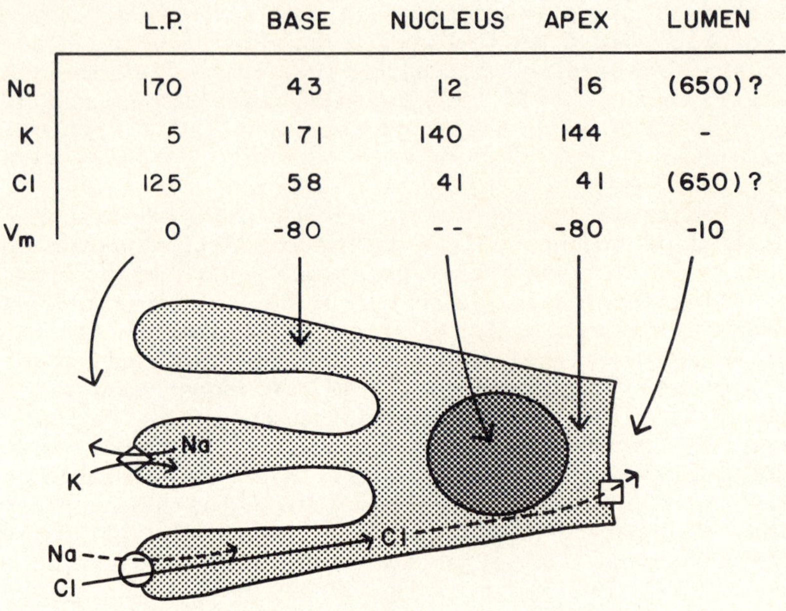

	L.P.	BASE	NUCLEUS	APEX	LUMEN
Na	170	43	12	16	(650)?
K	5	171	140	144	-
Cl	125	58	41	41	(650)?
V_m	0	-80	--	-80	-10

FIGURE 9. A schematic representation of a model for chloride secretion in the stressed duckling salt gland. The model indicates that net transcellular movement of chloride is effected by a neutral NaCl carrier (circle) followed by energetically-favored diffusion of Cl through an apical membrane channel or its equivalent (square). To maintain cell Na and to energize the transport process, Na is recycled out of the cell by the basolateral NaK-ATPase (diamond). The intracellular concentrations given in this table were calculated as mmol/L cell water (mM) from the microprobe data assuming that the electrolytes were totally dissolved in cell water. Concentrations (mM) in the lamina propria (L.P.) and the lumen were derived from chemical measurements of the plasma and from the literature, respectively. The uncertainty surrounding the composition of the actual luminal secretion is indicated by the question marks (see text). The predictions of membrane potential (V_m) are discussed in the text.

chloride movement in the stressed cell. This idea is further attractive because one major difference in electrolyte concentrations between stressed and unstressed cells was found in chloride analyses; the non-secreting cell had apical chloride concentrations 30% higher than the secreting cell and gave no evidence for non-uniform distribution of intracellular chloride. This difference would be expected if the regulation of secretion in this cell involved the control of chloride permeability at the apical membrane (Frizzell et al., 1979).

While the model just discussed is consistent with virtually all available data, there are a number of details and predictions for which there is no experimental support. More detailed electron microprobe analyses, along with electrophysiological and pharmacological studies, will at a minimum be necessary to test and refine this model. However, many such investigations are difficult or even impossible when the tissue must be manipulated in the _in vivo_ condition. We are, therefore, enthusiastic regarding the potential for an _in vitro_ system of isolated salt gland tubules that we have recently developed. (Figure 10); this preparation is particularly attractive not only because it relieves many experimental restrictions of _in vivo_ tissue, but also because it maintains the cell junctions and the polarity of the

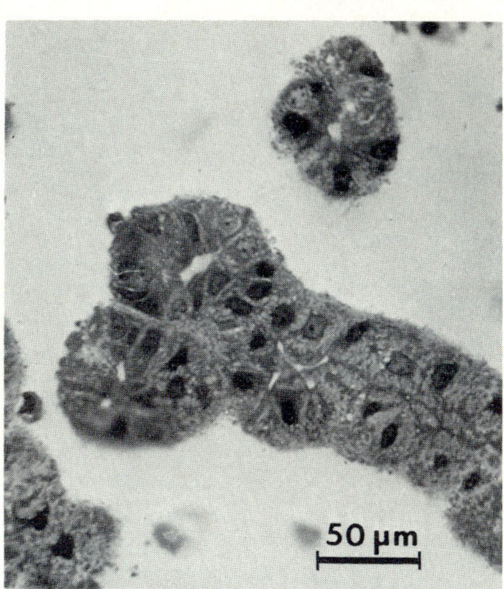

FIGURE 10. Light micrograph of a plastic section of secretory tubules isolated from the salt gland of a stressed duckling. The cell junctions and the tubular organization of the secretory epithelium have remained intact.

epithelium, and in this respect is superior to isolated single cells. Preliminary experiments have suggested that the cells of these tubules are capable of intracellular ion regulation and that the basolateral membrane potential does indeed follow the ratio of extra-/intracellular potassium. Thus, isolated tubules appear to offer the possibility of performing unambiguous electrophysiological studies on epithelia with compound tubular anatomy. Since the present studies have further indicated that microprobe results are likely to be useful in uncovering the subcellular details of ion transport, it appears that the necessary tools are now available to establish the mechanism and regulation of electrolyte transport in the salt gland.

ACKNOWLEDGMENTS

This work was supported in part by NIH Grants HL-25775, AM-21391, and S07-RR-05394, and by NSF Grant PCM-7725208. The authors also wish to thank Drs. H. Shuman, A. P. Somlyo, and A. V. Somlyo for sharing with us their algorithm for the multiple least-squares fit.

REFERENCES

Appleton, T. C. and Newell, P. F. (1977). Nature 266:854.

Bahr, G. F., Johnson, F. B., and Zeitler, G. (1965). Lab. Invest. 14:1115.

Binder, H. J. (1979). In "Mechanisms of Intestinal Secretion" (H. J. Binder, ed.), p. 1. Alan R. Liss, New York.

Dörge, A., Rick, R., Gehring, K., and Thurau, K. (1978). Pflügers Arch. 373:85.

Ellis, R. A., Goertemiller, C. C., Jr., and Stetson, D. L. (1977). Nature 268:555.

Ernst, S. A. and Ellis, R. A. (1969). J. Cell Biol. 40:305.

Ernst, S. A. and Mills, J. W. (1977). J. Cell Biol. 75:74.

Ernst, S. A. and Riddle, C. V. (1979). J. Membrane Biol. 45:21.

Frizzell, R. A., Field, M., and Schultz, S. G. (1979). Am. J. Physiol. 236:F1.

Gupta, B. L., Hall, T. A., and Moreton, R. B. (1977). In "Transport of Ions and Water in Animals" (B. L. Gupta, R. B. Moreton, J. L. Oschman, and B. J. Wall, eds.), p. 83. Academic Press, London.

Gupta, B. L. (1979). In "Microbeam Analysis in Biology" (C. P. Lechene and R. R. Warner, eds.), p. 375. Academic Press, New York.

Hall, T. A. (1971). In "Physical Techniques in Biological Research" (G. Oster, ed.), 2nd ed., Vol. 1A, p. 157. Academic Press, New York.

Hall, T. A. (1979). In "Microbeam Analysis in Biology" (C. P. Lechene and R. R. Warner, eds), p. 185. Academic Press, New York.

Hokin, M. R. (1967). J. Gen. Physiol. 50:2197.

Hossler, F. E., Sarras, M. P., and Barrnett, R. J. (1978). J. Cell Sci. 31:179.

Jones, R. T., Johnson, R. T., Gupta, B. L., and Hall, T. A. (1979). J. Cell Sci. 35:67.

Lechene, C. P., Bonventre, J. V., and Warner, R. R. (1979). In "Microbeam Analysis in Biology" (C. P. Lechene and R. R. Warner, eds.), p. 409. Academic Press, New York.

Peaker, M. (1971). J. Physiol. 213:399.

Peaker, M., and Linzell, J. L. (1975). "Salt Glands in Birds and Reptiles." Cambridge University Press, Cambridge.

Rick, R., Dorge, A., MacKnight, A. D. C., Leaf, A., and Thurau, K. (1978a). J. Membrane Biol. 39:257.

Rick, R., Dorge, A., von Arnim, E., and Thurau, K. (1978b). J. Membrane Biol. 39:313.

Schmidt, W. F., III, and McManus, T. J. (1977). J. Gen. Physiol. 70:59.

Schmidt-Nielsen, B. (1976). Am. J. Physiol. 230:514.

Schmidt-Nielsen, K., Jorgensen, C. B., and Osaki, H. (1957). Fed. Proc. 16:113.

Schmidt-Nielsen, K. (1960). Circulation 21:955.

Silva, P., Stoff, J., Field, M., Fine, L., Forrest, J. N., and Epstein, F. H. (1977). Am. J. Physiol. 233:F298.

Schamber, F. H. (1977). In "X-ray Fluorescence Analysis of Environmental Samples" (T. G. Dzubay, ed.), p. 241. Ann Arbor Science Publishers, Ann Arbor.

Shuman, H., Somlyo, A. V., and Somlyo, A. P. (1976). Ultramicroscopy 1:317.

Somlyo, A. P., Somlyo, A. V., and Shuman, H. (1979). J. Cell Biol. 81:316.

Somlyo, A. V., Shuman, H., and Somlyo, A. P. (1977). J. Cell Biol. 74:828.

DISCUSSION

SPEAKER: Brian Andrews.

FERRIER: On the question of the bremsstrahlung and the hole count, I think hole count is fine if you are looking at extraneous x-rays generated outside the main probe of your beam. But when you put that beam onto a specimen, you get significant elastic scatter through substantial angles; then the problem is that the backscattered electrons plus x-rays from below can dramatically influence the bremsstrahlung at high energies. For example, in some of the slides you were showing (Figure 3) the shape around 1 to 5 KeV seems to be rather different from what I would expect if the bremsstrahlung were essentially from the specimen itself.

ANDREWS: This spectrum (Figure 3) shows the peak of the continuum about where I expect it (1.5 - 2.0 KeV), but it is not pronounced enough. This is because the spectrum still has the copper grid contribution in it.

FERRIER: I am talking about the bremsstrahlung shape; you are making use of windows for the bremmsstrahlung, are you not?

ANDREWS: Yes, and the window is 2.89 - 3.19 KeV.

FERRIER: But even there you can find significant contributions from extraneous sources.

ANDREWS: I agree. I didn't emphasize this but this is a spectrum which has not been corrected yet for extraneous sources which we do using the techniques that Henry Shuman published in Ultramicroscopy (Volume 1, p. 317, 1976).

SOMLYO: In agreement with these remarks, I think this is why Henry (Shuman) was using a lower energy continuum region. Around 3 KeV there is a larger contribution from the true solid target, whereas the grid contribution is more like a solid target continuum showing self-absorption of the low energy x-rays. Moreover, this is different for every column. If you go to a beryllium grid or if you do something like Ted (Hall) does, use a support where there is no solid target contribution, then you do not have this problem. But the shape of the high-energy region (Figure 3) indicates that there is a relatively large contribution from extraneous continuum . . . more than if you were to take 1.34 - 1.64 KeV.

ANDREWS: You are entirely right; that error does exist. But I do want to emphasize again that this spectrum has not been corrected. So it is misleading to look at it and say it has the wrong shape. If you go the computer and subtract out the calculated extraneous contribution, you will get a spectrum that has a continuum shape that is appropriate for a low-Z target. The fact that this works so well gives us some confidence that we can correct for these contributions even though they may be significant.

FERRIER: We feel that it might be better to make sure you have some handle on the extraneous contributions. I mean, you can clean up the column until you think the hole count superb, but because of elastic scattering, you will still get significant extraneous contribution when the beam is on a specimen. Now what we would like to see is an arrangement of suitable materials so that you can get a handle on what these contributions are via characteristic peaks. As an example, you might use one element where you can get it at the higher end and aluminum so you can get it at the lower end. Even then, the corrections are only good to first order, although with thin films you are beginning to get a reasonable estimate. In any case, this is a difficult problem.

ANDREWS: I take it that you are not much convinced about the value of using the copper L-peak as that indicator?

FERRIER: No, I am not. We can have a long argument about it.

RICK: Do you think that the data shown in your last slide (Figures 7 and 8) actually reflect cytosolic concentrations or do they also contain intra-cytoplasmic organelles? I am not that familiar with the ultrastructural details but aren't these cells similar to oxyntic cells of the gastric mucosa, where you have quite a number of vesicles?

ANDREWS: In addition to the presence of extracellular spaces and the large number of mitochondria, there are, of course, other organelles including vesicles, golgi, and endoplasmic reticulum. However, none of the cellular structures is particularly well developed in this cell. Regarding the location of the probe, we tried to assess this by looking at random micrographs of conventional sections and examining likely probe areas for organelles. It turns out that we didn't pick up too much in the way of membrane-bound structures other than mitochondria. However, the mitochondrion is the only organelle we can identify for sure in cryosections; this is the only one that shows with definition. I am quite worried about the extent to which such contributions influence the analysis of basal regions of stressed cells. We just know that this region is ultrastructurally very complex.

ELECTRON MICROPROBE ANALYSIS OF Na TRANSPORTING EPITHELIA

R. Rick
A. Dörge
C. Roloff
K. Thurau

Department of Physiology
University of Munich
Munich, West Germany

INTRODUCTION

Assuming a transcellular pathway of active transepithelial Na transport, at least two different transport barriers can be distinguished, the apical and basolateral membranes of an epithelial cell. In the two-barrier model of transepithelial Na transport proposed by Koefoed-Johnsen and Ussing in 1958, Na entry into the epithelial cell from the outer medium is thought to be passive, while Na extrusion to the inner medium is active. The distributional space between the two membranes is called the Na transport compartment.

During the past twenty years, our understanding of the individual transport steps has been greatly expanded; however, still relatively little is known about the behavior of the Na in the transport compartment. This is mainly due to the fact that the methods employed to study the intracellular electrolyte concentrations provided only mean values for all epithelial cells. Since almost all Na transporting epithelia are composed of different epithelial cell types, with possibly different transport properties, methods are needed which allow the determination of the electrolyte concentrations in <u>individual</u> epithelial cells.

ISBN 0-12-362880-6

Electron microprobe analysis provides a method which is capable of detecting electrolyte concentrations in single cells or even in subcellular structures. Therefore, during recent years we employed this technique and applied it to various epithelial tissues (Rick et al., 1978a; Rick et al., 1978b; Beck et al., 1980; Rick et al., 1980). The present paper illustrates the use of this method for the analysis of transepithelial electrolyte transport mechanisms by reporting data on the active Na transport in the frog skin epithelium.

METHODS

The experiments were performed on isolated abdominal skins of frogs of the species Rana temporaria and Rana esculenta. The skins were cut into 3 to 4 pieces, which were incubated in Ussing-type chambers under different experimental conditions. In the control, both sides of the skin were bathed in normal NaCl frog Ringer's solution and the transepithelial potential difference was short-circuited by means of an automatic clamping device. Amiloride (10^{-4}M) and novobiocin (1,5 mM) were added to the outer bathing solution, while argininvasopressin (150 mU/Ml), ouabain (10^{-4}M), isoproterenol ($2x10^{-7}$M), and propranolol (10^{-4}M) were added to the inner bathing solution.

For electron microprobe analysis, small tissue samples were shock-frozen in liquid propane (-188° C) or in liquid propane/isopentane mixtures (-196° C). From the frozen material, cryosections of about 1 m thickness were cut at -100° C (Reichert OmU 2, Shandon FC 150), which were afterwards freeze-dried at -80°C and 10^{-6} Torr. The sections were mounted on thin collodion or formvar films, which were suspended over a large 1x2 mm wide opening of a nickel aperture.

Electron microprobe analysis of the freeze-dried cryosections was performed in a scanning electron microscope (Cambridge S150 or S4) which was equipped with an energy-dispersive x-ray detecting system (LINK). The conditions selected were 17-20 kV acceleration voltage and 2-5 x 10^{-10} A probe current. Measurements were performed by scanning approximately 1 μm^2 large areas of the specimen for 100 sec. Evaluation of the energy-dispersive x-ray spectra was performed by a computer program.

Quantification of the cellular element concentrations was achieved by comparing the characteristic x-ray intensities obtained in the cell directly with those obtained in an internal standard. This internal standard was produced by covering the specimen immediately prior to freezing with a thin layer of standard solution containing 20g% albumin and an extracellular composition of electrolytes. Further details of the methods have been published earlier (Bauer and Rick, 1978; Dorge et al., 1978; Rick et al., 1979).

RESULTS AND DISCUSSION

Tissue Preparation

Figure 1 shows a freeze-dried frog skin section as it is visualized during analysis in the scanning transmission mode. Though the cryosection was obtained from fresh, unfixed tissue and was neither stained nor coated with a conducting layer, the different epithelial cell types and layers can be easily distinguished. Furthermore, nucleus, cytoplasm, and intercellular spaces are discernible.

The preparation of the tissue by freezing, cryo-sectioning, and freeze-drying was chosen in order to minimize the movements of water-soluble elements. Figure 2 demonstrates that the original extra/intracellular concentration gradient for K is sufficiently well preserved. The different K $K\alpha$ -intensities obtained in the cellular and extracellular space reflect the known differences in the K concentrations in both compartments. From the S-shaped intensity profile across the cell membrane, using the 10% to 90% criterion, a spatial resolution of 0.6 μm can be calculated. The actual redistribution of K is probably much less, as oblique sectioning and the lateral scattering of the electrons within the section also contribute to the smearing of the signal. However, large movements of electrolytes may occur in tissues containing large watery spaces with no biological matrix content. In rat kidney, solutes of tubular fluid were found to be dislocated during freeze-drying (Dörge et al., 1975).

Quantification

Quantitative electron microprobe analysis of biological specimens is complicated by the sensitivity of biological specimens to electron bombardment which may lead to loss or

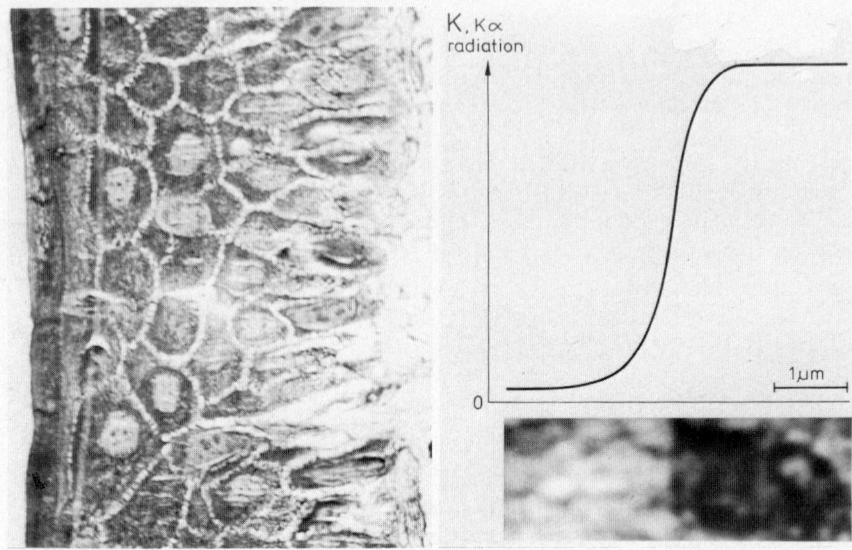

FIGURE 1. Scanning transmission electron micrograph of a 1μm thick freeze-dried cryosection of frog skin epithelium. The different epithelial layers are (from the left): stratum corneum, granulosum, spinosum, and germinativum. To the right, the underlying corial connective tissue is partly visible. The thickness of the epithelium is approximately 25 μm. Modified from Rick et al. (1978a).

FIGURE 2. K Kα -intensity distribution across a cell membrane. The detail of a scanning transmission electron micrograph shows the cytoplasm of a toad urinary bladder granular cell (right) and the adjacent extracellular albumin standard layer (left). From Dörge et al. (1978).

dislocation of elements and formation of a contamination layer. Fortunately, under the measuring conditions applied in this study, mass loss appears to affect only constituents of the biological matrix, as shown in Figure 3. The measurements were performed on freeze-dried albumin standard solutions prepared in the same way as the tissue samples. While the characteristic x-ray intensities of K, Cl, and Na remain unchanged at increasing current doses (the maximal dose applied corresponds to the normal measuring conditions), the S radiation and the white radiation, which in a freeze-dried biological soft tissue mainly reflects light elements such as C, H, O, and N, show a significant drop.

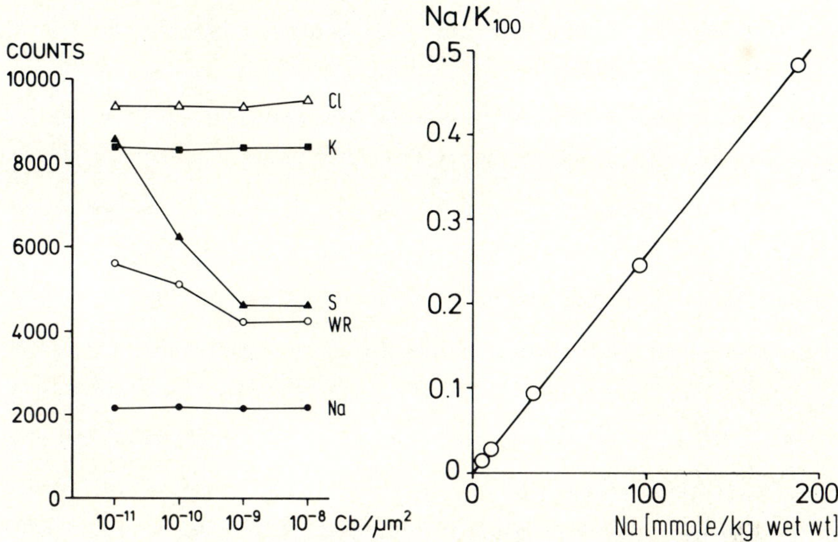

FIGURE 3. Net countrates of Cl, K, S, and Na and white countrate (WR) after administration of different current doses. The measurements were performed on freeze-dried cryosections of albumin standard solutions at room temperature. Modified from Rick et al. (1979a).

FIGURE 4. Calibration curve for Na obtained from dextran standards containing different Na concentrations and a constant K concentration of 100 mmole/kg wet weight. Na/K ratios were used to cancel out small differences of the section thickness. From Rick et al. (1979a).

Quantification of the cellular electrolyte concentrations was based on a direct comparison of the characteristic x-ray intensities obtained in the cell with those obtained in an albumin standard layer present in the same section. Compared to the continuum method (Hall, 1975), this procedure offers the advantage that in freeze-dried specimens the values represent concentrations per volume or, for a soft tissue with a specific weight of about 1, wet weight concentrations. The method presumes identical section thickness of cell and standard layer and that no differential volume changes occur during the preparation. Using an albumin standard solution, the small possible shrinkage during freeze-drying is most likely the same as in a biological soft tissue (Ingram et al., 1974).

A further precondition of this type of quantification is that the x-ray intensities are linearly dependent on the element concentrations. This can be expected to be the case if the specimen is sufficiently thin, so that deceleration of the exciting electrons and x-ray absorption and secondary x-ray fluorescence within the specimen itself can be neglected. Figure 4 shows that a linear calibration curve can be obtained even for Na which, because of its low energy radiation, is the most critical of all elements of interest.

Table 1 compares the cellular Na and K concentrations obtained by this method with those obtained by other techniques. The chemical analysis of isolated epithelial cells, which have been jet-washed to reduce the amount of adherent extracellular Na (Zylber et al., 1973), is in good agreement with the mean values obtained in the present study. Moreover, electron microprobe analysis of rat kidney proximal tubule revealed a total K concentration which is only about 20% higher than the minimal estimate of the free K concentration calculated from measurements with a K-selective microelectrode (Edelmann et al., 1978).

TABLE 1. Cellular electrolyte concentrations of frog skin epithelium and rat kidney proximal tubule.

	Na	K
	mmole/kg wet wt	
Frog skin		
X-ray microanalysis	9.4	118.4
Chemical analysis	12.1	109.1
Rat kidney		
X-ray microanalysis	20.2	139.5
K-selective electrode	--	113.0

Intraepithelial Element Distribution

Table 2 gives the element concentrations obtained in nucleus and cytoplasm of spiny cells under control conditions. For small cations, such as Na and K, the concentrations in both cellular compartments are virtually identical. This was found to be the case under all experimental conditions, supporting the notion that the intracellular space represents only one distributional compartment for small ions. In contrast, for mainly structurally bound elements, such as P and Ca, and for the dry weight content, significant differences exist.

The cells of the other living epithelial layers, granular and germinal cells, as well as further epithelial cell types, like gland cells and mitochondria-rich cells, showed very similar electrolyte concentrations in the control, only the Cl concentration in the mitochondria-rich cells, at 14 mmole/kg wet weight, being significantly lower. Under all experimental conditions, the electrolyte composition of the outer cornified cell layer approximated to that of the outer bathing solution, indicating that this layer functionally represents an extracellular space, freely equilibrating with the outside.

TABLE 2. Cytoplasmic and nuclear concentrations of Na, P, Cl, Ca, and dry weight.

	Na	K	P	Cl	Ca	dry wt
	mmole/kg wet wt					g/100g
Cytoplasm	7.4 ±4.5	111.8 ±12.1	98.8 ±19.2	36.5 ±5.0	1.1 ±1.2	25.4 ±2.3
Nucleus	5.1 ±3.9	115.0 ±11.4	144.4[a] ±19.1	32.9[a] ±3.7	0.3[a,b] ±0.6	2.18[a] ±1.9

Mean values ± SD, n = 19.

[a]Significantly different from the cytoplasmic value, 2P < 0.05.

[b]Not significantly different from zero.

Localization of the Na Transport Compartment

Previous chemical and radiochemical analysis of the frog skin epithelium often has demonstrated a behavior of the Na concentration inconsistent with the generally accepted two-barrier concept. To explain these findings, alternative transport models have been proposed, such as a double-pump model, according to which the Na influx is active (Cuthbert, 1972), or a non-cellular model in which Na is thought to be transported along the plasma membranes, never entering the intracellular space (Cereijido and Rotunno, 1968). Furthermore, it has been suggested that only one epithelial cell type or cell layer is involved in transepithelial Na transport (Voûte and Ussing, 1968).

The rationale underlying the present investigation was that the cells participating in transepithelial Na transport should display a characteristic behavior of their Na concentration. According to the two-barrier model, inhibition of the active transport step should result in an increase of the intracellular Na concentration, whereas inhibition of the passive transport step should lead to a drop. Figure 5 shows the result of such an experiment. In the control, the Na concentrations in all living epithelial layers are low, ranging from 10 mmole/kg wet weight in the granular cells to 6 mmole/kg wet weight in the germinal cells, whereas the K concentrations are high, on an average 116 mmole/kg wet weight. After inhibition of the active transport step by the cardiac glycoside ouabain, the Na concentration increased in all layers by about 100 mmole/kg wet weight, while the K concentration dropped by almost the same extent. The results of the two further experimental conditions depicted in Figure 5 demonstrate that the Na increase after ouabain originates mainly from the outer bathing solution. Under both conditions, when the Na influx is abolished either by using an Na-free outer bathing solution or by rendering the outer-facing membranes impermeable to Na with the diuretic amiloride, the Na concentrations in all epithelial layers remain low and the K concentrations high.

The fact that the cells of all living epithelial layers have shown the features typical of transepithelially transporting cells supports the notion of a syncytial Na transport compartment as originally proposed by Ussing and Windhaager (1964). The existence of cell-cell junctions with high Na permeability most easily explains the parallel

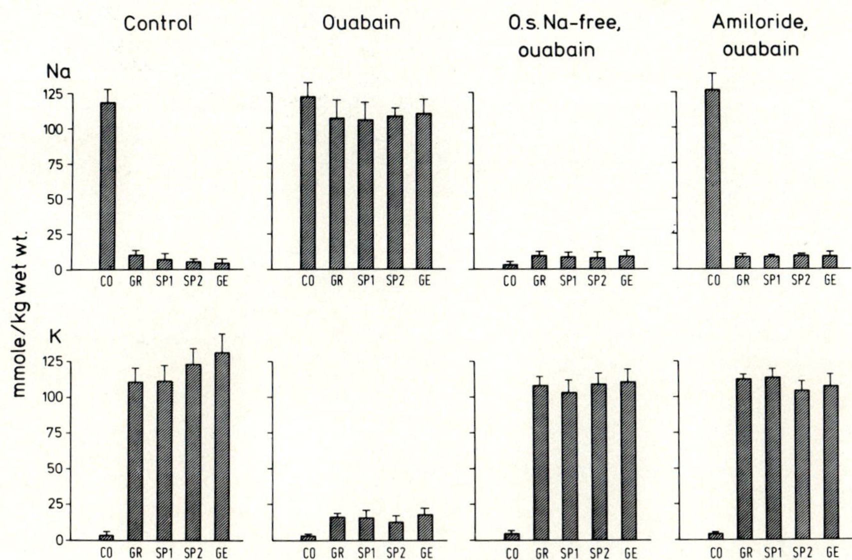

FIGURE 5. Na and K concentrations in the different epithelial layers of frog skin in control, after ouabain (90 min), and when simultaneously with the application of ouabain either the outside was incubated with a Na-free solution, or amiloride was added. The different layers are stratum corneum (CO), granulosum (GR), outer and inner spinosum (SP1, SP2), and germinativum (GE). Mean $\pm$ 2SEM, each bar representing about 10 measurements. From Rick et al. (1978a).

changes and the close similarity of the Na and K concentrations in all epithelial layers. Moreover, the finding that the Na increase even in the innermost epithelial cell layers was cancelled, when the Na influx from the outside was blocked, provides direct evidence for an intercellular Na exchange. Since in frog skin epithelium the shunt pathway between the cells is sealed at the level of the outermost living epithelial layer, the stratum granulosum (Erlij, 1971), only this cell layer is directly exposed to the outer medium, whereas all deeper cell layers could exchange their Na with the outside medium only via intercellular junctions.

However, apparently not all epithelial cell types share in this synctial Na transport compartment. Besides the cornified cells, functionally representing an extracellular space, further exceptions are the gland cells and the so-called mitochondrial-rich cells of the epithelium (Rick et al., 1978a). Figure 6 shows the Na and K concentrations in the mitochondria-rich cells. After ouabain, the Na concentration increases from 7 to 42 mmole/kg wet weight. This increase is much less pronounced than in neighboring granular and spiny cells, implying that this cell type is not coupled to its neighbors. Furthermore, it is evident that the Na increase after ouabain cannot be cancelled by amiloride. As in frog skin, both the Na influx across the outer membranes (Rick et al., 1975; Lindemann and van Driessche, 1977) as well as the transepithelial Na transport rate (Eigler et al. 1967; Dorge and Nagel, 1970) can be blocked almost completely by amiloride, this finding strongly argues against a significant contribution of this cell type to transepithelial Na transport.

Action of Modifiers of Transepithelial Na Transport

The rate of transepithelial Na transport can be modified by a large number of drugs and hormones. Analysis of accompanying changes in the Na concentration in the transport compartment should assist in further elucidation of their mechanisms of action.

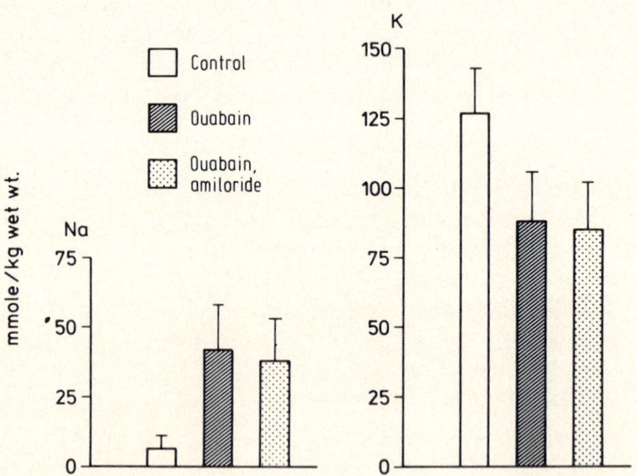

FIGURE 6. Na and K concentrations in the mitochondria-rich cells in control, after ouabain (90 min), and after simultaneous application of ouabain and amiloride. Data from Rick et al. (1978a).

Figure 7 shows an experiment in which the transepithelial Na transport was either inhibited with the diuretic amiloride or stimulated with the antibiotic novobiocin. In agreement with the view that amiloride blocks the Na influx from the outside, the Na concentration in all epithelial layers is lowered to about 50% of the control. In contrast, after novobiocin, the Na concentrations are increased by a factor of two to three, lending further support to the view that novobiocin stimulates the Na influx across the outer barrier (Cruz and Biber, 1976). The effect of novobiocin on both the intracellular Na concentrations and the transepithelial transport rate can be cancelled by simultaneous application of amiloride, demonstrating that novobiocin acts on an amiloride-sensitive Na transport step. The changes in the K values are small, only the drop after novobiocin being statistically significant.

Figure 8 reflects the changes in the intracellular Na and K concentrations after stimulating the Na transport with the anti-diuretic hormone, argininvasopressin (AVP). Similar to novobiocin, the Na concentrations in all epithelial layers

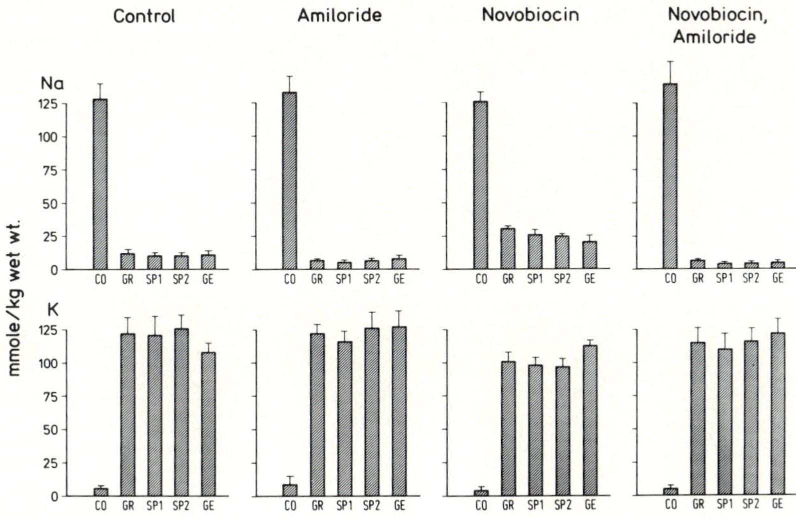

FIGURE 7. Na and K concentrations in the different epithelial layers of frog skin in control, after amiloride (30 min), after novobiocin (30 min), and after simultaneous application of amiloride and novobiocin. From Rick et al. (In press).

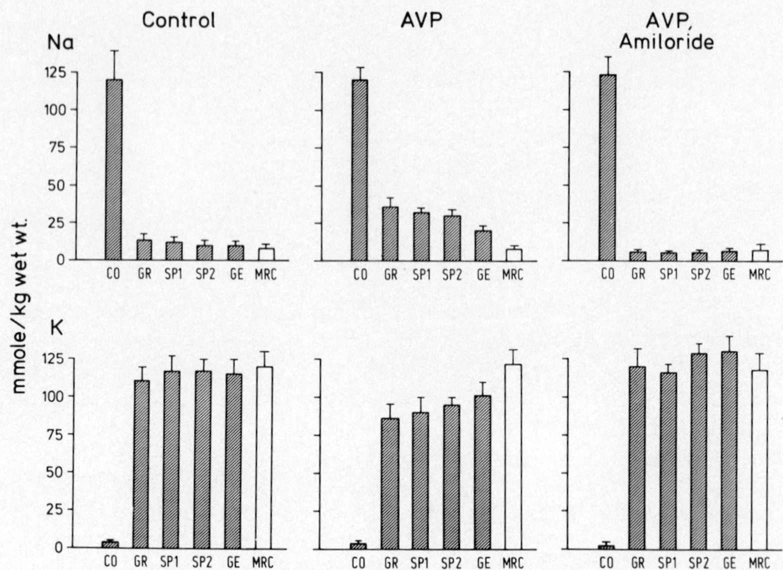

FIGURE 8. Na and K concentrations in the different epithelial layers of frog skin and in mitochondria-rich cells (MRC) in control, after AVP (60 min) and after simultaneous application of AVP and amiloride. From Rick et al. (In press).

are significantly increased, on an average from 11 to 30 mmole/kg wet weight, and the Na increase can be completely abolished when amiloride is applied simultaneously. Thus it appears that vasopressin also acts primarily by stimulating an amiloride-sensitive Na influx across the outer membranes. As the Na increase after vasopressin is sufficiently high to account for the enhanced Na extrusion at the inner barrier, there is no need to assume a direct stimulatory action on the active transport step (Aceves, 1977). It is further evident from Figure 8 that the Na and K concentrations in the mitochondria-rich cells are virtually unchanged, when argininvasopressin or amiloride are added, providing additional support to the view that this cell type is not involved in transepithelial Na transport.

A further hormone, which is known to stimulate the transepithelial Na transport, is epinephrine (Tomlinson and Woods, 1978). However, unlike vasopressin, epinephrine also effects a transepithelial Cl secretion in frog skin (Koefoed-Johnsen et al., 1953). A less-complicated situation exists, when a specific β-stimulator, such as isoproterenol, is used at relatively low concentrations. Under this condition, the resulting increase in the short-circuit current can almost completely be attributed to a stimulation of the amiloride-sensitive Na transport (own unpublished results). Figure 9 shows the effect of isoproterenol on the intracellular Na concentrations. Again, in all layers of the epithelium, an increase in the Na concentration can be detected, which is cancelled when amiloride is applied simultaneously. Addition of propranolol, a β-blocker, prior to isoproterenol, results in a complete blunting of the isoproternenol response.

According to these results also the effects of isoproterenol can be referred to a stimulation of the

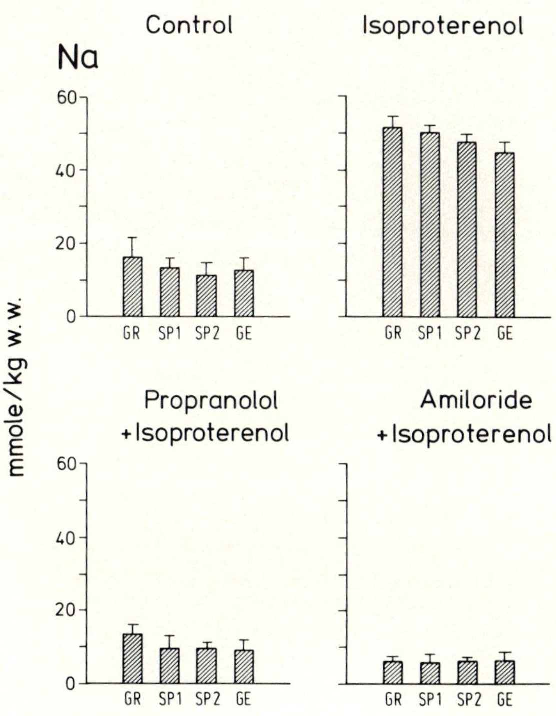

FIGURE 9. Na concentrations in the different epithelial layers of frog skin in control, after isoproterenol (20 min) and when 20 min prior to the addition of isoproterenol propranolol was applied, or when amiloride was added simultaneously.

amiloride-sensitive Na influx. Since vasopressin and isoproterenol are known to produce an increase in the intracellular c-AMP levels, one is tempted to speculate that the action of both hormones on the Na influx is mediated by the same intracellular messenger.

CONCLUSIONS

All findings of the present study are compatible with the two-barrier concept of transepithelial Na transport involving a passive Na uptake and an active Na extrusion. The Na transport compartment comprises the cells of all living epithelial layers. Exceptions are the cornified cells, mitochondria-rich cells, and gland cells. The transepithelial Na transport pathway can be described as shown in Figure 10: Na enters the epithelium across the outer-facing cell membranes of the outermost living epithelial cell layer, the stratum granulosum, from where it can readily diffuse via low-resistance intercellular junctions to all deeper epithelial layers. Na is extruded from this syncytial Na transport compartment across the basolateral membranes into the intercellular spaces. The Na uptake across the outer membranes can be blocked by amiloride and stimulated by novobiocin, vasopressin, and isoproterenol.

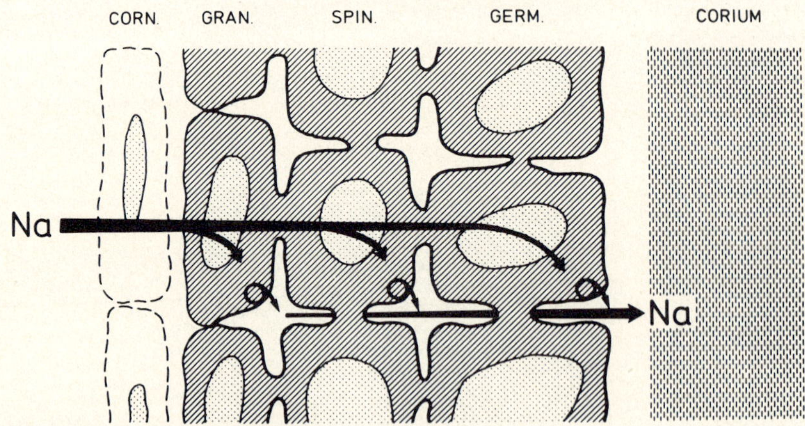

FIGURE 10. Proposed pathway of transepithelial Na transport. From Rick et al. (In press).

ACKNOWLEDGMENTS

This research was supported by a grant from the Deutsche Forschungs-Gemeinschaft.

REFERENCES

Aceves, J. (1977). Pflugers Arch. 371:211-16.

Bauer, R., and Rick, R. (1978). X-ray Spectrum. 7:63-69.

Beck, F., Bauer, R., Bauer, U., Mason, J., Dorge, A., Rick, R., and Thurau, K. (1980). Kidney Int. 17:756-63.

Cereijido, M., and Rotunno, C.A. (1968). J. Gen. Physiol. 51:S280-89.

Cruz, L.J., and Biber, T.U.L. (1976). Am. J. Physiol. 231:1866-74.

Cuthbert, A. W. (1972). J. Theor. Biol. 36:555-68.

Dörge, A., and Nagel, W. (1970). Pflügers Arch. 321:91-101.

Dörge, A., Rick, R., Gehring, K., Mason, J., and Thurau, K. (1975). J. Microsc. Biol. Cell. 22:205-14.

Dörge, A., Rick, R., Gehring, K., and Thurau, K. (1978). Pflugers Arch. 373:85-97.

Edelmann, A., Curci, S., Samarzija, I., and Fromter, E. (1978). Pflügers Arch. 378:37-45.

Eigler, J., Kelter, J., Renner, E. (1967). Klin. Wschr. 14:737-38.

Erlij, D. (1971). Philos. Trans. R. Soc. London, Ser. B. 262:69-89.

Hall, T.A. (1975). J. Microsc. Biol. Cell. 22:271-81.

Ingram, F.D., Ingram, M.J., and Hogben, C.A.M. (1974). In "Microprobe Analysis as Applied to Cells and Tissues" (T. Hall, P. Echlin, and R. Kaufmann, eds.), pp. 119-46. Academic Press, New York.

Koefoed-Johnsen, V., Ussing, H.H., and Zerahn, K. (1953). Acta Physiol. Scand. 27:38-48.

Keofoed-Johnsen, V., and Ussing, H.H. (1958). Acta Physiol. Scand. 42:298-308.

Lindemann, B., and Van Driessche, W. (1977). Science. 195:292-94.

Rick, R., Dörge, A., Nagel, W. (1975). J. Membrane Biol. 22:183-96.

Rick, R., Dörge, A., von Arnim, E., and Thurau, K. (1978a). J. Membrane Biol. 39:313-31.

Rick, R., Dörge, A., Macknight, A.D.C., Leaf, A., and Thurau, K. (1978b). J. Membrane Biol. 39:257-71.

Rick, R., Dörge, A., Bauer, R., Gehring, K., and Thurau, K. (1979a). SEM II:619-26.

Rick, R., Dörge, A., Gehring, K., and Bauer, R. (1979b). In "Microbeam Analysis in Biology" (C.P. Lechene and R.R. Warner, eds.) pp. 517-34. Academic Press, New York.

Rick, R., Dörge, A., Katz, U., Bauer, R., and Thurau, K. (1980). Pflügers Arch. 385:1-10.

Rick, R., Dörge, A., von Arnim, E., Weigel, M., and Thurau, K. (In press).

Tomlinson, R.W.S., and Wood, A.W. (1978). J. Membrane Biol. Special Issue:135-50.

Ussing, H.H., and Windhager, E. (1964). Acta Physiol. Scand. 61:484-504.

Voûte, C.L., and Ussing, H.H. (1968). J. Cell Biol. 36:625-38.

Zylber, E.E., Rotunno, C.A., and Cereijido, M. (1973). J. Membrane Biol. 13:199-216.

DISCUSSION

SPEAKER: R. Rick

SOMLYO: Have you estimated the difference between dry mass of a mitochondria-rich cell and the other cells?

RICK: Yes. The mitochondria-rich cells have somewhat more water. The measurements in the mitochondria-rich cells were performed only in the nucleus, since when you do measurements in the cytoplasm, almost a third of the excited volume consists of mitochondria.

SOMLYO: That is why I say that the mitochondria generally contain less water than the cytoplasm over the whole cell.

RICK: In Figure 8, we compared nuclear measurements with nuclear measurements. But as mentioned before, in the normal epithelial cells the nuclear concentrations of sodium and potassium are identical to the cytoplasmic concentrations.

WARNER: Roger, is the gradient you see there a believable gradient?

RICK: Which one?

WARNER: On AVP set.

RICK: Yes. You can see that, very often, compared to control, this gradient is increased after hormones or drugs which increase the sodium transport. That would be consistent with the assumption that sodium is actually diffusing from one cell to the other. If there is an increased sodium flux from one layer to the next, assuming that sodium permeability of the cell-cell junction remains fairly constant, then you would expect an increased gradient.

WARNER: Have you statistically validated that?

RICK: Yes, there is a fairly good correlation. The model is further supported by the fact that after inhibition of the transepithelial sodium transport, the concentrations in all cells are practically identical. As there is nothing moving through the system, there is no reason for building up a gradient.

SOMLYO: Is it just possible that some of the cellular gradient that you express on the basis of a computed per wet weight concentration could be due to small changes in the hydration of the different cells going across the cell layer? If I calculate it, you have a thirty mmole concentration in a given region and the hydration is 78% water; your dry weight concentration will be 106 mmoles per kilogram dry weight. If it is 75% water, then the dry weight concentration is going to be 90 mmoles per kilogram dry weight. Can one be certain enough of having really measured cell hydration to within the nearest 3% with your estimate, to be sure that a 10% difference in dry mass concentration that you are really measuring is not due to differences in hydration?

RICK: I really do not see your point. Because whether you have 78% or 75% water content, that would change the sodium values only by 3%.

SOMLYO: Let's say, if you have 30 mmolar sodium in 78% water, what you have effectively then (.78 x 30 /.22), the dry mass concentration you are effectively measuring, your measurement is really a concentration per dry weight. You are not <u>looking</u> at hydrated; you are <u>computing back</u> to hydrated, aren't you?

RICK: No. This is directly the way the data come out. Actually, they are concentrations per volume.

SOMLYO: But you are measuring frozen dry sections, aren't you?

RICK: Usually we do not calculate concentrations per dry weight. But if we would do that, the data would come out identical.

LECHENE: But in your volume, the difference of dry sodium which exists, depending upon if you have 75% or 78% hydration, would be very high.

RICK: If you have 22% dry matter or 25% dry matter and a sodium concentration of, let us say, 30 mmolar on the basis of volume, then the recalculation for the water would result in an error of about 1.5 mmolar (38.5 versus 40).

HALL: Yes, but I think there was a misunderstanding that the calculation of the original measurement of sodium per unit volume is, not to me, close to any estimation of dry weight.

RICK: I do not care about dry weight.

HALL: But when you convert to amount of element with water?

RICK: But in this case, I guess I am right that this would only make a difference of a few percent. But I am actually presenting the data on the basis of wet weight concentrations which is fairly safe, because that is almost identical to volume concentrations.

THE USE OF X-RAY MICROANALYSIS IN STUDIES OF THE ACROSOME REACTION IN SEA URCHIN SPERM

Marie Cantino
Center for Bioengineering
University of Washington
Seattle, Washington

INTRODUCTION

The aim of this project has been to determine quantitatively what elemental changes are associated with the acrosome reaction in sea urchin sperm. Although some results will be summarized at the end of the paper, the emphasis will be on the techniques employed to deal with the question, on several technical problems encountered, and on a comparison of the microprobe results with those obtained by chemical analysis. Another paper describing the results in more detail is in preparation.

BACKGROUND

The acrosome reaction consists of several morphological changes which occur just prior to fertilization in the sperm of most vertebrates and invertebrates. Characteristics of the acrosome reaction and its role in fertilization have been discussed in a number of comprehensive reviews (e.g., Bedford, 1970; Shapiro and Eddy, 1980), and only a few relevant details are mentioned below.

Although there is some variation between species, common features of the acrosome reaction include: fusion of the acrosomal and plasma membranes; release of the contents of the acrosome; and, in many invertebrates, polymerization of actin to form an acrosomal rod. Figure 1 illustrates these

ISBN 0-12-362880-6

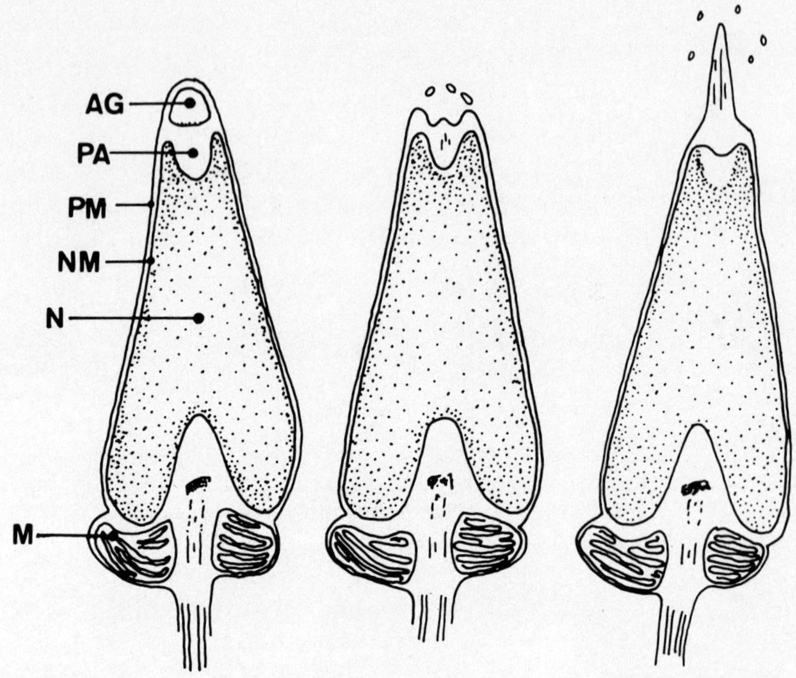

FIGURE 1. Schematic representation of the changes associated with the acrosome reaction in sea urchin sperm, beginning with the unreacted sperm (left), proceeding to fusion of the outer acrosomal and plasma membranes (center) and polymerization of actin to form an acrosomal rod (right). AG = acrosomal granule; PA = profilamentous actin; PM = plasma membrane; NM = nuclear membrane; N = nucleus; M = mitochondrion.

changes as they are thought to occur in the *Strongylocentrotus purpuratus* sperm used in this study.

The acrosome reaction, which is a prerequisite for fertilization, occurs *in vivo* when the sperm come into contact with the egg. *In vitro*, it can be induced in sea urchins by components of the egg jelly coat known as "egg jelly."

A number of investigators have implicated ion movements and dependencies in the acrosome reaction (Collins and Epel, 1977; Singh et al., 1978; and others). Results obtained at the University of Washington using radioactive tracers and ion-sensitive electrodes have demonstrated that movements of sodium, calcium, and potassium ions accompany the reaction in sea urchin sperm (Schackmann et al. 1978, 1981).

The role of x-ray microanalysis in these studies was, first, to determine whether these represent net elemental changes, and, second, to establish whether such changes are localized in subcellular regions.

SPECIMEN PREPARATION

The procedures used to collect sperm and induce the acrosome reaction were similar to those described by Schackmann et al. (1978). Undiluted ("dry") sperm, collected by intracoelomic injection of KCl, were diluted into artificial sea water (A.S.W.) containing egg jelly. Control samples (unreacted sperm) were prepared by diluting cells into A.S.W. without egg jelly. The medium consisted of 360 mM NaCl, 50 mM $MgCl_2$, 10mM KCl, 10mM $CaCl_2$, 2.5 Mm Tris, and 5.0 mM Hepes adjusted to pH 8.0. Five to ten percent dextran T_{10} was added prior to freezing, and final cell concentrations were 10^8-10^9 cells/ml.

Two methods for preparing cell suspensions for microanalysis were tested: direct application of cells to formvar-coated grids (wholemount preparation) followed by freezing and freeze-drying; and cryoultramicrotomy of frozen cell suspensions (section preparation) followed by freeze-drying. Each of these methods, along with the freeze-drying procedure, is described below.

Wholemount Preparation

Unfixed sea urchin sperm are only about two microns in diameter (as measured by phase-contrast microscopy) and undergo 30-40% shrinkage during freeze-drying. They are, therefore, well suited for wholemount preparation. Placement of live cells on the support film presented two difficulties, however. If the medium was left in large quantities surrounding the cells, it obscured the sperm and contributed error to the analysis; if, on the other hand, grids were blotted until they were nearly dry, the cells became visibly damaged. In order to reproducibly prepare samples in which both of these factors were minimized as much as possible, the suspension was placed on the back of the formvar-coated grids, so that the medium was trapped in the wells formed by the grid bars and the support film. The grids could then be blotted on a piece of filter paper without completely removing the medium. This procedure resulted in heavy salt deposition close to the grid bars, with salts gradually thinning towards the center of the grid square. Although some cell damage still occured in thin regions, undamaged sperm could usually be found elsewhere on the grid. Since the entire procedure was performed in a 10°C coldroom, and since grids were frozen within two seconds of the blotting step, drying was negligible. Figure 2 illustrates the wholemount preparation and freezing procedure. The micrographs in Figure 3 show a comparison of the cell morphology in two types of preparation: a glutaraldehyde-fixed, freeze-dried cell (A) and a cryofixed, freeze-dried cell (B).

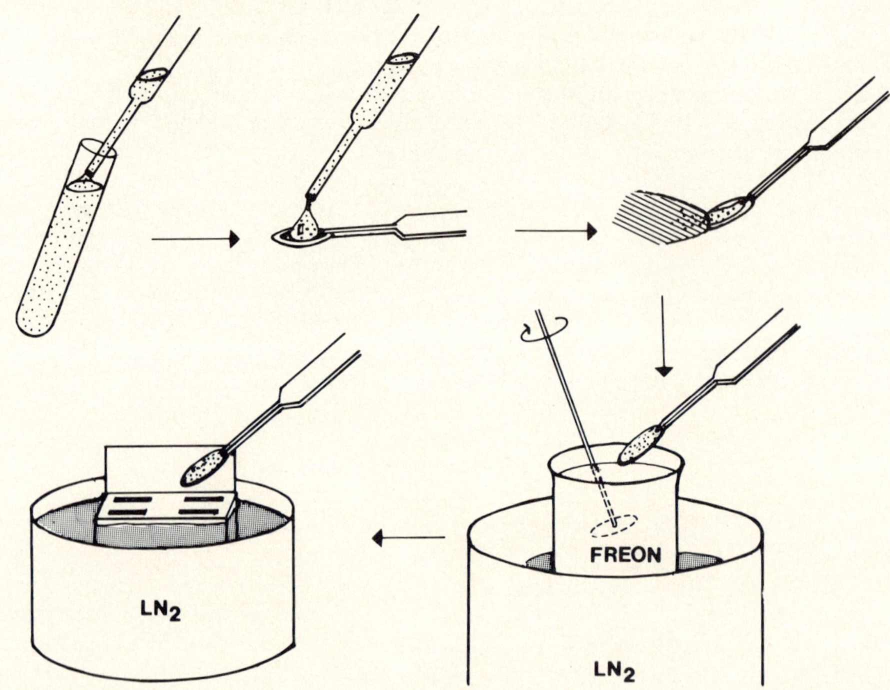

FIGURE 2. Schematic representation of the wholemount preparation procedure. Cells in suspension are pipeted onto a grid, the excess is blotted off with filter paper, the grid is plunged into Freon cooled in liquid nitrogen, and the grids are stored in a brass box under liquid nitrogen until they are freeze-dried.

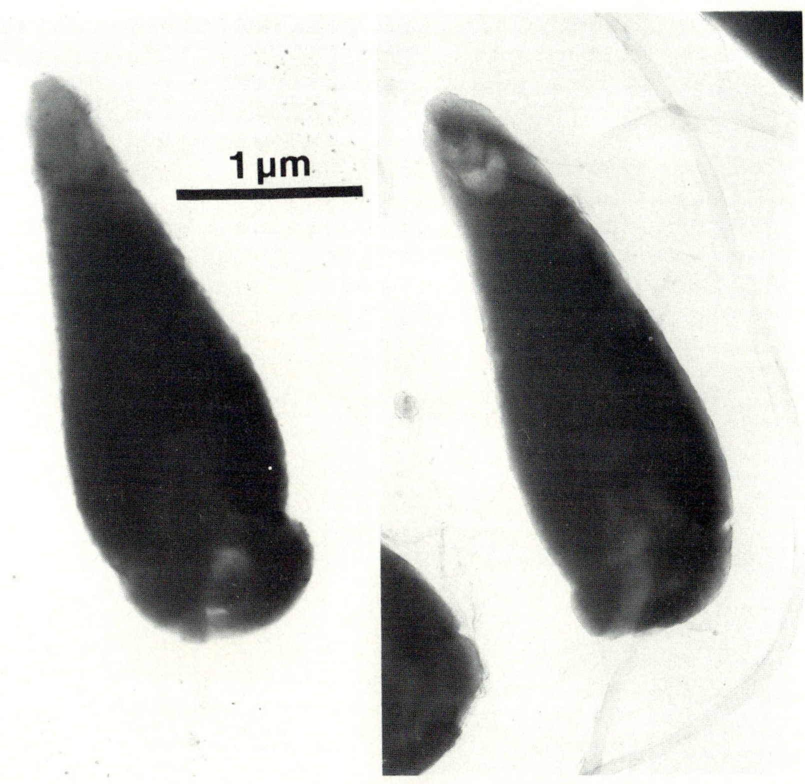

FIGURE 3. Micrographs showing unreacted sperm, prepared as wholemounts, which were glutaraldehyde-fixed prior to freezing and freeze-drying (left) or cryofixed only (right).

Section Preparation

The advantages of the wholemounts are that they are easy to prepare and freezing rates can be very high because the cryogenic liquid comes in contact with all of the cells. However, a second preparation method was sought to determine the contribution of two possible sources of error in the analytical results; namely: the contribution of salts deposited on top of the cells after drying, and possible damage incurred by the cells when they are placed on a support film. Although the cells analyzed appeared to be morphologically intact, it is possible that rapid elemental changes could have occurred without visible changes.

In order to investigate these possibilites, cryosections of the cell suspensions were prepared using a technique described by Tormey (1978). In this procedure small strips of millipore filter were dipped into the cell suspension, frozen, mounted in a vise chuck, and sectioned at -85° C in a Sorvall MT2 cryoultramicrotome. Sections were placed between two formvar films in a folding grid and were stored in a brass box for freeze-drying.

The medium used to suspend the cells was identical to that used for preparing wholemounts, except that 10% dextran (rather than 5%) was added. Figure 4A shows a freeze-dried cryosection. Freeze-damage was a more serious problem in

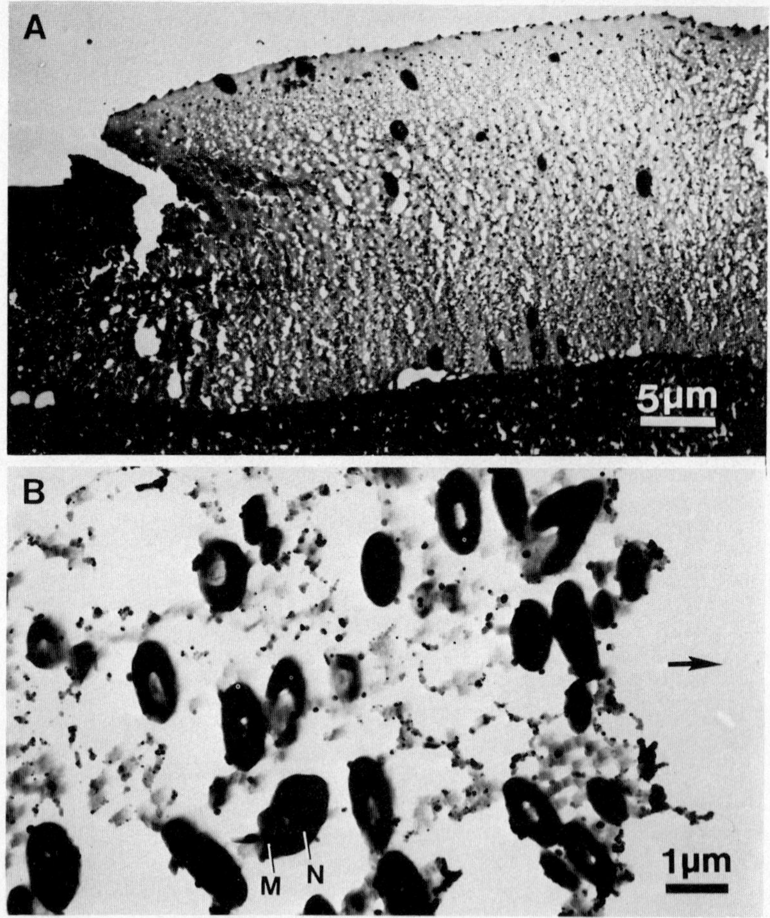

FIGURE 4. Cryosections of diluted, unreacted sperm (A) and undiluted ("dry") sperm (B). The dark region in the lower portion of (A) is the filter paper. M = mitochondrion; N = nucleus.

sections than in wholemounts, although profiles close to the surface could sometimes be found in which no ice crystal damage was visible. Knife compression of the cells is evident in Figure 4B, a cryosection of undiluted ("dry") sperm. The arrow in Figure 4B indicates the direction of motion of the knife relative to the block.

Freeze-Drying

Two methods of drying the frozen samples were compared. The simpler procedure was to place the cold brass box containing the samples in a desiccator over silica gel or molecular sieve. A rotary pump was used to evacuate the jar, and samples were permitted to warm up (2-4° /min) in the dry environment. The second procedure was somewhat more rigorous: the brass box was placed on a much larger precooled brass block in a carbon evaporator and dried at 10^{-5} to 10^{-6} torr. Warming occured at 0.2°C/min.

Figures 5 and 6 show a comparison of samples freeze-dried by the two techniques. In the case of whole sperm, the differences are dramatic. In addition, Figure 6 illustrates that in other types of preparations (i.e., tissue cryosections) the dessicator drying method did not preserve the tissue structure as well as the vacuum freeze-drying method.

Once the samples were freeze-dried, care was taken to ensure that rehydration did not occur. When the grids were still under vacuum in the freeze-dryer, they were brought to five or ten degrees above room temperature, and dry nitrogen was admitted to the evacuated area initially to ensure that no condensation of water occurred when they were finally brought out into room air. The samples were immediately transferred to a desiccator and stored under vacuum. Subsequent exposures to room air were minimized as much as possible.

In spite of these precautions, signs of rehydration were sometimes seen. One point at which rehydration may occur is when samples are withdrawn from the microscope column vacuum to room air, as shown by Figure 7. A section of the suspending medium (artificial sea water, described above) which was viewed and photographed for the first time (A) is compared to the same section after it had been removed from the microscope for approximately 30 seconds, then reloaded (B). Careful inspection reveals that salt crystals have

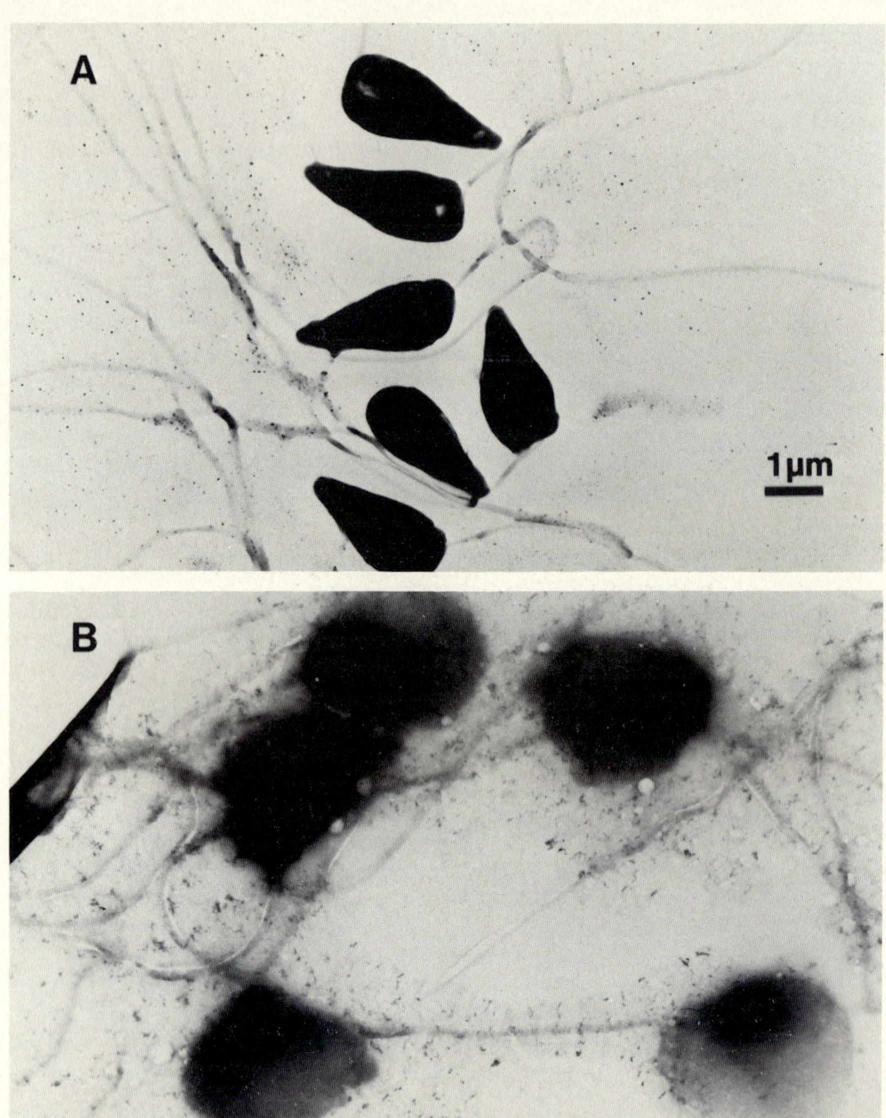

FIGURE 5. Wholemount preparations of sperm which were dried in a vacuum evaporator (A), or dried in a desiccator (B).

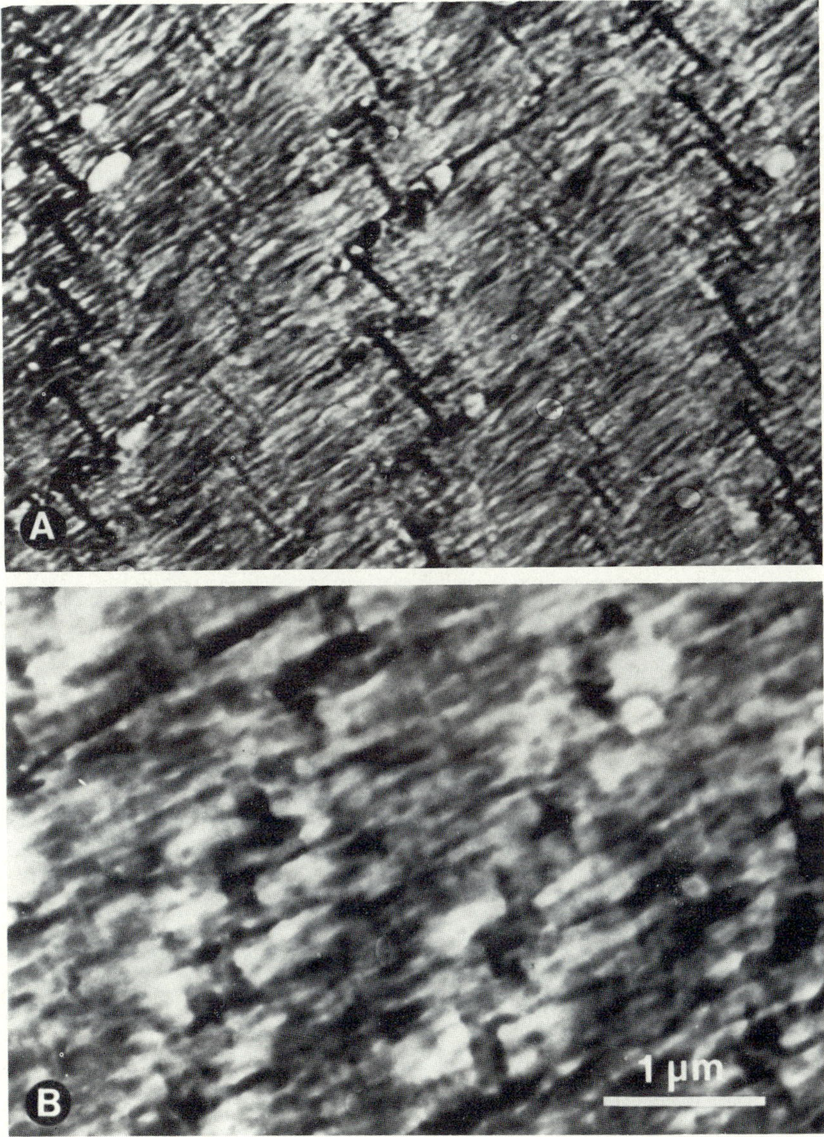

FIGURE 6. Cryosections of rat gastrocnemious muscle which were dried in a vacuum evaporator (A), or dried in a desiccator (B). (Micrographs kindly provided by E. Wang).

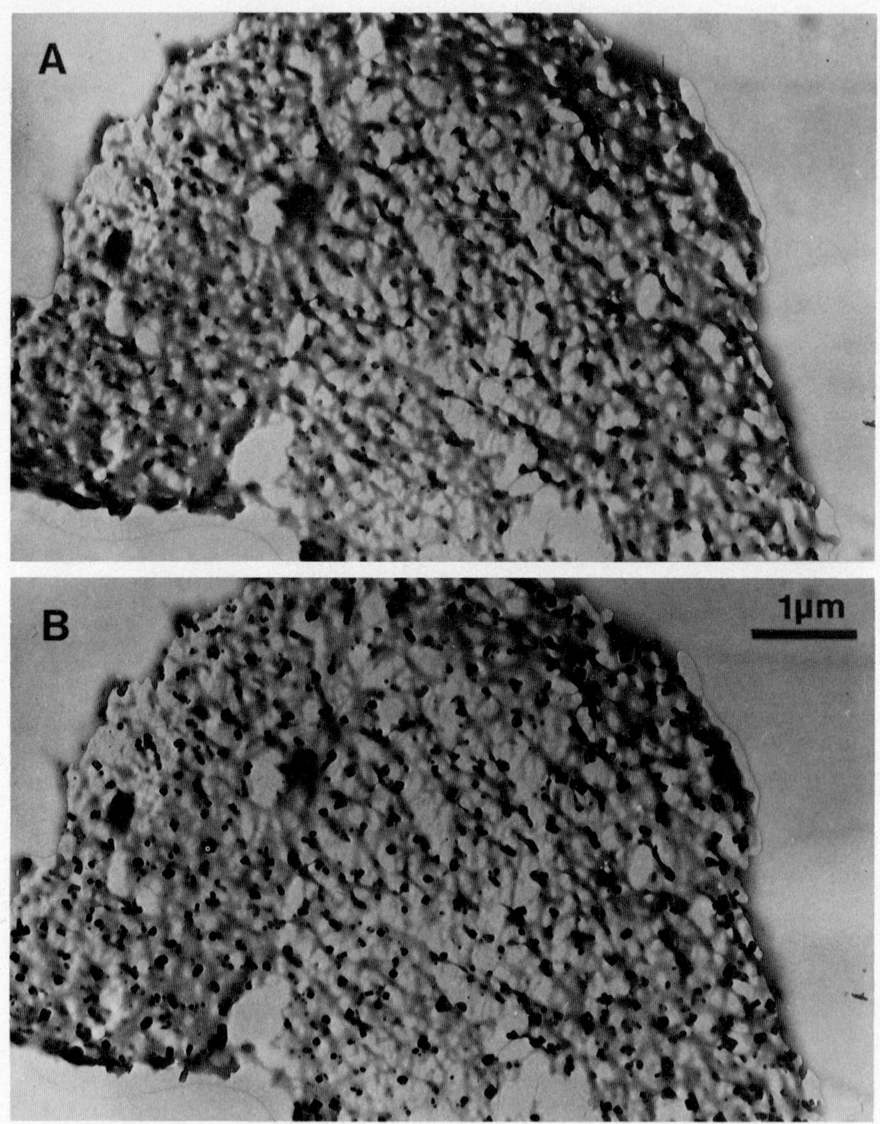

FIGURE 7. Freeze-dried cryosection of artifical sea water and 10% dextran before (A) and after (B) removal from the microscope vacuum.

coalesced and enlarged, indicating movement of ions. This suggests that even if exposure to room air is minimized, repeated transitions from the microscope vacuum to room air may result in significant changes in ion distributions.

Since the same quantitation and analysis routine (described below) was used for both types of preparation, an independent chemical technique, atomic absorption spectrometry, was used to measure total elemental concentrations of several elements. Cells were collected and suspended in sea water, as described above, then centrifuged through silicon fluid to remove the excess medium (2 minutes at 7700 g). The pellet was dried and organic matter was removed either by dry ashing at 500°C or by extraction in 0.2 N HCl solution followed by centrifugation (results obtained by the two methods were very similar and have been pooled). A correction for extracellular fluid remaining in the pellet was based on ^{14}C sucrose and tritiated water measurements on similar samples (Rottenberg, 1979).

ANALYSIS AND QUANTITATION

Spectra were gathered on a JEOL 100C transmission microscope with scanning attachment. The data presented here were collected in STEM or CTEM mode using a 0.25 μm^2 raster or spot.

It was not possible to collect the data in earlier experiments at cold temperatures to reduce mass loss; however, recent measurements using a cold temperature stage indicate that a 10 to 15% increase in the peak-to-continuum ratio occurs when samples are analyzed at room temperature as compared to analysis at -181° C. A correction has been applied to the data, based on these results.

Before analysis, the folding grids were opened to reduce the extraneous hard x-ray background and to remove one support film thickness. Corrections for the support film contribution to the mass thickness were made by analyzing an area of the film immediately adjacent to the cell. Peak integrals were calculated using a program provided by KEVEX which constructs gaussians to fit the data. An area of the modeled background from 1.0 to 3.9 KeV was used as a white region. The quantitation scheme used to convert peak-to-continuum ratios to absolute concentrations was based on that described by Shuman et al. (1976), where protein standards were used in combination with binary salt standards to measure the proportionality constants.

RESULTS

Table 1 shows the elemental composition of two subcellular regions in wholemounts. The designations "nucleus" and "mitochondrion" are used for convenience; however, in wholemounts, the actual volume excited by the beam includes two narrow bands of cytoplasm, as well.

TABLE 1. Elemental composition of sperm in wholemount preparations; means ± S.D. in mmols/kg dry wt.

Region (sample)	Total No. of cells	Na	Mg	P	S	Cl	K	Ca
Mitochondrion								
(Unreacted)[a]	78	620 ±160	150 ±20	600 ±100	180 ±40	910 ±200	400 ±70	24 ±4
(Reacted)[b]	69	660 ±170	160 ±30	670 ±60	170 ±40	890 ±170	290 ±50	110 ±70
Nucleus								
(Unreacted)[a]	81	330 ±90	60 ±10	1550 ±50	25 ±10	440 ±150	550 ±60	7 ±5
(reacted)[b]	71	520 ±170	68 ±7	1510 ±120	25 ±7	580 ±300	320 ±70	7 ±4

[a]Data were collected on sperm from 6 animals.

[b]Data were collected on sperm from 5 animals.

A paired-sample t-test was used to test differences between elemental concentrations in unreacted and reacted samples; $p \leq .05$ was considered to be significant. Potassium was found to be significantly lower in reacted cells, over both the nucleus and mitochondrion, and calcium increased over the mitochondrion. In three of the experiments, increases in nuclear sodium and mitochondrial phosphorus were significant.

Table 2 shows the composition of nuclei in unreacted sperm, as determined on cryosections and wholemounts (from Table 1). Although the dispersion in data taken on cryosections is greater, the mean values are consistent with those measured on wholemounts. The differences in sulfur and potassium may arise from differences in the sampling procedure; in wholemounts, the same region of the nucleus was analyzed in each cell, while in sections profiles were from random cross-sections through the nucleus and cytoplasm.

One of the original incentives for undertaking the preparation of cryosections was to evaluate the contribution of extracellular salts in wholemount analyses. Contrary to prior expectations, however, Na, Mg, and Cl concentrations were as high or higher in some cell profiles in the cryosections. Sampling distributions were distinctly skewed, however, with modes of Na, Mg, and Cl distributions at 200, 64, and 200 mmols/kg dry wt. respectively, all considerably lower than the means of those distributions (see Table 2). Elevation of Na, Mg, and Cl concentrations in some cells may be due to freeze-damage, which was quite variable from one cell to another. The small size of the cells also makes it likely that some longitudinal profiles analyzed in cryosections included extracellular medium in the projected area of illumination.

Results of atomic absorption spectrometry are shown in Table 3. Each entry is the mean of four values: two parallel determinations in each of two experiments. For comparison, the average concentrations of the four elements were computed from x-ray microanalytical results as an average of the nuclear and mitochondrial compartments (weighted according to their relative volumes).

Concentrations of Ca and K were within the uncertainties of the measurement (most of that uncertainty is associated with the calculation of average composition by X.E.S., since the composition of the tail and other compartments is not known). Both Na and Mg concentrations were lower in A.A.S.

TABLE 2. Elemental Composition of the Nuclear Region in Cryosection of Unreacted Sperm and Wholemounts of Unreacted Sperm. Table entries are means ± S.D. in mmols/kg dry wt. of data collected on cells from one animal.

Preparation	Total No. of cells	Na	Mg	P	S	Cl	K	Ca
Wholemounts	81	330 ±90	60 ±10	1550 ±50	25 ±10	440 ±150	560 ±60	7 ±5
Cryosections	22	390 ±270	80 ±60	1600 ±240	12 ±10	370 ±190	500 ±180	6 ±6

TABLE 3. Average Composition of Sperm as Determined by Atomic Absorption Spectrometry (A.A.S.) and X-ray Microanalysis (X.E.S.) of Wholemounts; Means ± S.D.

Technique	Na	Mg	K	Ca
	mmols/kg dry wt.			
A.A.S	310 ±20	66 ±7	440 ±30	17 ±5
X.E.S.	460 ±130	100 ±20	490 ±70	15 ±5

measurements, consistent with the supposition that X.E.S. values are systematically raised by contamination from extracellular medium.

CONCLUSIONS

Of the two types of samples used for x-ray microanalysis, wholemounts were found to be the easiest to prepare and analyze, and the least plagued by problems of freeze-damage. The data collected on cryosections and by atomic absorption spectrometry serve to confirm that the techniques used in preparing and analyzing wholemounts do not alter the concentrations of diffusable elements, aside from the systematic elevation of Na, Mg, and Cl (probably due to extracellular salt deposition). In the light of these results, data showing net elemental changes accompanying the acrosome reaction (e.g., Table 1) can be interpreted with increased confidence.

Preparing biological specimens for electron probe microanalysis of diffusable ions is by no means routine. However, each new study which is undertaken adds to a growing body of methodology and contributes significant biological results. It is hoped that this study will provide relevant and useful information to other investigators undertaking projects involving electron probe microanalysis of cells in suspension.

ACKNOWLEDGMENTS

The author gratefully acknowledges the advice of Dr. R. Schackmann in these experiments and his assistance with the extracellular space measurements. She would also like to thank Drs. D. Johnson, B. Shapiro, and T. Hutchinson for valuable suggestions and the use of their laboratory facilities. This work was supported by NSF Grant # PCM 7921657, NIH Grant # GM23910, and an Annual Fund Doctoral Dissertation Fellowship from the University of Washington.

REFERENCES

Bedford, J. M. (1970). Biol. of Reprod. (suppl.) 2:128
Collins, F., and Epel, D. (1977). Expl. Cell. Res. 106:211.
Decker, G. L., Joseph, D. B., and Lennerz, W. J. (1976). Dev. Biol. 53:115.
Monson, K. L. (1980). Doctoral Dissertation, University of Minnesota.
Rottenberg, H. (1979). Methods in Enzymol. 55:547.
Schackmann, R. W., Eddy, E. M., and Shapiro, B. M. (1978). Dev. Biol. 65:483.
Schackmann, R. W., and Shapiro, B. M. (1981). Dev. Biol. 81:145.
Singh, J. P., Babcock, D. F., and Lardy, H. A. (1978). Biochem. J. 172:549.
Shapiro, B. M., and Eddy, E. M. (1980). Int. Rev. of Cytol. 66:257.
Shuman, H., Somlyo, A. V., and Somlyo, A. P. (1976). Ultramicroscopy 1:317.
Tormey, J. M. (1978). In: "Scanning Electron Microscopy," Vol. 2, p. 259. SEM, Inc., Chicago.

DISCUSSION

SPEAKER: Marie Cantino

SOMLYO, Avril: Are the glutaraldehyde-fixed and the fresh cryo-fixed cells the same size?

CANTINO: Yes. The size of the cells is the same in both cases.

LECHENE: (Regarding rehydration.) There are a lot of phenomena with crystals. First, large crystals grow at the expense of smaller crystals because of radius of curvature. Vapor pressure of the crystal is a function of the radius of curvature. Second, we know with the drop technique that when you remove from your dryer, it is enough that you breathe a little on your sample in order to rehydrate it. You will have enough to wet and redry with formation of a few nucleation centers and formation of the large crystals where we had small crystals at the starting point. So, for example, what we do currently to keep our dry droplets with small crystals is to keep them in a desiccator where they are at slightly higher temperatures than room temperature, so that when you remove them there will not be a thermal gradient which could trap a little frost, and we avoid breathing on them.

CANTINO: When I take my samples out of the freeze-dryer, I usually take them up to higher than room temperature and do store them desiccated, but I do not keep the desiccator heated. I am sure that the types of media that I use are more hydroscopic than normal tissue because of the higher magnesium chloride concentration, but I do find that just the act of taking them out of the vacuum in the microscope into fairly humid room air will also rehydrate to some extent and this is what concerns me.

SOMLYO: Are they carbon-coated?

CANTINO: The support film is carbon-coated but not the sections or the whole cells on top of the film.

LECHENE: On your last slide (Table 1) weren't all the values increased by the same fraction? Oh, I see, the potassium is decreased.

CANTINO: Phosphorus is probably the best one to look at because it generally does not change and there is no phosphate in the medium. I use that as a standard for judging whether everything remains constant. As you can see, phosphorus remains constant here.

LECHENE: Did your atomic absorption calibration for potassium have the same composition as sea water?

CANTINO: The dilutions of the unknowns and standards for potassium were made using 1 mg/ml cesium chloride to reduce the problem of ionization interference with the sodium, but I didn't make the standards up with the same composition as sea water. I did run a known "artificial tissue" sample which I made up myself, having a composition similar to my unknown. The results were correct to within 5%, so I don't believe there was an interference problem.

MONSON: Do you think it might work if you would use one of the buffers that are purely hydrocarbons, having none of the salts that you are looking at. You can make that isotonic.

CANTINO: I have done experiments with choline chloride but I always hesitate to change the medium because it is so hard to interpret what is happening with the cells if you change anything in the medium. That is why, for example, I try to keep dextran as low as possible. If you took away all the normal ions, I do not know what the cells would do. For one thing you could not induce an acrosome reaction because the cells will not react without calcium, so that would certainly change things.

SOMLYO: You would probably lose cell potassium. When you try to wash cultured cells for the same purpose using the sucrose or ammonium acetate buffer you find out that you wash out not only what you think is extracellular sodium chloride--certainly your potassium values go down, too.

CANTINO: The problem I had with choline chloride was that the cells seemed to rehydrate immediately since the choline chloride is so hydroscopic.

MEASUREMENT OF MASS LOSS DURING MICROANALYSIS: METHODS AND PRELIMINARY RESULTS

James G. Hecker
Thomas E. Hutchinson

Center for Bioengineering
University of Washington
Seattle, Washington

INTRODUCTION

X-ray microanalysis is unique in its ability to combine the high spatial resolution of the electron microscope for subcellular localization with the identification and quantification of elements using x-rays generated within that fine localized region. To take full advantage of this powerful measurement capability, mass loss occurring during irradiation must first be separated from the often accompanying problems of contamination and etching, and must then be measured, described, and minimized.

An analysis procedure and test system for determination of mass loss was developed using a JEOL 100C scanning transmission electron microscope (STEM) with variable temperature cold stage. An x-ray energy-dispersive analysis system (EDS) detected elements within the specimen. A computer program to control acquisition and analysis of sequential x-ray spectra has been developed to allow any number of elements or regions of x-ray energy to be monitored for changes in elemental composition with time. Multi-element determinations of mass loss during microanalysis are now made routinely. Data on element-specific specimen changes can be obtained by utilizing count rate information which is always available during analysis.

ISBN 0-12-362880-6

The time required is nearly the same as that required for analysis without monitoring mass loss. Mass loss information is not only essential in determining specimen count rate or composition before mass loss has occurred, but can also be used to draw inferences about the dynamic interactions between specimen, electron beam, and vacuum system. Preliminary data from biological specimens such as albumin, muscle, and sea urchin sperm indicate mass loss is element and specimen dependent and appears to depend on vacuum conditions. Significant mass loss can occur under actual analysis conditions.

Although considerable progress has been made in understanding the reactions which occur in polymers or biological specimens during electron beam irradiation, questions about the severity, mechanisms, and dependence on operating conditions still remain. Contamination, etching, and elemental mass loss all appear to be potentially significant or even controlling for a range of operating conditions found in most commercially available electron microscopes, unless steps are taken to minimize or eliminate these effects. Following the nomenclature used by Hren (1979) in his excellent review of the problems of contamination and etching, contamination will be defined as the unintentional addition of mass to specimen surfaces during electron beam irradiation. Etching is defined (Glaeser, 1979) as the removal of specimen material (or contamination) through the synergistic action of certain residual gases (primarily O_2 and H_2O) and electron irradiation. Elemental mass loss is then an element-dependent loss which may occur in addition to either detectable contamination or the rapid loss of organic matrix attributable to etching. The combined effects of contamination, etching, and elemental mass loss cause a loss of spatial resolution, add extraneous peaks or Bremsstrahlung, change the minimum analysis volume, change the peak-to-background ratio, and alter local composition. Specimen preheating or vacuum desorption, cryoshielding, and ultra-high vacuum (UHV) pumping systems are usually effective in preventing contamination. However, contamination is highly dependent on the vacuum system, specimen, and irradiation, with the result that a change in one operating system parameter, such as beam diameter or current density, may often lead to a dramatic change in contamination rate.

Etching may be described (Hren, 1979) as a series of chemical reactions between specimen (or contaminant) and adsorbed molecules or molecular fragments, activated or ionized during irradiation, which lead to small reaction end products (mainly CH_4, CO_2, CO, H_2, and NH_3). Although etching reactions seem to proceed exceedingly slowly at liquid helium temperatures, organic specimens from room temperature to approximately -130°C are most susceptible to etching because of the increased condensation of reactant gas species. This is the reason etching can completely reverse severe contamination as specimen temperature is reduced (Egerton and Rossouw, 1976). The same practices most readily available for alleviating contamination are also useful in reducing etching.

According to Glaeser (1979) the inelastic scattering of incident electrons, leading to molecular ionizations, is the primary mode of radiation damage in organic materials. Chemical bond dissociation, or radiolysis, occurs following molecular ionization, and the highly reactive-free radicals and ions formed react further in a variety of secondary reactions which depend on the specific system studied. Many similar reactions and reactive gas species are presumably found in both etching and elemental mass loss. Etching and specimen heating can cause significant radiation damage, but these effects occur long after severe ionization damage has already occurred (Glaeser, 1979). They are, however, well within the range of exposures found during most microanalysis. Glaeser reports that a wide variety of biological materials exhibit an exponential decay with cumulative electron exposure to a residual mass, with a decay constant of 1-10 electrons/$Å^2$. Particularly with hydrated material, but even with radiation resistant biological aromatic materials, radiation damage is usually severe long before the exposures used in routine analysis are reached. The present study was undertaken to develop methods to measure and account for mass loss which still occurs despite efforts to eliminate it, and which may still occur long after radiation damage to the specimen is severe.

METHODS

An interactive program for acquiring, analyzing, and plotting mass loss data has been developed. The mass loss information available during analysis and the use and flexibility of the program are illustrated by preliminary data for three systems: sea urchin sperm, rat soleus muscle, and polyvinyl chloride. Mass loss measurements are made on a JEOL 100C electron microscope with scanning attachment (STEM) and a Kevex 7000 energy-dispersive Si(Li) detector and analyzer (EDS). A FORTRAN program controls acquisition of spectra and analysis using FORTRAN callable Kevex library subroutines. This program is run under the RT-11 operating system on a dedicated PDP 11/03, rather than under software from Kevex. Data storage is on single density floppy disks.

Specimens used for preliminary mass loss determinations were chosen both for convenience and as representative biological samples. Sea urchin sperm were rapidly frozen in Freon slush and freeze-dried as wholemounts or cryosectioned before freeze-drying (Cantino, 1981). Cryosectioned and freeze-dried rat soleus muscle was supplied by Eunice Wang (University of Washington). PVC films were cast onto glass from a 1 wt% solution of PVC in tetrahydrofuran and were 100-200 nm in thickness. PVC films were chosen to duplicate the results of mass loss work previously done examining the effect of operating conditions on chlorine loss (Delgado, 1976; Delgado and Hutchinson, 1978; Griffin, 1978). Every effort was made to minimize exposure before beginning data collection. Copper grids covered with PVC film could be shifted to a new unexposed region and data collection started simultaneously, while the grids supporting sea urchin sperm and muscle were examined briefly using low beam current and low magnification to locate the specimens. A Keithley electrometer was used to measure the beam current above the specimen. Conditions for these specimens were chosen to duplicate operating conditions during routine microanalysis. The conditions used are given below.

Values to be used in calculations, such as beam current, magnification, area of analysis, and specimen tilt, are entered interactively into the computer. Windows for any number of peaks or regions are also marked before analysis begins. Currently a maximum of 16 windows is allowed, with atomic numbers from sodium (Z=11) to zinc (Z=30). This includes the elements of primary biological interest in this laboratory. There is no programming limitation, however, on the number of elements which may be examined, only the usual detection limitations. This allows the number and size of the windows to be set at the time of analysis. Normally 10 or 11 windows are set, all below 4 keV except for copper K and a Bremsstrahlung region set between 4-6 keV. Full width at half maximum, or approximately 130 to 150 eV, is usually chosen for the variable window width. This window width is occasionally set smaller and shifted from the peak centroid to avoid overlapping peaks. This is only a substitute for deconvolution, but the time now required for deconvolution makes it inappropriate for mass loss analysis.

The live time interval is also entered during this interactive phase, and acquisition begins. Live time is the time x-rays are actually collected and counted, while dead time is the time the pulse processor is handling detector pulses and is unavailable to process additional x-rays. The sum of live and dead time is real time. After every live time interval the counts per channel within each of the selected regions are recorded. Real time for each spectrum is also recorded for later calculation of actual specimen exposure. The channel-by-channel difference in count rate during each live time interval is obtained by subtracting from each region or window in each spectrum the preceding spectrum and corresponding window. Each region in each spectrum is summed, resulting in a data array containing the number of counts during each live time interval for each region or window in each spectrum.

Time of data acquisition is virtually the same using this mass loss program as during acquisition otherwise. The only additional time required is that needed to enter values interactively before collecting a spectrum. The spectrum at the end of acquisition can be processed and analyzed normally. Mass loss information obtained is always present while acquiring a spectrum. This program simply takes advantage of that information, while adding very little in additional time.

The data are fitted to a model, and results of the fit can be printed out immediately after analysis. A rough plot of mass loss for each element can also be printed out on a Teletype for an immediate confirmation of the severity of mass loss. To use the plotting facilities of the academic computer center, data stored on floppy disks are transferred via modem and further processed. The same calculations of exposure, flux, and fit to a model are again performed. The program written for the main computer (Cyber 170/750) allows flexibility in correcting values used in calculation which may not have been known accurately before analysis, such as magnification, used to calculate irradiation area, or beam current. These are usually measured after analysis, using a diffraction grating replica and Faraday cage, respectively. Plotting parameters and plotting titles for display of the results in a variety of formats are also entered interactively. Three- or five-point smoothing can be selected, and any number of curves can be displayed on the same plot. The program is designed to be flexible and self-explanatory.

Background counts are subtracted from each peak or region by using a collected Bremsstrahlung region. The ratio of the integrated Bremsstrahlung curve under the window selected for a peak to the collected Bremsstrahlung region, currently 4.00 to 6.00 keV, can be determined by modeling the total Bremsstrahlung curve for a series of spectra from a particular specimen type. The Bremsstrahlung in each region or window was found to be a relatively constant fraction of the collected Bremsstrahlung region. For a series of spectra from similar specimens, the ratio of Bremsstrahlung under a peak to a collected Bremsstrahlung region, or to the total Bremsstrahlung curve presumably, is constant. Within a series of spectra from similar specimens, a 5% variation in this ratio was found. To check the sensitivity of the mass loss results to the value determined for this Bremsstrahlung correction ratio, this calculated ratio was varied by $\pm$ 10% to 15% for three sets of data. The resultant mass loss curves are insensitive to changes of this magnitude and nearly indistinguishable from unchanged curves. The use of an average value for a series of like specimens is acceptable, and separate determinations of this ratio for each peak and each spectrum do not appear necessary. One important assumption implicit in this type of Bremsstrahlung subtraction is that of a relatively constant Bremsstrahlung curve with time. The general shape of the smoothly varying Bremsstrahlung does not appear to change significantly for spectra collected for different acquisition times or exposures, and as the mass loss curves are relatively insensitive to variations in the ratio of any one region to the Bremsstrahlung region, this assumption is probably acceptable.

RESULTS

Preliminary data collected are of two types: low exposure polyvinyl chloride (PVC) data, and high exposure data for sea urchin sperm or rat muscle, corresponding to actual analysis conditions. Exposure, E (Coulombs/cm^2), is defined as

$$E = f \cdot t \text{ (amps sec/cm}^2\text{)} \quad (1)$$

where

f = flux electrons to surface = I_b/A (amps/cm^2)

I_b = beam current (amps) measured with Faraday cage
A = area of irradiation (cm^2)
t = actual exposure time

The actual area of irradiation is found from the following and is dependent on the square of magnification.

$$A = SA/(\cos \Theta M^2) \qquad (2)$$

where

SA = area of scanning image, or beam spot area on the viewing screen in (cm^2)

Θ = specimen tilt angle
M = measured magnification

Cumulative exposures are reported up to 10^3 or 10^4 (Coulombs/cm^2). Although this is a high exposure, it is equivalent to the following conditions:

Θ = 30° to 45° = .87 to .71, $\cos\Theta \sim 1$
10 nm spot diameter
$I_b = 1 \times 10^{-9}$ amps (1 na)
t = 0.5 seconds live time, ~1 second real time

or

$I_b = 1 \times 10^{-10}$ amps (0.1 na)
t = 5 seconds live time, ~10 seconds real time

or

$I_b = 1 \times 10^{-9}$ amps (1 na)
A = 0.1μm x 0.1 m scan
t = 500 sec live time, ~1000 sec real time

Results reported in recent literature often indicate cumulative exposures of 1 to 1000 Coulombs/cm^2, although for many specimens the onset of radiation damage occurs after an exposure of only 10^{-2} Coulombs/cm^2. Table 1 shows a range of normal operating conditions and calculated exposures. The conditions chosen for these preliminary data are within the range of conditions commonly used in microanalysis, although radiation damage, at least as determined by ultrastructure or crystallinity, may already be severe at an exposure several orders of magnitude less than the exposure reached during microanalysis.

TABLE 1. Examples of cumulative exposure during microanalysis. (NR = not reported; calculated values are in parentheses)

	Chandler 1976	Rick 1978	Somlyo 1974	Somlyo 1977	Shuman 1976	Monson 1980	Wendt-Gallitelli 1979-1980
Beam spot diameter	10 μm	NR	200-300 nm	50-14000 nm	60 nm resolution	20 nm	20 nm
or scan area	($8 \times 10^{-7} cm^2$)	($2 \times 10^{-8} cm^2$)	($5 \times 10^{-10} cm^2$)	NR	($3 \times 10^{-11} cm^2$)	($3 \times 10^{-12} cm^2$)	($3 \times 10^{-12} cm^2$)
I_B [Amps]	0.1 μA (100 nA)	0.5 nA	20 nA	0.4-3.5nA	NR	1 nA	4 μA
Live time [secs]	1000	200-400	300	100	NR	200	100
Real time [estim]	(2000)	(400-800)	(600)	(200)	NR	(400)	(200)
Exposure [$coul/cm^2$]	(250)	(15)	(2.4×10^4)	500	10^3	(1×10^5)	(2×10^8)
Exposure [$elec/Å^2$]	(2×10^5)	(9×10^3)	(1.5×10^7)	(3×10^5)	(6×10^5)	(6×10^7)	(4×10^{10})
Accel volt [keV]	80	15	40	NR	NR	80	NR
Comments	warm	warm 1 min prerad.	warm	-100°C 6 spec sum.	-100°C	-170°C	warm

Figure 1 is an example of chlorine loss from PVC film, which has been used previously as a model system for mass loss (Delgado, 1976, 1978; Griffin, 1978). The solid line represents the model used--a linear least squares fit to an exponential decay with offset. The fit to PVC data is quite good, although in some cases the initial decay suggests a sum of exponentials might give a better fit. This may be the result of the mode of analysis used. The horizontal ramp input used for scanning (STEM) analysis has a dead time of 10% to 50% of the horizontal sweep, during which time the beam is located at the left edge of the scan. The model used in fitting is contained within a subroutine and can be easily modified as the data warrant. Work by previous investigators (Delgado, 1976; Griffin, 1978; Delgado, 1978) has shown this to be radiation-limited damage, rather than a chain reaction. Figure 2 is the corresponding Bremsstrahlung curve for the PVC sample of Figure 1. Figures 1 and 2 are intended to illustrate the use of this mass loss determination procedure for the case of rapid loss of highly volatile components such as chlorine from a radiation-sensitive material such as PVC. The notation on the ordinate refers to the live time interval chosen for recording of successive

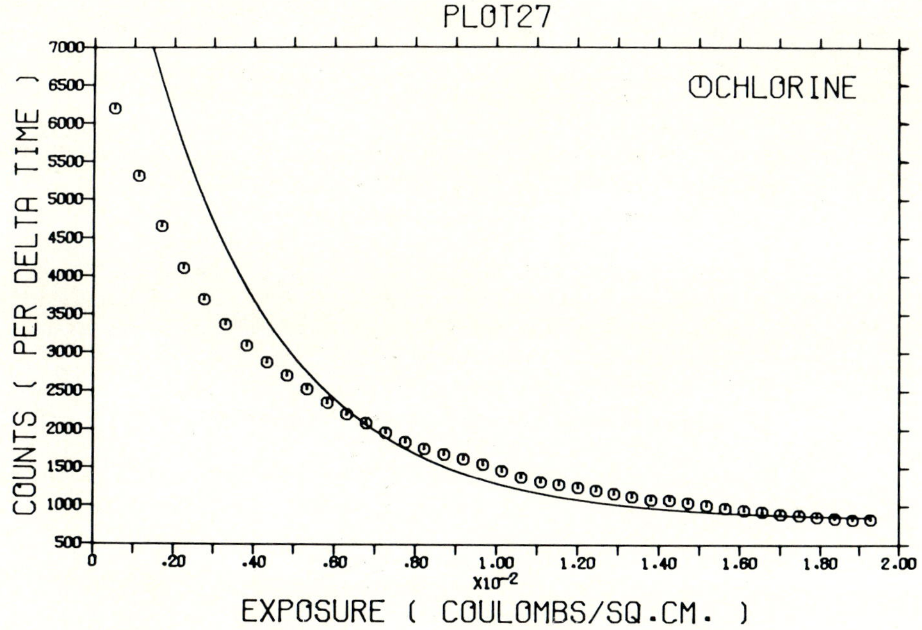

FIGURE 1. Chlorine loss from PVC.

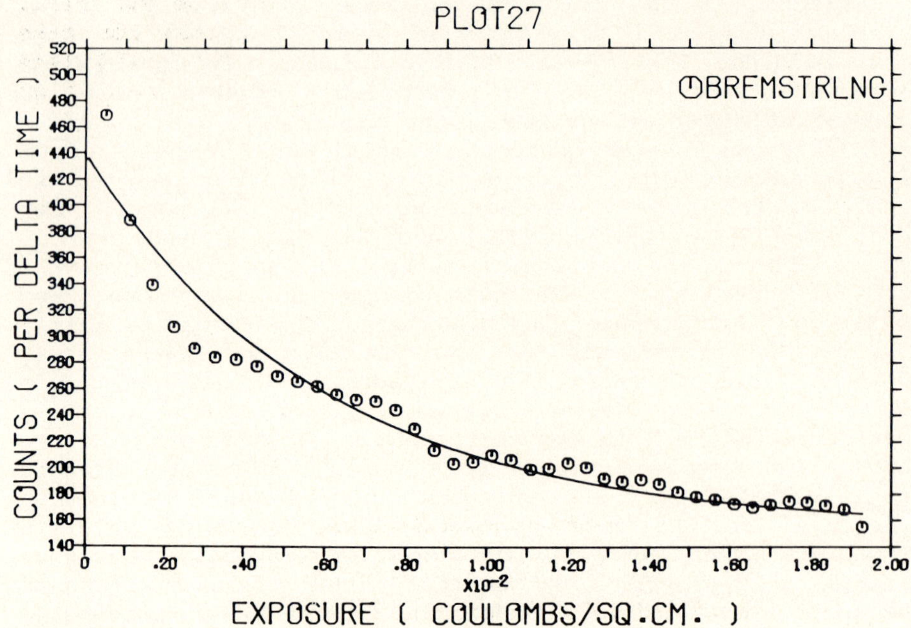

FIGURE 2. Bremsstrahlung loss from PVC.

spectra. The exposures plotted on the abscissa are calculated from the real time recorded at each interval.

Figure 3 indicates loss occurring over a long time period and high exposure from sea urchin sperm. Although the initial flat region of the phosphorus curve is not representative of all mass loss curves, the elemental mass loss of chlorine, potassium, and phosphorus is typical. Data are from a sea urchin sperm wholemount, uncoated and freeze-dried. Analysis was in STEM mode, 300 seconds live time, specimen tilt 45°, beam current 2.5×10^{-9} amps, 5 second live time interval, at a measured magnification of 72000X. The scan area measured on the viewing screen was 0.56 cm^2, and some specimen drift was noted during collection of data. Cumulative exposure calculated was 8.5×10^3 Coulombs/cm^2. A smooth steady Bremsstrahlung increase was noted during data collection--also a common occurrence--indicating either contamination or an increase in mass thickness due to drift. However, the increase in Bremsstrahlung is evidence for drift into the specimen and not drift off the specimen. The mass loss curves, therefore, are not artifacts of the drift seen occasionally during data

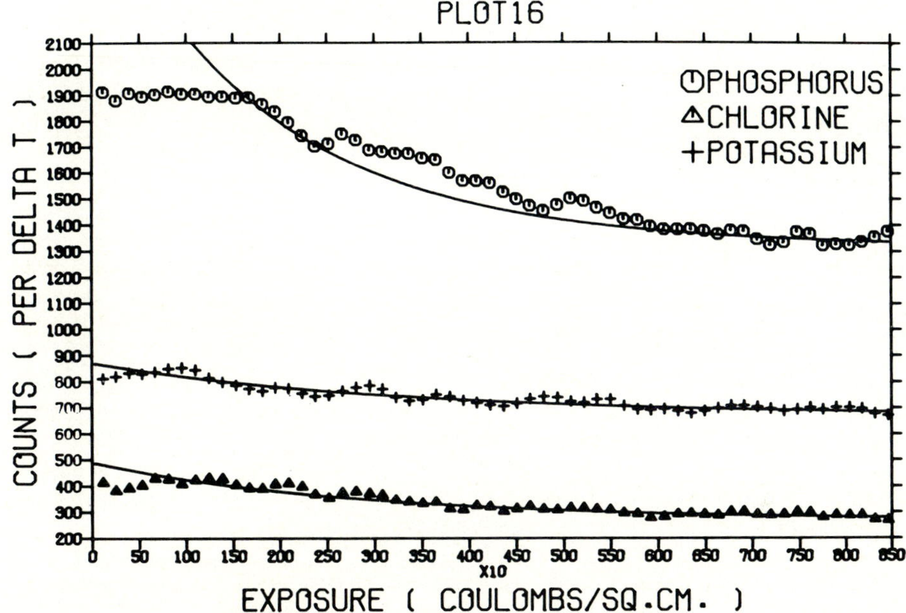

FIGURE 3. Elemental mass loss for muscle.

collection. Figure 3 is representative of elemental loss seen from the two types of biological specimens examined--muscle and sea urchin sperm--although this is one of the highest exposures used. The display of plotting results and option of plotting any number of curves on the same plot are also illustrated by Figure 3. Figure 3 is an example of qualitative mass loss behavior. Considerable variation in the extent of Bremsstrahlung and elemental mass loss was found.

In all, 22 experiments are reported in these preliminary results: 4 from PVC, 12 from sea urchin sperm and 6 from rat soleus muscle. (A series of spectra of calibration standards from sectioned Spurr resins have been collected at room temperature and at -176° C and processed, but analysis and plotting have not been completed.) For the biological specimens, windows were set corresponding to Na, Mg, Al, Si, P, S, Cl, K, Ca, Bremsstrahlung, and Cu. For PVC, only Cl, Bremsstrahlung, and Cu were set. Considerable mass loss, or apparent gain in some instances (to be discussed below), was noted consistently for P, S, Cl, K, Bremsstrahlung, and Cu. Significant changes were often seen for Na, Si, and Ca, but the lower count rate, high degree of scatter, and interfering overlap of the copper L and potassium K peaks makes interpretation of these data more difficult. In the case of PVC, chlorine makes up a substantial portion of total mass, and the rapid mass loss of chlorine results in a simultaneous loss of a large portion of specimen mass. Although the initial portion of the chlorine loss curve was fit poorly by the model chosen--a single exponential decay with offset--the rest of the Cl curve and the Bremsstrahlung and Cu curves were fit quite well by this model, were reproducible, and showed little scatter. The biological specimens are primarily a matrix of C, H, N, and O, and are not as consistent.

The collected Bremsstrahlung region (4.0-6.0 keV for these data) represents irradiated mass thickness of specimen, support film, coating if present, and any extraneous sources due to scattering or secondary excitations. For these uncoated specimens, assuming negligible support film and extraneous sources, the Bremsstrahlung is a measure of specimen mass thickness. For very thin specimens, the support film may also make a sizeable contribution. Because the physiologic ions of interest are present in low concentrations, the Bremsstrahlung is indicative, at least initially, of the scattering of the organic matrix. Changes in the Bremsstrahlung count rate reflect changes in mass thickness due to contamination, rapid organic loss due to etching, and perhaps elemental mass loss also. Bremsstrahlung count rate exhibited both increases, up to +60% (+275% in one case), and decreases, to -30%, of the initial count rate. The range for most determinations was -30% to +35%, and most of these were -10% to +25%. Although the system is a JEOL 100C with appropriate diffusion pump traps and anticontaminator, the behavior of the Bremsstrahlung results under routine analysis conditions shows dramatic changes, seeming to reflect etching and contamination interaction. A smooth linear increase in Bremsstrahlung was seen on 7 out of 22 occasions. Smooth exponential increases and decreases were also seen on some runs, as well as sharp changes in the slope of Bremsstrahlung count rate versus time during a very few runs. Scatter for the Bremsstrahlung and other regions with a high count rate was light.

Table 2 summarizes the preliminary data. No Al was present, except in the few analyses with the cold stage. Mg was also not present in consistently measurable quantities. Scatter in the mass loss curves for Na and Ca and in some of

TABLE 2. Percent change from initial count rate, after exposure E.

Run #	Na	Si	P	S	Cl	K	Ca	Brem	Cu	E $\frac{coul}{cm^2}$	Comments
PVC											
10					-80			-59	-31	2×10^{-1}	
11					-83			-65	-34	4×10^{-1}	
27					-85			-67	-52	2×10^{-2}	
25					-67			-42	-35	9×10^{-1}	
Mass Gain											
4	0	-10	0	+18	+33	+ 3	+33	+30	+43	1×10^{5}	low counts
8	+ 6	- 4	-21	- 3	+38	+22	+ 7	-10	- 2	4×10^{4}	high Cu, drift
15	+68	+67	+34	0	+10	+33	+50	+31	+32	1×10^{3}	
20	+18	+220	+42	+28	+36	+15	+59	+23	+ 6	1×10^{3}	high Cu
21	0	+16	-12	+13	0	+ 2	+16	+14	+ 8	2×10^{2}	
22	- 6	+ 7	+ 7	+ 6	-10	-10	+13	+12	+10	2×10^{2}	
24	-60	+100	- 8	+14	+ 2	+ 6	+ 4	+19	+15	6×10^{1}	
Mass Loss											
2	-41	+41	-35	-32	-45	-28	+14	-35	-29	7×10^{4}	
3	+ 5	- 3	- 7	-10	-32	- 9	- 3	0	+ 3	1×10^{5}	
5	-12	+ 4	- 7	- 1	-10	-10	-24	- 7	- 3	2×10^{3}	
6	- 3	- 8	- 8	-11	-14	- 7	-11	- 8	- 6	9×10^{3}	
7	- 4	+ 2	-10	0	-12	-10	- 5	+18	+12	1×10^{5}	
14	0	-38	-12	-63	-11	-10	+36	+20	+13	3×10^{3}	
16	-37	-67	-32	-80	-42	-21	-10	+24	+20	5×10^{3}	
17	-14	-25	- 5	-51	-11	- 6	-11	+19	+20	4×10^{3}	
18	-17	-10	-13	-100	-18	-17	-26	+81	+80	6×10^{3}	
19	-25	+107	-11	- 2	0	+ 4	0	+15	+ 1	1×10^{3}	
23	-	-	-64	-51	-38	- 4	-16	+275	+263	3×10^{4}	

the Si and S curves makes the values listed for these subject to some question. In addition, counts for run number 4 were unusually low; run number 20 had unusually high Cu and Si count rates; and run number 8 had both unusually high Cu and substantial specimen drift during analysis. These three runs are all in the subgroup showing elemental mass gain. Of the elemental curves with moderate to light scatter, P, S, Cl, and K usually showed substantial mass loss. Just as Bremsstrahlung usually increased, P, S, Cl, and K usually decreased with exposure. The loss was neither consistent nor the same for all elements. There were also several cases where Bremsstrahlung decreased and one case showing no change. Assuming for the moment that runs 4, 8, and 20 are too suspect to be considered further, still in several other cases elemental mass gain was also shown. Possible explanations for this apparent behavior will be discussed below. Silicon may be present as a contaminant from a silicon-based diffusion pump oil used previously in the instrument, from the Si(Li) detector, or possibly from glass knives used in cryosectioning the muscle sections (numbers 19-24).

SUMMARY

Although, in general, elemental mass loss for P, S, Cl, and K was observed and Bremsstrahlung mass gain was seen, too few analyses have been completed for this preliminary discussion of contamination, etching, and elemental mass loss to reach a conclusive description of specimen reactions during microanalysis. Clearly, a variety of combinations of elemental mass loss and apparent gain and total mass thickness gain and loss is possible. The varied results possible with the high quality commercially available microscope used under routine analysis conditions suggest the problems of contamination, etching, and elemental mass loss are not completely understood. This is one of the first attempts to determine simultaneously the rates of mass loss for all elements of interest in biological or polymeric specimens. In addition, it is possible to use information from the Bremsstrahlung radiation to monitor during microanalysis the severity of etching and contamination. Results presented above tend to support Glaeser's statements (1979). The dynamic interaction of contamination and etching is complex and subtle enough that a simple change in operating conditions, beam spot diameter, or magnification, for example, can cause a dramatic change in the severity of either. It may be possible, and seems necessary, to separate elemental mass loss from these contributing and simultaneous difficulties, as microanalysis results represent the cumulative record of specimen reactions and changes during irradiation. On the basis of these preliminary results, we think simultaneous determination of rates of contamination, etching, and multi-element mass loss is needed during microanalysis.

The loss or gain of mass thickness, as represented by the Bremsstrahlung radiation, may be interpreted as the combination of contamination and etching discussed earlier. It appears elemental mass loss can be both substantial and element dependent. The secondary reactions occurring after primary excitation or ionization of vapor phase molecules or of molecules within the specimen are probably complex and varied, leading through many different reaction sequences depending on specimen, conditions, and activated primary species. For loss to occur, however, volatile products must be formed. If significant loss is occurring at the specimen surface, due to organic matrix reactions, then the local specimen concentrations of the nonvolatile remaining elements may be increased. Qualitatively, then, preliminary data can be described by a model of contamination, etching, and elemental mass loss. When contamination is more severe than etching, increases in Bremsstrahlung and decreases in elemental mass thickness are seen. If etching is more severe than contamination, then perhaps the preferential rapid loss of organic matrix via etching reactions leads to an increase in average atomic number and a resulting increase in the Bremsstrahlung. This can occur because fewer reactions leading to volatile products are available to the higher atomic number physiologic ions. A local increase in elemental composition of these physiologic ions occurs, thereby providing a possible explanation for the apparent mass gain seen occasionally.

The methods and results presented both indicate the type of count rate information present during microanalysis and illustrate the use of that information in monitoring specimen-beam interactions which can interfere with quantitative analysis. The necessity of determining these interactions is emphasized by the variety of elemental and Bremsstrahlung count rates shown during routine biological microanalysis. Preliminary results show the methods used for determination of mass loss are capable of following changes seen during rapid chlorine loss from PVC and from biological specimens.

REFERENCES

Cameron, I. L., Smith, N. R., and Pool, T. B. (1978). In "Ninth International Congress on EM," Vol. 2, p. 120. Toronto.

Cantino, M. E. (1981). In this volume.

Delgado, L. A., and Hutchinson, T. E. (1979). Ultramicroscopy 4:163-8.

Delgado, L. A. (1976). Doctoral Dissertation, University of Minnesota.

Egerton, R. F., and Rossouw, C. J. (1976). J. Phys. D: Appl. Phys. 9.

Glaeser, R. M. (1979). In "Introduction to Analytical Electron Microscopy" (J. J. Hren, J. I. Goldstein, and D. C. Joy, eds.). Plenum Press, New York.

Hren, J. J. (1979). In "Introduction to Analytical Electron Microscopy" (J. J. Hren, J. I. Goldstein, and D. C. Joy, eds.), Chapter 18. Plenum Press, New York.

Monson, K. L. (1980). Doctoral Dissertation, University of Minnesota.

Rick, R., Dorge, A., MacKnight, A. D. C., Leaf, A., and Thurau, K. (1978). J. Membrane Biol. 39:257-71.

Shuman, H., Somlyo, A. P., Somlyo, A. V. (1976). Ultramicroscopy 1:317-39.

Somlyo, A. P., Somlyo, A. V., Devine, C. E., Peters, P. D., and Hall, T. A. (1974). J. Cell Biol. 61:723-42.

Somlyo, A. V., Shuman, H., and Somlyo, A. P. (1977). J. Cell Biol. 74:828-57.

Wendt-Gallitelli, M. F., Wolburg, H., Schlote, W., Schwegler, M., Holubarsche, C., and Jacob, R. (1980). Basic Research Cardiology 75:66-72.

Wendt-Gallitelli, M. F., Wolburg, H., Schwegler, M., and Schlote, W. (1979). Experientia 35:1591-93.

DISCUSSION

SPEAKER: James G. Hecker.

HUTCHINSON: The first sample is after 10×10^2 coulombs per square centimeter.

HECKER: Yes.

RICK: That is already a fairly high dose (exposure) so that much of the mass loss would not be seen in your measurements because, at least to our experience, the mass loss of white radiation and sulfur is already ended when you have exposed specimens with such a dose.

HECKER: That is what I found for that run for the white count region. I found the continuum radiation to be decreasing.

RICK: Yes. Well, then this is a secondary mass loss which has nothing to do with this past mass loss that occurs before you have the first measurement.

ANDREWS: You want us to move the decimal point two places to the left?

HECKER: That should be multiplied by 100 so that the middle of the scale is 40,000 (this is one of the highest exposures used for mass loss determination).

FERRIER: You are talking in coulombs per square meter in biological material; you might be talking about, at most, one coulomb per square meter to really get significant data. There is something wrong with the scale. I think you surely must be wrong.

HECKER: (The scale is correct. See Table 1 in text.)

RICK: How long do you irradiate to get to that dose?

HECKER: This is 300 seconds live time which corresponds to something like 500 seconds.

HALL: What kind of vacuum do you have?

HUTCHINSON: Usually 2×10^{-7} torr.

HECKER: Contamination can be a problem and it seems to vary during the day. There seem to be several competing effects, such as etching, contamination, and mass loss, and there is also a time dependence.

HUTCHINSON: We can operate comfortably, Ted, with very little contamination, and at other times it can be very substantial.

HALL: Have you checked the composition of any contamination?

HUTCHINSON: No. Well, let us say it does not contain a tremendous amount of silicon, if that is the point that you are questioning.

HALL: Oh. Sometimes we get chlorine. It depends on the type of pump oil you use.

HUTCHINSON: No. It looks like hydrocarbons.

Part II

ELECTRON PROBE ANALYSIS OF CARDIAC, SKELETAL, AND VASCULAR SMOOTH MUSCLE

Avril V. Somlyo
Henry Shuman
Andrew P. Somlyo

Pennsylvania Muscle Institute
University of Pennsylvania
Philadelphia, Pennsylvania

INTRODUCTION

Advances in rapid freezing techniques, cryoultramicrotomy, and electron probe x-ray microanalysis (Hall, 1971; Hall, 1979; Hutchinson, 1979; Lechene and Warner, 1979) have made it possible to quantitate the in situ elemental composition of cell organelles in a variety of systems, including muscle (Gonzalez-Serratos et al., 1978; Misra et al., 1980; Somlyo, Shuman, and Somlyo, 1977a,b; Wroblewski, Roomans, Jannson, and Edstrom, 1978). However, while the importance of Ca in regulating contraction and the primary role of the sarcoplasmic reticulum (SR) as a source and sink of Ca in fast skeletal muscle are now generally accepted (Ebashi and Endo, 1968; Endo, 1977), the relative contributions of, respectively, the SR and the mitochondria to regulating cytoplasmic Ca in smooth and in cardiac muscle have not been established with comparable certainty. Similarly, while a wealth of studies have dealt with the mechanism of Ca uptake by the SR of striated muscle, information bearing on the mechanism of, and ionic changes associated with, calcium release is lacking. The above areas of uncertainty are largely due to technical limitations on the information obtainable from studies of fractionated organelles. Electron probe analysis of ultrathin cryosections is a uniquely suitable approach, obtaining the required information about the composition of in situ

ISBN 0-12-362880-6

organelles at rest and during activation of muscle. In the following, we shall summarize some of our studies bearing on the apparent non-participation of mitochondria in the physiological regulation of cytoplasmic Ca++ in cardiac and in vascular smooth muscle, and on changes in the composition of the TC of the sarcoplasmic reticulum that occur during a tetanus. Some of these findings have been published in detail (Somlyo, Shuman, and Somlyo, 1977; Somlyo, Somlyo, and Shuman, 1979) and others have been presented at Symposia (Somlyo, Somlyo, Gonzalez-Serratos, Shuman, and McClellan, In Press).

METHODS

The experimental animals (frog and rabbit) and the preparation of the muscles have been described in our detailed publications (Somlyo, Shuman, and Somlyo, 1977; Somlyo, Somlyo, and Shuman, 1979). We wish to reemphasize the importance of careful dissection in preparing tissues for electron probe analysis, lest the use of cut or otherwise damaged surface fibers result in erroneously high mitochondrial calcium content (granules) and abnormally low cytoplasmic K and high NaCl. All tissues were frozen in supercooled Freon 22 (Somlyo, Shuman, and Somlyo, 1977a); the rates of freezing obtained by this method have also been published (Costello and Corless, 1978).

Smooth muscle cells were made hyperpermeable by incubation in 50 μg/ml saponin for 30 minutes followed by exposure to varying concentrations of free Ca using a Ca-EGTA buffering system (Endo et al., 1977). Contractile responses to Ca were measured prior to freezing. The method of obtaining hyperpermeable cardiac myocytes has been published previously (Reiser et al., 1979; Chiesi et al., submitted).

Tension was monitored from bundles of approximately 20 frog semitendinosus fibers while stimulating at 40 shocks per sec. At 1.2 sec from the onset of stimulation a beaker of supercooled Freon 22 was shot up to the muscle. This technique allows the flash-freezing of activated fibers while recording tension.

Sections ≃ 1000-2000Å thick were cut on an LKB cryo-ultramicrotome (LKB Productor, Bromma, Sweden) modified to maintain an ambient temperature in the cryochamber of -130 °C at a specimen temperature of -110 °C and knife temperature of -100 °C. Methods for removal of the sections from the knife and freeze-drying at or below 5 x 10^{-6} Torr have been previously published (Somlyo, Shuman, and Somlyo, 1977; Somlyo and Silcox, 1977). All of the tabulated electron-probe data on frozen-dried sections were obtained on unstained specimens.

Some frozen samples were processed by freeze-substitution (Feder and Sidman, 1958; Van Harreveld and Crowell, 1964; Heuser and Reese, 1976; and Franzini-Armstrong et al., 1978). The water in the frozen material was substituted with acetone at -80 °C for 3 days and then gradually warmed up to room temperature at which time osmium tetroxide crystals were added to the acetone to a concentration of approximately 5%; the cells were subsequently embedded in Spurr's resin. Sections for electron probe analysis were cut on anhydrous glycerol (Ornberg and Reese, 1978) or, for structural studies, on water. While monovalent ions are lost during freeze-substitution, large deposits of divalent cations such as mitochondrial granules, platelet-dense bodies, and calcium oxalate deposits are maintained. Their localization and concentrations are similar to those found in cryosections of the same samples. Freeze-substitution can be readily used in the development of freezing techniques, as sections taken at right angles to the specimen surface monitor the gradient of ice crystal size.

Analyses were done on a Philips EM400 electron microscope interfaced with a 30mm^2 Kevex Si (Li) energy dispersive detector and Kevex 7000 multichannel analyzer, using Hall's approach (Hall, 1971) of determining concentrations in ultrathin sections through the relationship of the characteristic x-ray/continuum ratio. Details of the multiple least square fitting technique and the sensitivity of our instrumentation have been published (Shuman, Somlyo, and Somlyo, 1976, 1977). Two modifications have recently been added to the instruments previously described: (1) the microscope is now fitted with a field emission gun normally operated at 80 Kv and separated from the main column by a differentially pumped aperture, and (2) the multiple least squares fitting routine has been rewritten in our laboratory (by Mr. K. McGinnis) and is now executed in less than 5 seconds on a PDP11/34 computer equipped with hard wired multiplier and dual RK05 discs, as compared to the approximately 200-second computation required by the software multiplication-based Tracor Northern NS880 system.

X-ray mapping was performed with the field emission source by scanning the beam; x-ray counts from four appropriate characteristic windows were entered into the computer while simultaneously accumulating the scanning transmitted electron image. The STEM images and x-ray maps were stored on disc.

RESULTS AND DISCUSSION

Composition of the Cytoplasm and Organelles in Vascular Smooth Muscle

A typical unstained cryosection of rabbit portal anterior mesenteric vein smooth muscle is shown in Figure 1, which illustrates "vitreous" freezing while in Figure 2 a small amount of ice crystal damage is observed. There was no difference in the elemental composition of such preparations, when the electron beam was focused to ≧ 0.5 μm for analysis of the cytoplasm in the smooth muscle cells. The mean

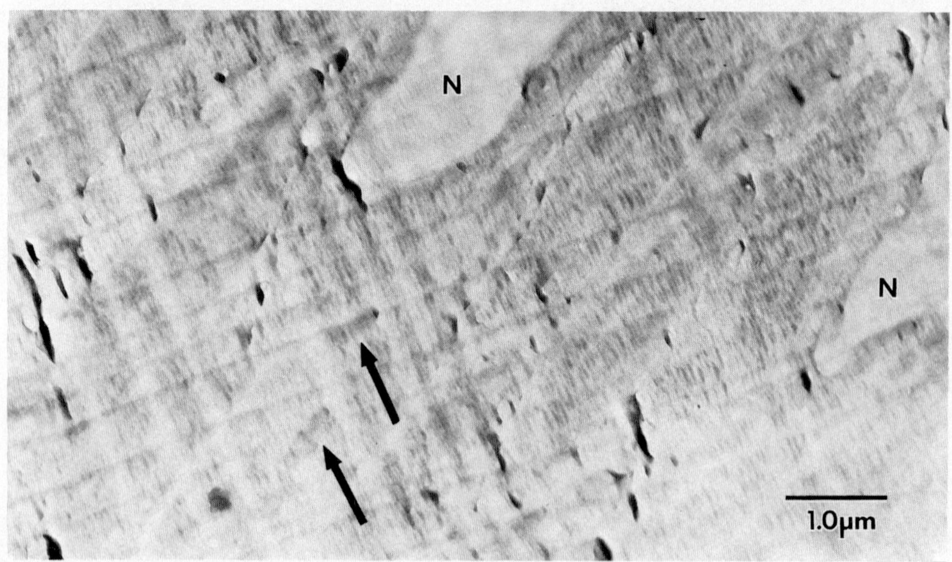

FIGURE 1. Unstained cryosection of PAMV smooth muscle showing vitreous freezing. Knife marks run diagonally from left to right. The long, narrow, closely-packed cells are from the media of the vessel. N, nucleus. Arrow, mitochondrion. x14000. (Somlyo, Somlyo, and Shuman, 1979).

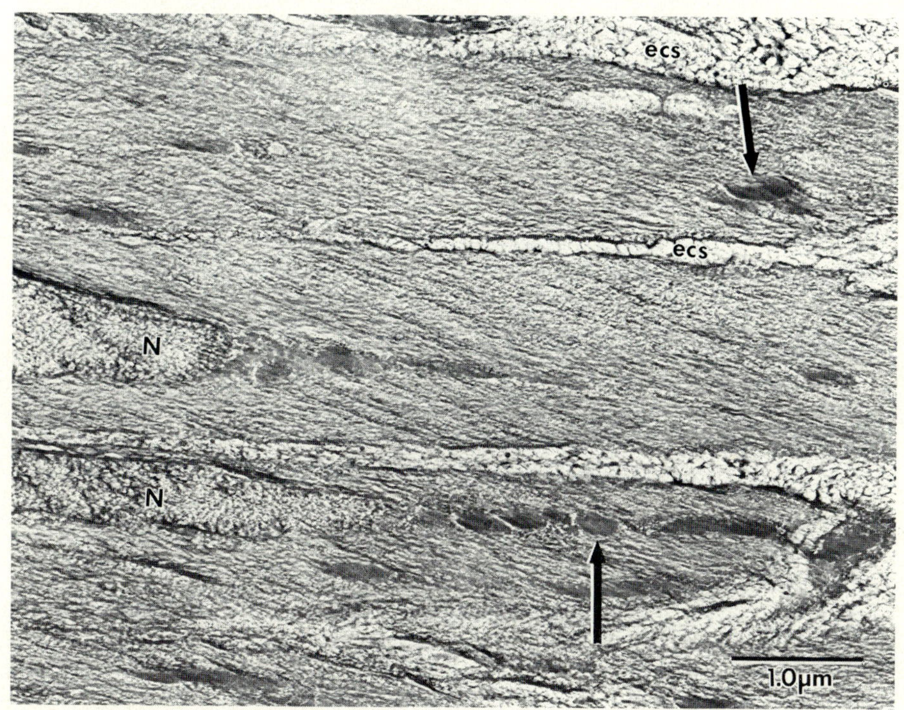

FIGURE 2. Unstained cryosection of PAMV smooth muscle showing some increase in contrast in the presence of small ice crystals. The cell outlines, mitochondria (arrow), nucleus (N), and the extracellular spaces (ecs) are clearly discernible. x14,400. (Somlyo, Somlyo, and Shuman, 1979).

cytoplasmic Cl content measured in 168 fibers from 8 rabbits was 278mmoles/Kg dry wt. or 78mmole/l cell H_2O based on a cell H_2O of 78% (see Table 2 in Somlyo, Somlyo, and Shuman, 1979) and is far in excess of the concentration predicted on the basis of a Gibbs-Donnan distribution (Boyle and Conway, 1941) but agrees with chemical analysis and flux studies (Casteels, 1971; Jones and Miller, 1978; Jones, Somlyo, and Somlyo, 1973; and Kao and Nishiyama, 1969). The electron probe studies show that this extra Cl is not compartmentalized, but is distributed throughout the cytoplasm. This could not be determined by chemical analysis and flux techniques. The possibility of an inward Cl-pump as well as the existence of unidentified non-diffusible anions in smooth muscle have been discussed elsewhere (Somlyo, Somlyo, and Shuman, 1979). Values for cytoplasmic K were in excellent agreement with flame photometric measurements on this same tissue (Jones, 1979).

Mitochondrial Ca was low and did not differ significantly from cytoplasmic values, both in control (Table 1) as well as in muscles frozen after a 30-minute contracture produced either with 80mM KCl alone or with added (supramaximal concentration) norepinephrine. The absence of mitochondrial Ca loading over the 30 minutes is consistent with the high apparent K_m of 17 μm measured in isolated vascular smooth muscle mitochondria (Vallieres, Scarpa, and Somlyo, 1975). If the apparent K_m <u>in situ</u> were lower, i.e., 10^{-7} to 10^{-6}M, as required of a physiological relaxing system, then massive mitochondrial loading would have occurred within the 30 minutes of maintained contracture (estimated Ca^{++} $\sim 10^{-5}$). Therefore, mitochondria do not play a major role in lowering cytoplasmic-free Ca during the normal contraction-relaxation cycle.

Mitochondrial CaP granules up to 2 mmole/kg dry wt. are observed only in damaged cells where cytoplasmic Na is increased and K decreased (Figure 3, Table 2). An association of mitochondrial granules with damaged cells has also been observed in epithelial cells (Gupta, Hall, and Naftalin, 1978).

Ca deposits are localized to the sarcoplasmic reticulum of normal undamaged smooth muscle (Table 2, Figure 3).

Mitochondrial Ca Uptake in Hyperpermeable Smooth and Cardiac Muscle Cells

In cryosections of rabbit portal anterior mesenteric vein made hyperpermeable with saponin and exposed to various concentrations of free Ca for 30-minute periods, mitochondrial Ca loading occurred at 10^{-5}M but not at 10^{-6}M Ca pH 6.6 (preliminary experiments done in collaboration with Professor Makoto Endo). The conclusion that these cells were in fact chemically skinned is based on their contractile responses to Ca and on electron probe analysis of the cytoplasmic Ca. The total Ca bound to the EGTA buffer (ethyleneglycol-bis-(β -amino ethylether)- N-N'tetraacetic acid) was 8mM which would be concentrated to approximately 24mmole/kg in the dried sections. The finding of 24mmoles Ca/Kg dry wt. in the cytoplasm of these cells confirmed that

TABLE 1 - Paired comparison of mitochondrial and cytoplasmic composition in rabbit portal-anterior mesenteric vein smooth muscles.

	K	Na	Cl	Mg	Ca	P	S	Continuum
Mitochondria	464±1.7	193±3.5	220±1.1	38±1.5	0.8±0.5	571±2.1	142±1.0	1789±7
Cytoplasm	565±2.6	246±4.6	308±1.7	43±1.9	0.7±0.7	261±1.7	125±1.2	1071±6
$\frac{[X_i]\ \text{mito}}{[X_i]\ \text{cyto}}$	1.2	1.1	1.1					

68 pairs of mitochondria and adjacent cytoplasm (each) from 9 animals were analyzed; values are mmol/kg dry weight except continuum, given as the number of counts.

From A. P. Somlyo, A. V. Somlyo and H. Shuman, 1979.

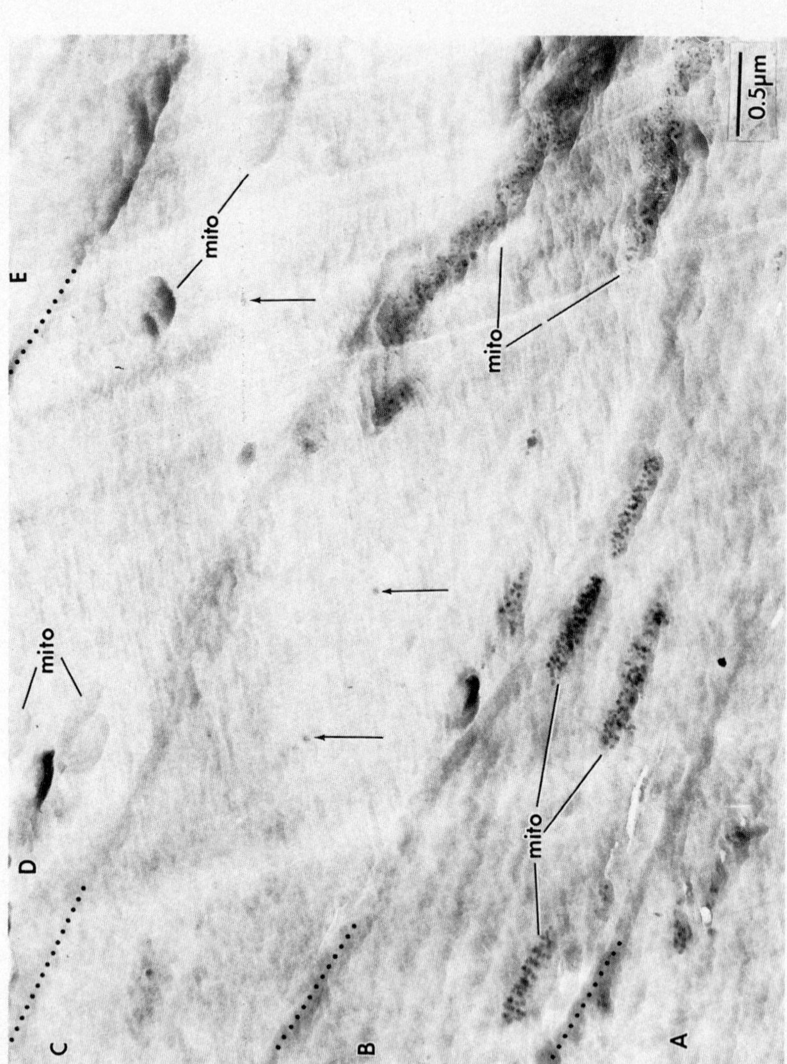

FIGURE 3. Unstained cryosection of PAMV smooth muscle including damaged fibers containing mitochondria with Ca- and P-containing granules. Cells A, B, C, and E contain mitochondrial granules. Fiber C also contains Ca deposits (arrows) in the SR. Fiber D does not contain granules, but one small deposit of Ca was present in the SR. Fibers containing granules had high cytoplasmic Na and low K concentrations. Dots=cell borders. X27,500. (Somlyo, Somlyo and Shuman, 1979.)

TABLE 2. Examples of cytoplasmic elemental concentrations of rabbit portal vein smooth muscle incubated in Kreb's + 80mM KCl.

	Na	P	Cl	K
	mmol/kg dry wt $\pm$ SD			
Cell w/o Ca^{++} granules	80$\pm$34	249$\pm$15	378$\pm$17	728$\pm$29
Cell w/Ca^{++} granules "only" in SR	67$\pm$60	198$\pm$25	363$\pm$29	731$\pm$50
Cell with mitochondrial and SR Ca^{++} granules	554$\pm$69	145$\pm$22	748$\pm$52	502$\pm$36

From A. P. Somlyo, A. V. Somlyo, and H. Shuman, 1979.

they were in fact hyperpermeable. These *in situ* measurements of mitochondrial Ca in cells, in which cytoplasmic-free Ca is directly controlled, are in agreement with the apparent K_m for Ca uptake measured in isolated smooth muscle mitochondria and with the electron probe measurements of the intact, non-skinned cells.

Hyperpermeable cardiac myocytes isolated enzymatically from rabbit ventricle cells are in a relaxed state at free Ca concentrations of less than 10^{-7}M but undergo spontaneous rhythmic contractions of increasing frequency as the free Ca concentration is raised towards 10^{-6}M (Chiesi et al., submitted; Reiser et al., 1979). Ca uptake into the SR of such cells can be augmented by oxalate, and the Ca oxalate crystals in the *in situ* SR have been identified by electron microscopy and electron probe analysis (Figures 4 and 5). Mitochondria did not accumulate Ca in these oxalate loading experiments (Figure 5) or in the spontaneously rhythmically contracting myocytes as determined by analysis of cryosections (Chiesi et al., submitted). At high Ca concentrations ($> 10^{-6}$) the myocytes undergo contracture, associated with distortion of the filament lattice and the appearance of mitochondrial granules containing Ca, P and Mg. Therefore, our studies of intact, resting, contracted, and hyperpermeable vascular smooth muscle and rabbit cardiac myocytes suggest that mitochondria do not play a major role in physiological Ca transport in these tissues and that Ca granules occur only in damaged or pathological cells.

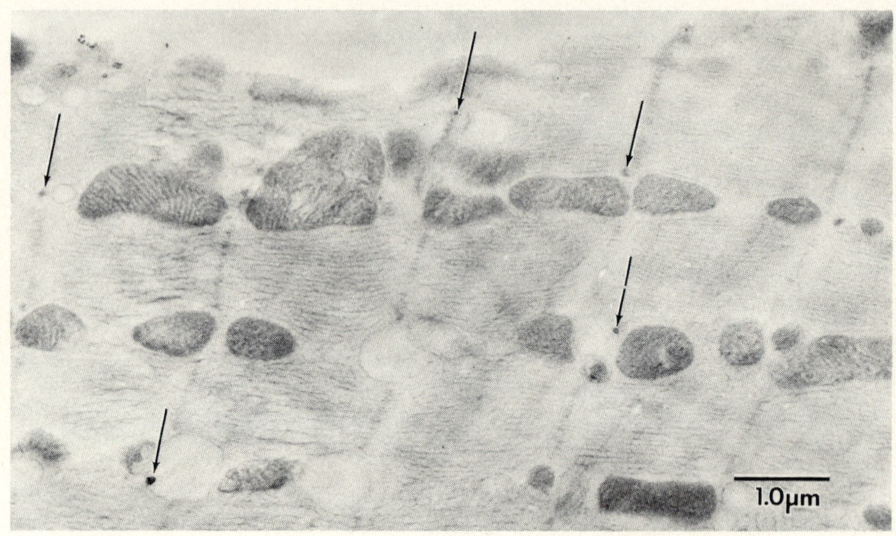

FIGURE 4. Calcium oxalate-loaded isolated cardiac cell, frozen, processed by freeze-substitution and sectioned on glycerol. Large Ca signals were obtained in the regions of SR indicated by the arrows. The majority of the deposits occurred in the SR in the Z band region. The only stain present is the osmium included in the acetone. At higher magnification the deposits were typical "bubbly" Ca oxalate deposits. (Chiesi et al., submitted).

Composition of the Sarcoplasmic Reticulum in Frog Muscle Fibers at Rest and During a Tetanus

The quality of freezing achievable in frog muscle by flash-freezing during monitored tension development is illustrated in the freeze-substituted sample shown in Figure 6. This muscle was frozen when tension development had reached 50% of maximal tetanic tension. After cryosections were obtained, the remainder of the specimen was freeze-substituted. There is no evidence of ice crystal damage in these sections cut parallel to the surface. The Z bands and actin and myosin filaments are crisp. In sections cut more deeply into the tissue, the first regions to show ice crystal damage are the Z lines, outer mitochondrial membrane-space, and the region of A-I filament overlap. A graded increase in damage due to ice crystals occurs towards the center for the specimen.

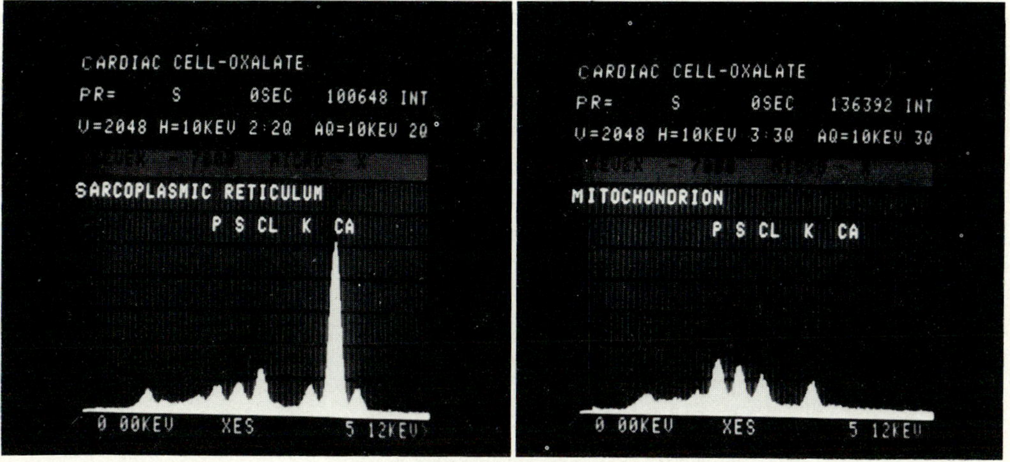

FIGURE 5. Spectra from analysis of SR containing a Ca oxalate deposit and of a mitochondrion in the same region of a rabbit cardiac myocyte in a cryosection. Note the presence of a large Ca peak in the former and the absence of detectable Ca in the mitochondrion. The Ca peak in the SR is not associated with P, which is compatible with the deposit being Ca oxalate. (Chiesi et al., submitted).

The cryosections used for electron probe analysis included both regions free of ice crystals and others with small ice crystal damage such as Figure 7. For analysis the electron beam was focused to 500Å over a terminal cisterna (TC) region of the SR. A paired analysis using the same probe parameters was done on a region of cytoplasm in the I-band next to the TC. Approximately 5 muscle fibers were examined from each of 8 pairs of tetanized and control muscles. The TC and adjacent cytoplasmic concentrations of both Na and Cl were not significantly different, ruling out the hypothesis based on earlier flux studies of whole muscles that the SR is an extracellular compartment (Somlyo, Shuman, and Somlyo, 1977). Recent flux studies done on single fibers, confirming the electron probe results, failed to show a component of Na efflux that could represent an internalized extracellular volume of approximately 10% representing the SR (Neville, 1979).

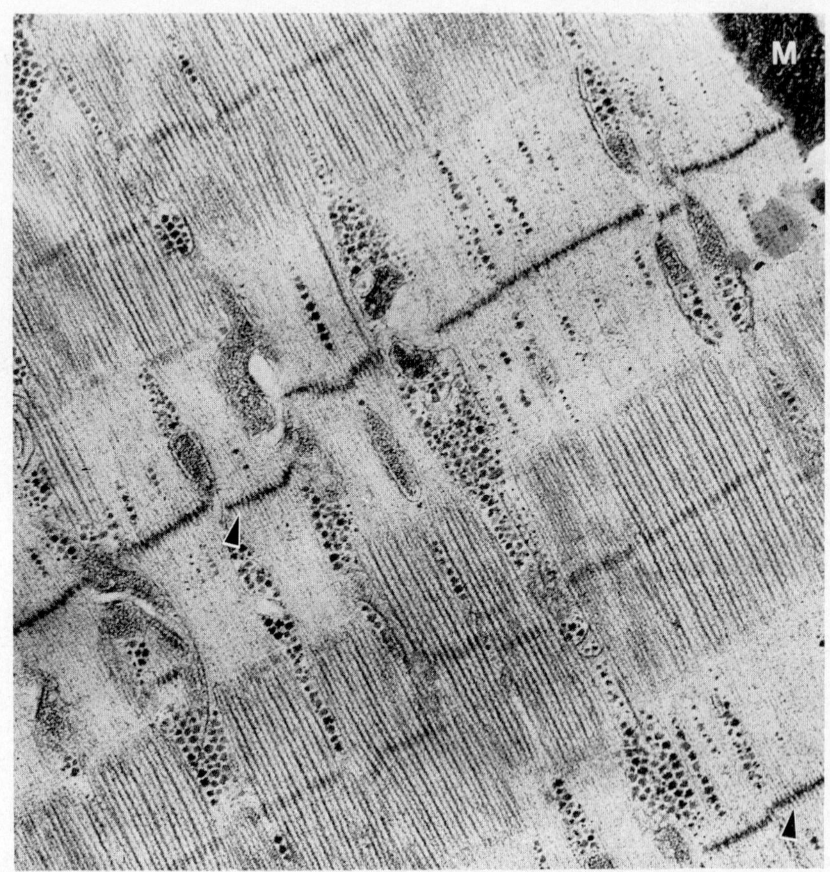

FIGURE 6. Section from a freeze-substituted frog semitendinosus muscle flash frozen during tension development at the point of 50% full-tetanic tension. Note the absence of ice crystal damage, including at the Z band regions (arrowheads) and the outer mitochondrial membrane. M=mitochondrion Pb stained section. x30,000.

The mean elemental concentrations of 192 control and 179 tetanized terminal cisternae are shown in Table 3. Ca is localized to the TC and during a 1.2 sec. tetanus approximately 63mmole/Kg dry TC wt. is released. This amount of Ca released into the cytosol would increase the cytoplasmic Ca by approximately 0.8mmole/l fiber volume, far in excess of the approximate 0.2mmole/l fiber volume required to saturate the 2 Ca sites on troponin necessary to activate the actin-myosin interaction. However, the concentration of the Ca binding protein, parvalbumin, would

TABLE 3 - Elemental composition of the terminal cisternae of frog muscle. ($\bar{x}$ ± SD mmol/kg dry wt.).

	n	Na	Mg	P	S	Cl	K	Ca
Control	192	54±35	58±21	413±78	216±39	42±19	549±128	108±46
Tetanus	179	52±32	70±21	427±69	212±55	42±20	595±109	45±19

Effect of tetanus (mEq): - 126 Ca + 46 K + 24 Mg = -56mEg

From A.V. Somlyo et al., 1980a,b.

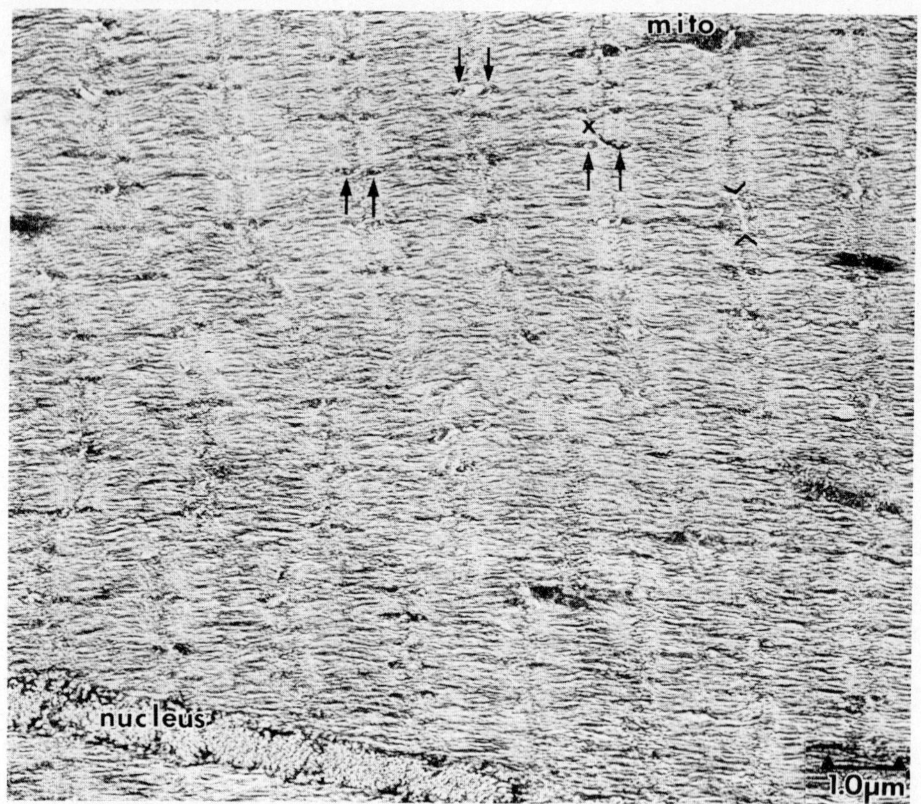

FIGURE 7. Longitudinally oriented, unstained, dried cryosection from a control bundle of frog semitendinosus fibers. A nucleus is shown in the lower portion. Pairs of terminal cisternae (arrows) occur at the Z lines. An en face view of a triad is indicated by the brackets. A small amount of ice crystal damage is present. (Somlyo, Somlyo, Shuman, and Stewart, 1979).

account for an additional, approximately 0.7mmole Ca bound/l fiber volume (Gosselin-Rey and Gerday, 1977). Thus, the predicted amount of Ca bound to parvalbumin and to troponin is in very good agreement with the measured amount released from the TC. The Ca-Mg binding sites on parvalbumin are presumed to be occupied by Mg at the free Ca and Mg concentrations present in the resting muscle. The slow Mg off time $t_{1/2}$ = 230-550msec (Haiech, Derancourt, Pechere, and Damaille, 1979; Potter, Robertson, Mandel, and Johnson, submitted) would preclude Ca being bound to these sites during a twitch but not during a 1.2 sec. tetanus.

A further objective of these experiments was to determine the net ion movements which take place across the SR membrane during Ca release. There was no significant difference in the concentrations of Na and Cl between the resting and tetanized TC (Table 3). While there was a highly significant increase in K and Mg (P 0.001) in the tetanized TC, these changes were not large enough to compensate for the change in charge of 126meq/kg dry TC due to the 63mmole Ca/Kg dry TC released. It is possible that protons and or organic ions account for the approximately 56meq charge deficit. It should be noted that the above measurements represent the net change of both Ca release and uptake processes.

Analyses utilizing astigmated probes over elements of longitudinal SR and adjacent cytoplasm showed no significant difference in the Ca concentrations of these regions at the approximately $\pm$ 3mmole/kg dry wt level. Cytoplasmic Ca was significantly increased during the tetanus and accounted for the majority of the Ca released from the TC, in agreement with the proposed Ca binding to cytoplasmic parvalbumin.

Even in fibers that were Ca loaded due to damage during dissection, Ca was not localized in the elements of longitudinal SR, although the terminal cisternae contained molar concentrations of Ca and P. This is well-demonstrated in the Ca and P x-ray maps (Figure 8) from one of these Ca-loaded cells. The spatial resolution using a field emission source is sufficient (10-20nm) for distinguishing two adjacent terminal cisternae of a triad separated by a T-tubule.

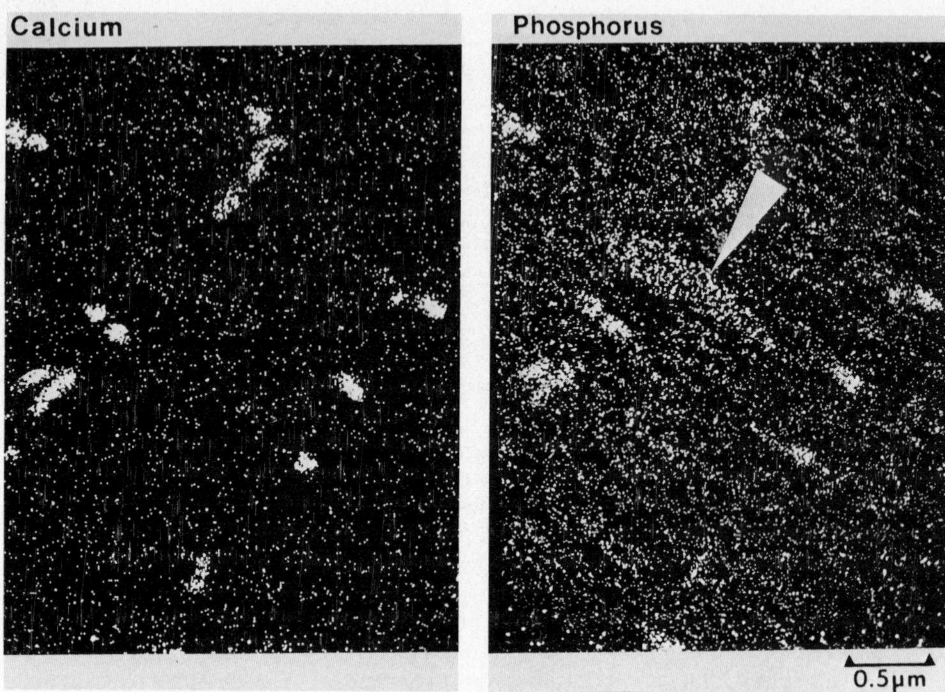

FIGURE 8. X-ray maps (calcium and phosphorus) from a longitudinally oriented cryosection of a frog muscle fiber that was "calcium-loaded" due to damage during dissection. X-rays in the calcium energy window (centroid 3.69kV) were collected without subtraction of the adjacent K Kβ peak. This and x-ray background contribute to the counts observed over regions of the fiber outside the SR. Maps obtained with field emission gun. The very high calcium content of the SR in this heavily calcium loaded muscle provides a "test object" to demonstrate the spatial resolution instrumentally obtainable with electron probe analysis. Note the presence of the mitochondrion (arrow) in the phosphorus, but not in the calcium x-ray map. (Somlyo, Somlyo, Gonzalez-Serratos, Shuman, and McClellan, 1980).

SUMMARY AND CONCLUSIONS

1. Electron probe analysis of muscles flash-frozen with supercooled Freon 22 is a uniquely suitable method for quantitation of the elemental composition of in situ organelles. The spatial resolution of the method determined in tropomyosin paracrystals is at least 10-20nm; in X-ray maps of Ca-loaded fibers, two terminal cisternae (TC) separated by a T-tubule can be readily resolved.

2. High concentrations of sequestered calcium in the SR have been observed in normal vascular smooth muscle and in isolated rat cardiac cells at physiological cytoplasmic-free Ca concentrations, when mitochondrial Ca was low. Significant Ca uptake by mitochondria, often in the form of granules, was only observed at abnormally high free cytoplasmic Ca levels, either in experimentally produced hyperpermeable cells or in cells damaged during dissection. These findings suggest that the mitochondria do not play a significant role in the physiological regulation of free Ca-levels.

3. The ionic composition of the SR of frog skeletal muscle indicates that this compartment is not in ionic communication with the extracellular space. Furthermore, the NaCl concentrations of the terminal cisternae (TC) of the SR are not detectably changed during a 1.2 sec. tetanus.

4. The amount of Ca released during a 1.2 sec. tetanus is sufficient to raise the total cytoplasmic Ca concentration by approximately 0.8mmol/liter fiber volume. This concentration is equivalent to that required to saturate the Ca/Mg sites on both troponin and parvalbumin.

5. Ca release during tetanus is accompanied by an increase in the Mg and K content of the TC. This increase is significantly less than the amount of charge (126m equivalents/kg dry TC) lost through Ca release. It is suggested that proton movement into the SR or the movement of organic ions may compensate for the charge deficiency observed.

ACKNOWLEDGMENTS

This work was supported by National Institutes of Health grant HL 15835 to the Pennsylvania Muscle Institute.

REFERENCES

Boyle, P. J. and Conway, E. J. (1941). J. Physiol. 100:1-63.

Casteels, R. J. (1971). J. Physio. 214:225-43.

Chiesi, M., Ho, M. M., Inesi, G., Leung, J., Somlyo, A. P., and Somlyo, A. V. (Submitted, 1980).

Costello, M. J., and Corless, J. M. (1978). J. Microsc. 112:17-38.

Ebashi, S., and Endo, M. (1968). In "Progress in Biophysics and Molecular Biology" (J. A. V. Butler and D. Noble, eds.), Vol. 11, pp. 123-83. Pergamon Press, Oxford and New York.

Endo, M. (1968). Physiol. Rev. 57:71-108.

Endo, M., Kitazawa, T., Yagi, S., Iino, M., and Kakuta, Y. (1977). In "Excitation Contraction Coupling in Smooth Muscle" (R. Casteels, T. Godfraind, and J.C. Ruegg, eds.), pp.199-209. Elsevier/North-Holland Biomedical Press, Amsterdam.

Feder, N., and Sidman, R. C. (1958). J. Biophys. Biochem. Cytol. 4:593-602.

Franzini-Armstrong, C., Heuser, H. E., Reese, R. S., Somlyo, A. V., and Somlyo, A. P. (1978). J. Physiol. 283:133-40.

Gonzalez-Serratos, H., Somlyo, A. V., McClellan, G., Shuman, H., Borrerro, L. M., and Somlyo, A. P. (1978). Proc. Natl. Acad. Sci. 75:1329-33.

Gosselin-Rey, C., and Gerday, C. (1977). Biochem. Biophys. Acta. 492:53-63.

Gupta, B. L., Hall, T. A., and Naftalin, R. J. (1978). Nature 272:70-73.

Haiech, J., Derancourt, J., Pechere, J., and Demaille, J. G. (1979). Biochemistry 13:2752-58.

Hall, T. A. (1979). J. Micros. 117:145-63.

Hall, T. A. (1971). In "Physical Techniques in Biological Research" (G. Oster, ed.), Vol. 1A., pp.158-275. Academic Press, New York.

Heuser, H. E., and Reese, R. S. (1976). J. Cell Biol. 70:357.

Hutchinson, T. E. (1979). Biochem. Biophys. Acta. 58:115-58.

James-Kracke, M. R., Sloane, B. F., Shuman, H., and Somlyo, A. P. (1979). Proc. Natl. Acad. Sci. 76:6461-65.

Jones, A. W. (1980). In "Handbook of Physiology. II. Cardiovascular System" (D. F. Bohr, A. P. Somlyo, and H. V. Sparks, eds.), pp.253-300.

Jones, A. W., and Miller, L. A. (1978). In "Blood Vessels, Molecular and Cellular Aspects of Vascular Smooth Muscle in Health and Disease" (D.F. Bohr and F. Takenaka, eds.), pp. 83-92. S. Karger AG, Basel, Switzerland.

Jones, A. W., Somlyo, A. P., and Somlyo, A. V. (1973). J. Physiol. 232:247-73.

Kao, C. Y., and Nishiyama, A. (1969). Am. J. Physiol. 217:525-31.

Lechene, C. P., and Warner, R. R. (1979). In "Microbeam Analysis in Biology" (C.P. Lechene and R. R. Warner, eds.). Academic Press, New York.

Misra, L. K., Smith, N. R. R., Chang, D. C., Sparks, R.L., Cameron, I. L., Beall, P. T., Harrist, R., Nichols, B. L., Fanguy, R. C., and Hazelwood, C. F. (1980). J. Cell Physiol. 103:193-200.

Neville, M. C. (1979). J. Physiol. 288:45-70.

Ornberg, R., and Reese, R. S. (1978). J. Cell Biol. 79:257.

Potter, J. D., Robertson, S. P., Mandel, F., and Johnson, D. (submitted). J. Biol. Chem.

Reiser, G., Sabbadini, R., Paolini, P., Fry, D., and Inesi, G. (1979). Am. J. Physiol. 236:70-77.

Shuman, H., Somlyo, A. V., and Somlyo, A. P. (1976). Ultramicroscopy 1:317-39.

Shuman, H., Somlyo, A. V., and Somlyo, A. P. (1977). Scan. Elect. Microsc. 1:663-72.

Somlyo, A. P., Somlyo, A. V., and Shuman, H. (1979). J. Cell Biol. 81:316-35.

Somlyo, A. P., Somlyo, A. V., Shuman, H., and Stewart, M. (1979). Scan. Elec. Micros. IIT Res. Inst. 2:711-22.

Somlyo, A. P., Somlyo, A. V., Gonzalez-Serratos, H., Shuman, H., and Clellan, G. (1980). In "Oji International Seminar on Regulatory Mechanisms of Muscle Contraction" (S. Ebashi, K. Maruyama, and M. Endo, eds.), pp.421-33. Japan Sci. Soc. Press. Springer Verlag, Tokyo.

Somlyo, A. V., and Silcox, J. (1979). In "Microbeam Analysis in Biology" (C. P. Lechene and R. R. Warner, eds.), pp. 535-55. Academic Press, New York.

Somlyo, A. V., Somlyo, A. P., Gonzalez-Serratos, H., Shuman, H., and McClellan, G. (1979). Symp. on Regulation of Muscle Contraction: E-C Coupling. In press, Los Angeles.

Somlyo, A. V., Shuman, H., and Somlyo, A. P. (1977a). J. Cell Biol. 75:828-57.

Somlyo, A. V., Shuman, H., and Somlyo, A. P. (1977b). Nature 268:556-58.

Vallieres, J., Scarpa, A., and Somlyo, A. P. (1975). Arch. Biochem. Biophys. 170:659-69.

Van Harreveld, A., and Crowell, J. (1964). Anat. Rec. 149:381-386.

Wendt-Gallitelli, M. F., Wolburg, H., Schwegler, M., and Schlote, W. (1979). Experientia 35:1591-93.

Wroblewski, R., Roomans, G. M., Jansson, E., and Edstrom, L. (1978). Histochemistry 55:281-92.

DISCUSSION

SPEAKER: Avril Somlyo

HUTCHINSON: Avril, is there a good reason for the difference in scale between those two? You seem to have collected very much less in the cytoplasm. Do you feel you were getting mass loss in that area and decided to get out?

SOMLYO, Avril: No. The large deposits contain molar concentrations of Ca and P in mitochondria of either damaged cells or in "skinned" cells when cytoplasmic Ca is raised to ∿ 10^{-5}M. The beam was focused on the granule and not the entire mitochondrion. Large Ca and P peaks build in a few seconds due to the high mass. Analysis over adjacent cytoplasm was run for 100 seconds in order to accumulate sufficient K, Na, and Ca counts to determine whether the cell was damaged ($\uparrow [Na]_i$ $\downarrow [K]_i$) or "skinned" (Ca-EGTA = 10mM). Generally, in the absence of these kinds of deposits we use the same dose for paired analysis which allows comparison of the absolute counts as well as the concentrations of the respective elements.

RICK: You would have used the same scale once the potassium concentration is in both problems?

SOMLYO, Avril: I doubt it because you are also introducing a great deal of scatter and I do not think that the quantitation of the potassium in the presence of these huge granules is accurate. One could analyze mitochondrial K by focusing the beam on the mitochondrial matrix between the granules and paired analysis over adjacent cytoplasm using the same dose.

TORMEY: By oxalate loaded, you mean there is oxalate in the external bathing solution or did you do a substitution?

SOMLYO, Avril: In the isolated hyperpermeable single cardiac cells, Ca oxalate loading of the SR was done during the incubation prior to freezing. After freezing, the cells were processed by freeze-substitution. Oxalate was not included in the acetone.

RICK: From the cytoplasm in the sarcoplasmic reticulum, don't you assume that in order to make that balance, that the volume stays identical? Is there any evidence for that?

SOMLYO, Avril: Yes. One of the reasons we did the freeze-substitution studies on the remainder of the samples after the cryosections were removed was to look more carefully for ultrastructural changes. Inspection of these pictures shows that the control and tetanized muscles from the ten frogs indicate no obvious change in reticulum volume. There are some published stereology measurements of SR volumes of fatigued frog muscles where no volume change was observed (Eisenberg and Gilai, 1979, J. Gen. Physiology, 74:1-16). I plan to do the stereology on these. Plus, there was no significant difference in the continuum measurement in the control and tetanized terminal cisterna operating at the same "probe parameters" and averaging about 200 analyses in each group. If you can accept that variations in the probe parameters would average out, the continuum in the terminal cisternae in the control and the tetanized muscles are also compatible with an absence of change in the terminal cisterna volume during a tetanus.

LECHENE: On the calcium translocation: the distance goes with the diffusion constant, are you close enough to diffuse fast? Will you move fast enough in the twitch?

SOMLYO, Avril: Yes.

LECHENE: What are the distances involved?

SOMLYO, Avril: The sarcomere length used was generally 2.8 μm. The actin filaments where the Ca binds and switches on the muscle, are 1 μm in length. The reticulum actually forms an entire collar around the cylindrical fibrils of approximately 0.5 μm diameter with the Ca storage sites (TC) adjacent to the actin filaments. As the Ca sensitive regulatory sites are on the actin filaments, the Ca has only to diffuse over the 0.2 μm radius. The time to peak twitch tension is 20 msec at 20°C and the diffusion coefficient at 20°C for Ca in the cytoplasm is approximately $1 \times 10^{-5} cm^2 sec^{-1}$. Therefore, tension development is not diffusion limited. (Hill, A. V., 1948, Proc. Roy Soc. B. 135:446-51).

HUTCHINSON: You were speaking of something on the order of 24 millimolar in the I-band, does that include the SR?

SOMLYO, Avril: Yes. In those particular experiments, an astigmated spot was focused over the I-bands of several fibrils including several terminal cisternae.

HUTCHINSON: So, what do you figure the sort of volume, volume/ratio would be as far as SR vs. non-SR within the I-band?

SOMLYO, Avril: If you take a terminal cisternae volume of about 5% of the entire fiber volume, then the TC volume would be about 10-20% of the I-band, depending on sarcomere length.

RICK: If you assume that a large part of the calcium is bound to the calcium sequestering protein, one should expect that the concentration of that protein would be different in different areas in a highly organized tissue so that you would see within the cytosol the difference in the total calcium concentration. Can you detect that?

SOMLYO, Avril: At rest, the cell Ca is localized to the I-band, specifically to the terminal cisternae which contain the low affinity Ca binding protein, calsequestrin. During activation, 60% of the cisternae Ca is now in the cytoplasm of the fiber presumably on paralbumin (during a tetanus) and troponin. I do not have enough data to make a rigorous comparison of the small cytoplasm analysis over the A-band excluding the SR and over the I-band region, although we have a great number of measurements in the I-band region. In the few measurements that I have over only the A-band there is, in fact, slightly less calcium than in the I-band cytoplasm, and it sort of fits with what one would predict from the differences in hydration of the two regions.

SOMLYO, Andrew: What was the question about the parvalbumin distribution or the calsequestrin distribution?

RICK: Whether the sequestrin protein is in the cell.

SOMLYO, Avril: Calsequestrin is localized to the terminae cisternae. There is good evidence that the parvalbumins are in the cell water and diffuse freely.

LECHENE: You have published that when the cell was damaged, you had a small change in the ratios in potassium, like sodium enters and potassium exits. This was the case where you were looking at mitochondria calcium. But you say in this cell you do not have this load of mitochondria yet. You have an extreme change in sodium and potassium ratio. How do you explain it? Are they totally dead?

SOMLYO, Avril: With regard to the isolated cardiac cells, they are chemically skinned and are hyperpermeable to ions and low molecular weight fluorescent markers, in fact, the molecular weight cutoff is around 500. They are in a solution containing 120 mM sodium chloride which has distributed across the membrane.

LECHENE: This I quite understand, but this was the condition in which you were finding a calcium in mitochondria in cardiac cells.

SOMLYO, Avril: The difference between Ca loaded and non-loaded mitochondria is in the free calcium concentration which is buffered with EGTA. These isolated cells will beat rhythmically at a [Ca] of $5x10^{-7}$ but will undergo an irreversible contracture and the mitochondria will load with calcium if we shift the calcium concentration just above a [Ca] of 6. The data you refer to applies to normal (not skinned) tissue where mitochondrial CaP granules are correlated with a reversal of the cytoplasmic K:Na due to cell damage or pathology, presumably extracellular Ca also leaks in.

SOMLYO, Andrew: Conversely, in a normal ringer solution when you make the membrane leaky, you have a 1 mmolar calcium concentration outside and that goes in and you get rocks, but not at a [Ca] of 6.

TORMEY: Also, in the cardiac cells where you find calcium deposits around the T system in the presence of oxalate, what is your feeling about whether they are localized in the T or whether they are localized in the SR?

SOMLYO, Avril: My feeling, with considerable bias, is that it is in the SR. Structures that look like T-tubules do not contain Ca oxalate. The difficulty arises in imaging the terminal cisternae in cardiac muscle in cryosections. Even when looking at conventionally fixed cardiac muscle, the lateral sacs of the cisternae are so flattened that they are often difficult to image under the best of circumstances. Ca fluxes have also been studied in these disassociated cell preparations. ATP dependent Ca uptake is enhanced with oxalate and prevented with Ca ionophores or methylxanthines.

ELECTRON PROBE X-RAY MICROANALYSIS OF NORMAL AND INJURED MYOCARDIUM: METHODS AND RESULTS

H. Hagler
K. Burton
L. Buja

Department of Pathology
Southwestern Medical School
The University of Texas
Health Science Center
Dallas, Texas

INTRODUCTION

Analytical electron microscopy has important applications for the study of many aspects of cell injury in a variety of tissues, including our particular area of interest, cardiac muscle cell injury (Buja et al., 1976, 1977; Burton et al., 1977, 1980a,b; Hagler et al., 1977, 1979a,b, 1980a,b,c). Successful use of analytical electron microscopy has required development of appropriate methods of tissue preparation and analytical electron microscopic techniques. In this presentation, we will describe the methods we have developed or applied to studies of cardiac muscle cell injury and the results obtained. The presentation will focus on two main areas: namely, studies of pathological calcification in ischemic myocardium and studies of membrane permeability alterations and electrolyte shifts which lead to the development of irreversible injury and pathological calcification.

Analytical Electron Microscopic Methods

Our ultrastructural and analytical electron microscopic studies were performed with a JEOL 100C electron microscope

ISBN 0-12-362880-6

equipped with a free lens control for condensor lens CL1, a high resolution scanning attachment, and a Kevex 30 mm2 energy dispersive x-ray detector with 158eV resolution. Ultrathin sections stained with uranyl acetate and lead citrate were used for transmission electron miscroscopy (TEM) and unstained, epoxy or freeze-dried sections were used for analytical studies. For analytical electron microscopy, the microscope was operated in the scanning transmission mode (STEM) at 80kV, 50 microamps emission current (above a dark current of 52-54 microamps), beam diameter of 40 to 80 nm, 30-degree specimen tilt, and specimen-to-detector distance of 15mm. X-ray spectra were collected at 20 eV per channel, using a Tracor Northern TN 2000 multichannel analyzer.

One method of data collection involved generation of a summed spectrum from 10 ultrastructurally similar inclusions in the same section using a brief 20-to-30-second analysis time for each inclusion with the beam focused to a small spot (spot mode) (Hagler et al., 1977). A second method of data acquisition involved collection of spectra in the raster mode over an area of 0.2 μm^2 for 100 seconds (Hagler et al., 1980a). Both of these methods served to optimize detection of elements in biologic thin sections and to reduce the problems of progressive mass loss, specimen contamination, and reduction in peak-to-background ratio as analysis time increases (Hall and Gupta, 1974).

Analysis of the spectra was based on the thin film criteria described by Hall et al. (1973), and applied by Shuman et al. (1976, 1977). Elemental reference spectra were collected from thin crystals of various salts (NaCl, $CaCl_2$, etc.) which had elements of interest with well-separated peaks. The unknown spectra were analyzed using the multiple least squares fitting routine available on the TN multichannel analyzer (Schamber, 1976). The elemental nature of the various peaks was identified, based on the characteristic x-ray energies of each element (Johnson and White, 1970). Elemental peaks were considered significant in the summed spectra when they were demonstrable at a two-sigma (standard deviation) confidence limit. A one-sigma confidence limit was used with the raster scan mode to avoid values of 0 when statistics were computed on the peak-to-continuum ratios of elements. The analysis and comparison of thin biological specimens was based on the peak-to-continuum ratio being proportional to concentration, as described by Hall et al. (1973). The continuum used for the calculations was from 5.5 to 6.5 keV. For absolute quantitation of calcium (Ca) concentration in test spectra, these spectra were compared with data obtained from thin plastic and cryo standards of varying elemental concentration, described below.

STUDIES OF PATHOLOGICAL CALCIFICATION

These studies were directed to an analysis of different types of mitochondrial inclusions known to develop with acute cell injury induced by a variety of insults, including hypoxia and ischemia (Hagler et al., 1979b; Jennings and Ganote, 1976; Trump et al., 1976, 1978; Ashraf and Bloor, 1976; Ashraf et al., 1976; Jennings et al., 1978) in order to define the nature of these inclusions and to characterize different stages in the process of pathological calcification.

Tissue Preparation

Tissue samples were obtained from mongrel dogs following 40 to 60 minutes of temporary coronary occlusion, followed by 20 minutes of reflow, or following 24 to 48 hours of permanent coronary ligation. The tissue was prepared for analytical electron microscopy, as described by Hagler et al. (1979b). Ultramicrotomy was performed with an LKB Ultratome III or a Sorvall MT 5000. Water was usually used as a trough liquid, but some sections were cut on glycerol to evaluate the effects of an anhydrous versus aqueous medium on preservation of calcific inclusions (Hagler et al., 1980b; Ornberg, 1979; Landis et al., 1977).

Preparation of Metallic Carboxylate Plastic Standards

The use of metallic carboxylate paint dryers as a source of plastic-soluble standards for calcium and other metals was suggested by Ornberg, 1979. Calcium (Ca) naphthenate (5 gm%) was obtained courtesy of Witco Chemical Organics Division (New York, New York). From this compound, sets of standards were prepared by mixing with Mollenhauer's epon-araldite mixture number one. Several series of epoxy standards were prepared to contain varying percentages from 0.05 to 0.70 gm% of Ca naphthenate. It was found that a maximum of 0.8% by weight of Ca naphthenate could be mixed with the epoxy so that the resin would still cure. Initial studies indicated similar retention of Ca naphthenate with sections cut on water and glycerol. Therefore, sections were routinely cut on water using an LKB Ultratome III and picked up on formvar-coated 50-mesh copper grids. The grids were dried and coated with 30 nm carbon prior to analysis. The sections were analyzed using the STEM raster mode at 10,000X magnification.

Optimal Preservation of Early Calcific Inclusions

Our analytical electron microscopic studies were directed to an evaluation of the various mitochondrial inclusions from normal and damaged cardiac muscle cells shown diagrammatically in Figure 1. Initial studies of fixed tissue suggested that differences existed in the calcium content of the various inclusions and that the small granular inclusions represented an early, relatively labile form of calcification (Table 1). Further studies were performed to determine optimal conditions for preservation of these inclusions (Hagler et al., 1979a,b) (Table 2). Direct

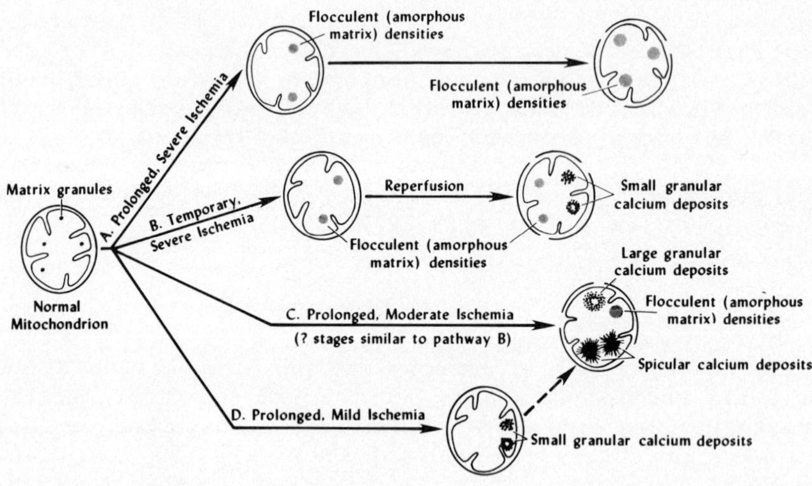

FIGURE 1. Diagram of various mitochondrial inclusions showing their relationship to the duration and severity of ischemia.

TABLE 1. Identification of Calcium in Various Mitochondrial Inclusions from Osmium-Phosphate-Fixed, Infarcted Myocardium Using the STEM Spot Mode and Spectral Summing Technique.

Mitochondrial inclusion types	Analysis time	Ca peak/ continuum ratio
Confluent dense masses	50 sec	6.6
Large granular-to- spicular inclusions	10 x 30 sec	1.07
Small granular inclusions	10 x 30 sec	0.16
Amorphous matrix densities	10 x 30 sec	ND[a]

[a]Not detected.

TABLE 2. Effect of Tissue Preparation on the Preservation of Calcium in Early Mitochondrial Calcific Inclusions Following Temporary Coronary Arterial Occlusion.

Tissue preparation	Ca peak/continuum ratio[a]
Glu-PO_4 and Os-PO_4	ND[b]
Os-PO_4	0.14 ± 0.43
Alcohol	1.15 ± 0.04
Unfixed cryosections	0.91 ± 0.05[c]

[a]Each value represents the peak intensity and its confidence limit (1 sigma) from a summed spectrum of 10 microareas using the spot mode for analysis.

[b]Not detected.

[c]Relatively reduced due to increased continuum from copper grid compared to nylon grid for other sections.

fixation in phosphate-buffered osmium was the only technique of aqueous fixation which resulted in ultrastructurally demonstrable early granular inclusions with small calcium peaks. Analytical electron microscopy of the unstained, alcohol-fixed samples of ischemic myocardium revealed prominent Ca and P peaks in the mitochondrial inclusions (Table 2). Similar inclusions with prominent Ca and P peaks also were observed in cryosections of ischemic myocardium but not of control myocardium (Table 2).

Although thin sections of alcohol-fixed ischemic tissue exhibited calcific inclusions after routine cutting on water, the mitochondria in these sections had many holes and relatively few intact inclusions (Figure 2). When glycerol was used as the trough liquid, the mitochondria of alcohol-fixed ischemic tissue showed no holes and had no apparent loss of inclusions (Figure 2). Table 3 shows the

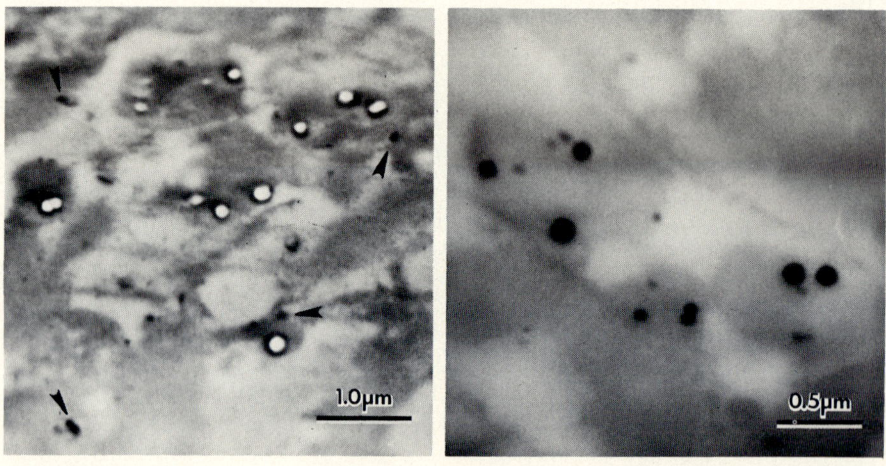

FIGURE 2. Left panel. Scanning transmission electron micrograph of alcohol-fixed, plastic-embedded heart tissue after 40 minutes temporary coronary occlusion and 20 minutes reperfusion. Thin epoxy section cut on water shows dissolution of many mitochondrial inclusions and preservation of some small inclusions (arrowheads). No stains were used for this micrograph. Right panel. Scanning transmission electron micrograph of the same tissue as shown on the left except that this tissue was sectioned on glycerol. No dissolution of mitochondrial granules was noted in these sections.

TABLE 3. Comparative Preservation of Calcium in Early Mitochondrial Calcific Inclusions from Alcohol-Fixed Tissue Sectioned on Water Versus Glycerol.

Trough liquid	Ca P/C ratio[a]
Water (n = 10)	.85 $\pm$ .38 (SD)
Glycerol (n = 20)	2.11 $\pm$ .97

[a]P/C (peak-to-continuum) ratios obtained from spectra collected in STEM raster mode.

results of x-ray microanalysis of mitochondrial inclusions from alcohol-fixed tissue sectioned on water versus glycerol. The spectra from tissue sectioned on water were collected only from those inclusions which remained in the tissue. The mean Ca peak-to-continuum ratio was 2.1 for inclusions sectioned on glycerol, while the mean Ca peak-to-continuum ratio was 0.85 for inclusions cut on water (Table 3). Thus, sectioning on water resulted in at least a 50% loss in Ca from those granules which remained intact and a complete loss from those that dissolved out and left holes (Hagler et al., 1980b). With tissue fixed in phosphate-buffered osmium, sectioning on glycerol resulted in little improvement in the Ca peak-to-continuum ratio for the smaller granular inclusions, indicating that the major loss of calcium occurred during aqueous fixation.

Quantification of Pathological Calcification Using Calcium Naphthenate Standards

The thin epoxy sections of calcium naphthenate standards were uniform and homogenous under STEM up to 30,000 magnification, which was the highest magnification used. Analysis of the standards showed a linear relationship between mean Ca peak-to-continuum ratios and Ca concentrations between 0.70 and 0.10 gm% with increased uncertainty and nonlinearity when the concentration was lower than 0.1% (Figure 3). Similar results were obtained with different sets of Ca naphthenate standards (Hagler et al., 1981).

An additional set of analyses of calcific inclusions was performed for purposes of comparison with the calcium

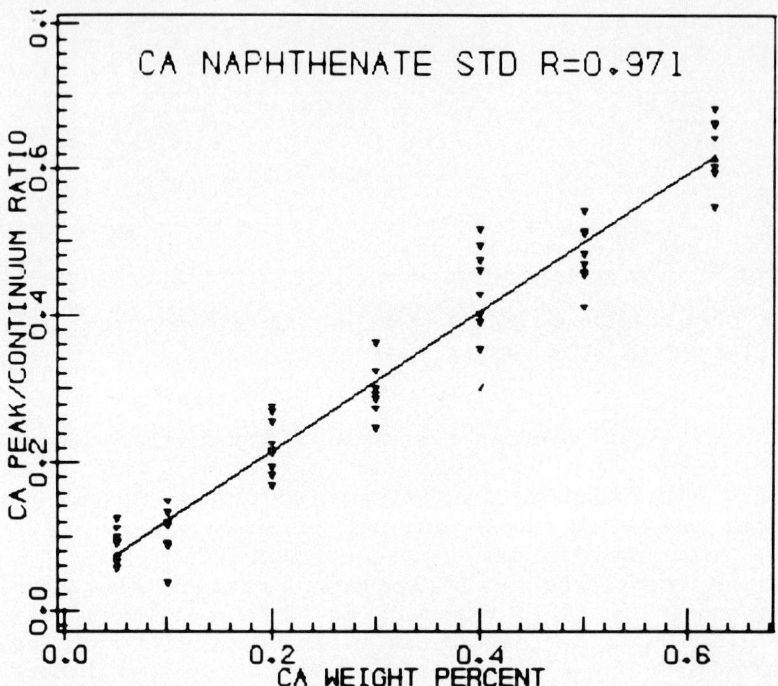

FIGURE 3. This graph shows the weight percent of Ca naphthenate in epoxy versus the Ca peak-to-continuum ratio.

naphthenate standards. For this analysis, unstained epoxy sections of alcohol-fixed tissues were cut on glycerol and mounted on 50-mesh formvar-coated copper grids which then were carbon coated. As shown in Table 4, peak-to-continuum ratios for Ca and P of early inclusions were 2.11 ± 0.97 (SD) and 1.83 ± 0.84 and for advanced discrete inclusions 7.09 ± 2.47 and 4.50 ± 1.47. The Ca-to-P peak intensity ratios were 1.16 ± 0.13 for the early inclusions and 1.57 ± 0.08 for advanced inclusions. Calcium peak-to-continuum ratios were converted to concentrations using the data from the Ca naphthenate standards. Ca concentration (weight %) for the early inclusions was 2.20 ± 1.00 and for the advanced inclusions 7.46 ± 1.22.

TABLE 4. Quantitative Analysis of Calcific Inclusions in Alcohol-Fixed Tissue Sectioned on Glycerol.

Type of Inclusion	Ca P/C ratio[a,c]	Ca concentration[b] (Wt. %)	P P/C ratio[c]	Ca/P peak[c] intensity ratio
Small granular incl.	2.11 ± 0.97 (SD)	2.20 ± 1.00	1.83 ± 0.84	1.16 ± 0.13
Large granular to spicular incl.	7.09 ± 2.47	7.46 ± 1.22	4.50 ± 1.47	1.57 ± 0.08

[a]Ca and P P/C ratios represent mean values from multiple individual inclusions analyzed with the STEM raster mode.

[b]Ca P/C ratios were converted to concentration by comparison with the data from the Ca naphthenate standards.

[c]P/C ratio = peak-to-continuum ratio; Ca/P ratio = calcium-to-phosphorus ratio.

Significance of Analytical Electron Microscopic Studies

Analytical electron microscopy has provided direct information regarding the elemental content of various mitochondrial inclusions observed in normal, ischemic, and infarcted myocardium. The weight percent of Ca of mitochondrial inclusions in plastic sections can now be computed, based on the Ca naphthenate standards. Our quantitative data confirm the concept that the small granular inclusions formed during initial stages of calcification represent a readily soluble, chemically amorphous (noncrystalline) form of calcium phosphate with a relatively low Ca/P ratio, and that the larger inclusions of more advanced calcification exhibit higher Ca/P ratios, consistent with formation of partly crystalline, hydroxyapatite-like material (Posner et al., 1969; Landis and Glimcher, 1978).

Our findings indicate that the method of tissue preparation significantly affects the amount of calcium preserved in calcific inclusions. Since the general preservaton of tissue with alcohol fixation is poor, direct fixation in phosphate-buffered osmium appears to provide a reasonable compromise for preservation of general ultrastructural features and ultrastructural identification of calcific inclusions. Anhydrous preparation of the tissue can then be used for study of the elemental content of the inclusions. The calcification process also appears to involve binding of calcium phosphate to an organic matrix (Bonucci et al., 1973). Further work is needed to define the nature and role of various organic species in the calcification process (Lehninger et al., 1978; Levy et al., 1980; Gallop et al., 1980; Means and Dedman, 1980).

Studies in our own and other laboratories of the flocculent (amorphous matrix) densities of severely damaged cells have shown that these inclusions are osmiophilic rather than inherently electron dense, have low or nondetectable calcium content, are rich in lipid and protein, and form in mitochondria of severely damaged cells, regardless of the mechanism of injury and level of perfusion (Buja et al., 1976; Hagler et al., 1979a,b; Jennings et al., 1978). These findings suggest that the amorphous matrix densities represent aggregates of denatured organic material formed as a result of mitochondrial damage. Although the flocculent densities and calcific inclusions develop as independent processes, there may be some secondary binding of calcium to the organic deposits, as suggested by some studies (Ashraf and Bloor, 1976; Ashraf et al., 1976).

Both the flocculent (amorphous matrix) densities and granular calcifications of injured cells differ from the normal mitochondrial matrix granules (Burton et al., 1977, 1980b; Trump et al., 1976). Several lines of evidence indicate that the normal granules are composed primarily of osmiophilic organic material with a high phospholipid content (Hagler et al., 1979b; Barnard and Ruska, 1979; Erkocak, 1977). Although these granules have some affinity for calcium binding, they do not ordinarily have a demonstrable calcium content. In some microprobe studies of fresh-frozen cryosections of control tissues, calcium phosphate granules have been identified. The relationship of these granules to the normal osmiophilic matrix granules is unclear. Considerable evidence suggests that formation of calcium phosphate granules in control tissues results from movement and aggregation of ions when tissue sampling, freezing, and cryosection preparation are less than optimal (Saetersdal et al., 1977; Somlyo, 1975; Séveus et al., 1978). The above considerations apply specifically to mitochondria from soft tissues. In contrast, mitochondria from cells in calcifying tissue contain multiple small calcium phosphate deposits (Landis et al., 1977; Sutfin et al., 1971). Progressive mitochondrial calcification in ischemic myocardium is associated with membrane alterations and excess calcium influx and is not simply the result of redistribution phenomena (Jennings et al., 1976; Burton et al., 1977, 1980a,b).

Thus, when using plastic-embedded tissue, methods are available to quantitate the amount of Ca retained in the sections after processing. Nevertheless, with plastic-embedded tissue, one is left with using anhydrous fixation and sectioning to fully document calcium phosphate deposition associated with relatively advanced stages of injury. In addition, there is little or no retention of sodium (Na), chlorine (Cl), and potassium (K) wth these techniques, so the important issue of studying early alterations in diffusible ions cannot be approached with these techniques. For the study of these initial and more subtle stages of cell injury, we have explored other approaches, described below.

STUDIES OF ALTERED MEMBRANE PERMEABILITY AND ELECTROLYTE SHIFTS

Models and Tissue Preparation

These studies were performed with two isolated heart muscle models: (1) a superfused, isometrically contracting papillary muscle preparation using right ventricular papillary muscles from cats (Burton et al., 1977, 1980b), and (2) a perfused interventricular septal preparation using male New Zealand rabbits (Langer and Brady, 1968; Burton et al., 1980a). These preparations were superfused or perfused with physiological salt solutions maintained at 29° to 30° C and electrically paced at constant rates. The muscles were attached to a force transducer for measurement of muscle contractile parameters. Some muscles were maintained under control-oxygenated conditions. After an initial stabilization period, other muscles were subjected to varying periods of hypoxia (medium bubbled with 95% N_2-5% CO_2 instead of 95% O_2-5% CO_2) with and without reoxygenation or to ischemia (perfusion of septal preparation stopped) with or without reperfusion.

For the evaluation of membrane permeability alterations, an ionic lanthanum (La) probe technique was developed (Burton et al., 1977, 1980a,b). Control and hypoxic papillary muscles were exposed to media containing 5 mM $LaCl_3$ for the last hour of the experiments prior to fixation under tension. Control and experimental septa were perfused with media containing 2mM $LaCl_3$ for 30 minutes prior to perfusion fixation. La^{+3} is an electron-dense ion with properties and size similar to that of Ca^{+2}. Several ultrastructural studies have shown that La selectively binds to the basal lamina-plasma membrane complex and is not visualized within cardiac muscle cells (Langer and Frank, 1972; Martinez-Palomo et al., 1973; Burton et al., 1977, 1980a,b). As an extension of these previous studies, Burton et al. (1977, 1980a,b), have shown that abnormal intracellular deposition of La can serve as a cytochemical marker of altered membrane permeability associated with cell injury. The ionic La probe technique appears to represent a more sensitive method for the detection of membrane injury than the exposure of tissue to fixatives containing colloidal La (Hoffstein et al., 1975; Ashraf et al., 1979).

We have also developed a method for the study of the subcellular distribution and concentration of soluble electrolytes in sections of fresh frozen myocardium (Hagler et al., 1980a). We have found that typical biopsy approaches to a solid organ, such as the heart, are poorly suited for x-ray microanalysis of frozen sections because optimal freezing is essentially limited to the cut edges of the biopsy. To circumvent this problem, we have turned to the perfused interventricular septa model. For these studies, the septa were prepared with the small right ventricular papillary muscles, normally attached to the septum, left in place for subsequent sampling (Figure 4a). After the desired period of in vitro perfusion, a papillary muscle was clamped in a specially designed muscle holder and the base of the papillary muscle was cut away from the septum. A silver tissue mounting pin was placed between the muscle and a small retaining clip (Figure 4b). The whole assembly was then plunged into a double-freezing dewar of propane slush, cooled with liquid nitrogen. The time from entering the chamber to

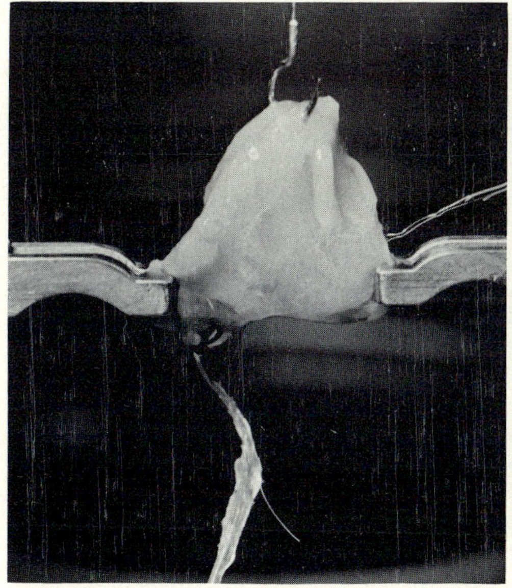

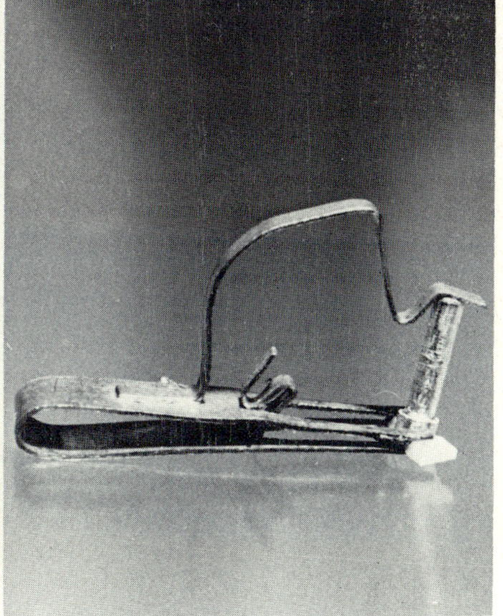

FIGURE 4. Left panel (a). The rabbit interventricular septal preparation with attached papillary muscles is shown here. The base of the septum is shown clamped between two opposed forceps; the force transducer is attached to the apex; a thermocouple for monitoring the muscle temperature is on the right; and the perfusion cannula enters the muscle from the bottom. Right panel (b). The muscle clamp, papillary muscle, and silver mounting pin are illustrated in this photograph. This assembly is snap frozen in liquid nitrogen-cooled propane slush.

freezing was typically 30 seconds. Propane cooled over liquid nitrogen has proven to be the best coolant for freezing this tissue (Hagler et al., 1980a). The papillary muscles were typically 0.5 mm in diameter and 6 mm in length. When positioned over the pin in the holder, the muscles presented a thin, flat surface for freezing and subsequent cryosectioning (Hagler et al., 1980a). The muscle sampling assembly was transferred from the propane to liquid nitrogen, and the pin and attached central portion of the papillary muscle were removed from the clamp by rotating the pin about the long axis of the papillary muscle, thereby fracturing the center of the muscle from the ends held in the clamps. The pins with attached muscles were stored under liquid nitrogen prior to cryoultramicrotomy. Three types of papillary muscles from the rabbit septa were studied: (1) nonperfused control papillary muscles taken immediately after excision of the heart; (2) perfused control papillary muscles taken after 1 hour of stabilization and 1 to 3 hours of perfusion; and (3) hypoxic muscles taken after 1 hour of stabilization and 1 to 1-1/2 hours of hypoxia.

Cryoultramicrotomy and Freeze Substitution

Cryoultramicrotomy was performed, using an LKB ultratome and cryokit. In addition to the modifications previously described by us and others (Hagler et al., 1979; Barnard and Séveus, 1978; Séveus, 1978), further improvements in the cryochamber environment have been made (Hagler et al., 1980a). Freeze substitution was used to check the adequacy of freezing and quality of ultrastructural preservation. Papillary muscles were secured in the muscle clamp and snap frozen in liquid nitrogen-cooled propane slush. The frozen tissue was transferred to a vial containing a mixture of 10% osmium in acetone which had been frozen in liquid nitrogen. The vial was placed in a liquid nitrogen reservoir and then allowed to warm to room temperature over 16 to 20 hours as the liquid nitrogen evaporated. The tissue was embedded in epon-araldite, thin-sectioned and stained with ethanolic uranyl acetate and lead citrate for routine transmission electron microscopy.

Preparation of Gelatin Standards for Use with Cryosectioning

Based on the results described by Roomans and Séveus (1977) and Roomans (1979), cryostandards were prepared using a mixture of various salts in a gelatin-glycerol matrix. A mixture of 20% gelatin and 5% glycerol was used as a base to which was added $CaCl_2$, KH_2PO_4, or NaH_2PO_4 in varying concentrations. The glycerol and salts were prepared and mixed thoroughly prior to the addition of the gelatin. When all solutions were prepared, the salt solutions were heated in a water bath to 50°C, the gelatin was added, and the mixture was vortexed for 30 seconds. The mixture was returned to the water bath for another 30 seconds of heating. This was repeated three times and then a drop was placed on a silver pin and snap frozen in propane slush at -187°C. The gelatin standards were then stored under liquid nitrogen until they were transferred to the cryoultramicrotome. The standards were found to section best at a knife temperature of -80°C and a specimen temperature of -90°C. At these temperatures, flat cellophane sections were obtained. These sections were transferred to formvar-coated 50-mesh copper grids, pressed, and freeze-dried, as described in the previous section on cryoultramicrotomy.

Ultrastructure and X-Ray Microanalysis of Cryosections

The modified LKB cryokit gave improved performance for low temperature cryoultramicrotomy of unfixed tissue. Cryosections of the propane-frozen tissue had only a narrow band of tissue which had no or minimal ice crystal damage and was acceptable for x-ray microanalysis (Figure 5a) (Hagler et al., 1980a). Unfixed, frozen-dried sections of normal myocardium exhibited myofibrils and dense ovoid structures with the distribution of mitochondria (Figure 5b). The sections showed some variation in contrast which appeared to be related to the quality of freeze-drying with increased visualization of ultrastructure in areas with slight ice crystal damage (Séveus, 1978; Séveus and Barnard, 1978). Some perfused control muscles exhibited evidence of damage with formation of multiple calcium phosphate granules in the mitochondria. This appeared to be related to poor perfusion and/or problems with freezing and cryosectioning (Somlyo et al., 1975; Séveus et al., 1978; Barnard and Séveus, 1979). Most perfused controls, however, appeared similar to nonperfused controls and did not show mitochondrial granules (Figure 6a). Muscles subjected to 1 to 1-1/2 hours of hypoxic perfusion typically showed formation of mitochondrial granules in some muscle cells (Figure 6b).

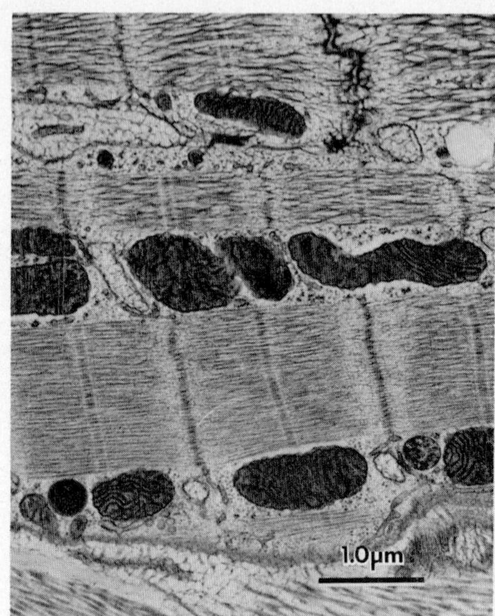

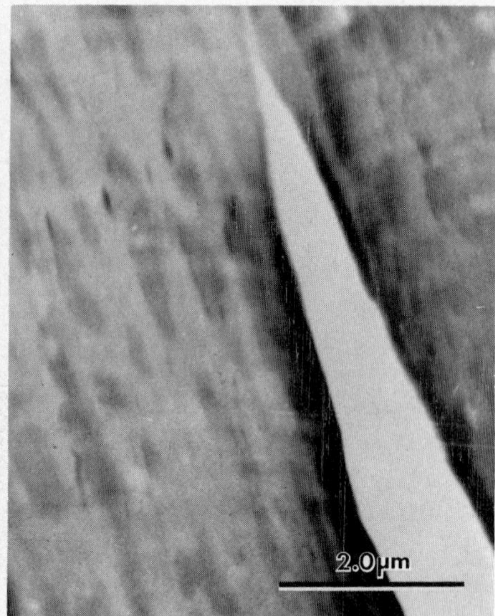

FIGURE 5. Left panel (a). A transmission electron micrograph of a freeze-substituted, nonperfused papillary muscle demonstrating the graded ice damage as one moves from the surface (bottom) toward the center of the muscle (top). Freeze-substituted in 10% osmium in acetone, stained with uranyl acetate and lead citrate. Right panel (b). Scanning transmission electron micrograph of nonperfused cryosectioned papillary muscle showing the low contrast of well-frozen tissue with minimal to no ice crystal disruption. The dense ovoid structures have the shape of mitochondria.

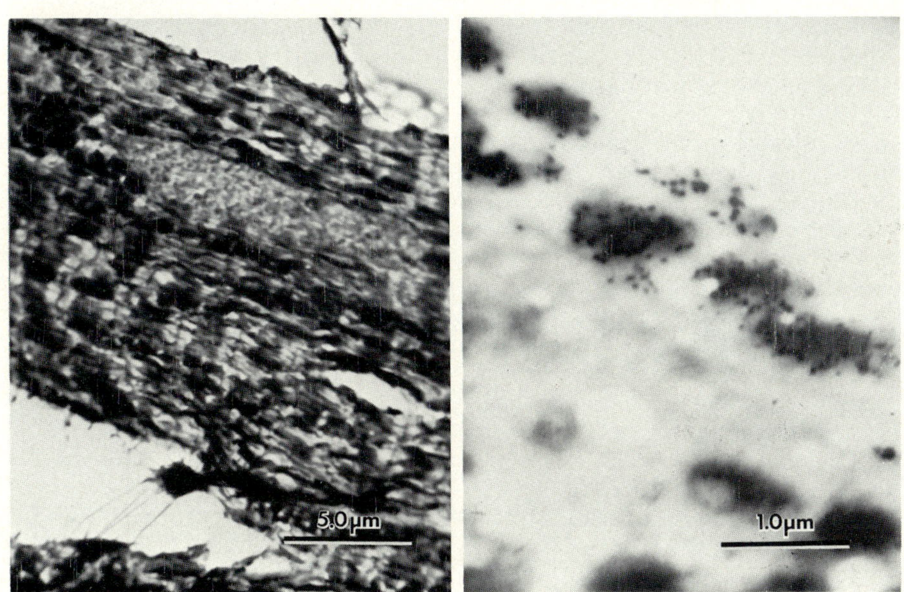

FIGURE 6. Left panel (a). Scanning transmission electron micrograph of a perfused control papillary muscle showing good ultrastructural definition. The enhanced contrast is due to small ice crystal damage which has caused slight separation of structures, thus allowing better visualization. This degree of ice damage has the advantage of allowing better identification of subcellular structures. (Right panel) (b). Scanning transmission electron micrograph of a cryosection from a one-hour hypoxic muscle. The mitochondria have become loaded with calcium phosphate granules. These cells have increased Na and decreased K concentrations.

X-ray analysis data from the cryosections are shown in Tables 5 and 6. Typical x-ray spectra from damaged and control mitochondria are shown in Figure 7. Nonperfused controls showed low Na and high K and P levels and equivocal Ca peaks in cytoplasm (primarily myofibrillar regions) and mitochondria. Well-perfused, oxygenated control muscles had similar levels of the various elements. Although the mean values for some elements were slightly different for nonperfused and perfused controls, the standard deviations suggested considerable overlap in the values. In hypoxic muscles (no reoxygenation), myocytes without mitochondrial inclusions showed marked electrolyte alterations in cytoplasm and mitochondria. The changes consisted of increases in Na and Cl and decreases in K and P without convincing changes in Ca or Mg. Hypoxic muscle cells with mitochondrial inclusions exhibited marked alterations in elemental content of cytoplasm and mitochondria. The mitochondrial inclusions were associated with a prominent mitochondrial Ca peak and a relative increase in P.

Quantification of Calcium in Cryosections

Figure 8 shows the standard curve obtained by microanalysis of the calcium standards prepared in a gelatin matrix. As with the Ca naphthenate standards, a linear relationship was found between peak-to-continuum ratio and concentration down to a concentration of about 0.1 gm%. Quantification of Ca in cryosections is obtained by comparing the Ca peak-to-continuum ratios in Tables 5 and 6 with the standard curve shown in Figure 8. Quantification of other elements and myocardium has recently been obtained (Burton et al., 1981).

In Vitro Lanthanum Probe Studies

Analytical electron microscopy has provided data regarding the localization of lanthanum in isolated muscle preparations after control and experimental conditions (Burton et al., 1977, 1980b). In control oxygenated muscles from the cat papillary muscle and rabbit perfused septal preparations, electron-dense deposits consistent with lanthanum accumulation were confined to extracellular regions involving the basal lamina-plasma membrane complex of muscle cells in both fixed epoxy-embedded sections and unfixed freeze-dried cryosections. Analytical electron microscopy confirmed the selectively extracellular localization of detectable lanthanum in control preparations. After 75

Table 5. Elemental Peak-to Continuum Ratios in Cytoplasm of Control and Hypoxic Rabbit Myocytes Prepared by Cryoultramicrotomy (STEM Raster Mode Analysis). (± s.d.)

	Na	Cl	S	K	Ca	Mg	P
Non-Perfused Controls (2 muscles, 13 spectra)	.30 ± .12	.79 ± .28	1.37 ± .25	2.22 ± .41	.03 ± .04	.11 ± .05	1.66 ± .40
Perfused Controls (3 muscles, 25 spectra)	.53 ± .24	1.33 ± .52	1.00 ± .22	1.75 ± .24	.08 ± .06	.09 ± .04	1.33 ± .35
1.0-1.5 Hour Hypoxia							
A. Cells with no mitochondrial inclusions (2 muscles, 10 spectra)	.77 ± .16	1.23 ± .30	0.90 ± .19	0.88 ± .40	0.10 ± .08	.06 ± .05	.88 ± .34
B. Cells with mitochondrial inclusions (1 muscle, 5 spectra)	.76 ± .07	.97 ± .13	.70 ± .14	.45 ± .03	.19 ± .07	.09 ± .03	.63 ± .11

Table 6. Elemental Peak-to-Continuum Ratios in Mitochondria of Control and Hypoxic Rabbit Myocytes prepared by Cryoultramicrotomy (STEM Raster Mode Analysis). (± s.d.)

	Na	Cl	S	K	Ca	Mg	P
Non-Perfused Controls (2 muscles, 13 spectra)	.35 ± .17	.88 ± .44	1.26 ± .27	1.73 ± .47	.03 ± .03	.08 ± .04	1.71 ± .28
Perfused Controls (3 muscles, 24 spectra)	.46 ± .31	1.00 ± .50	1.01 ± .23	1.58 ± .28	.07 ± .08	.09 ± .04	1.62 ± .42
1.0-1.5 Hour Hypoxia							
A. Cells with no mitochondrial inclusions (2 muscles, 12 spectra)	.60 ± .15	1.09 ± .26	.90 ± .21	.86 ± .44	.11 ± .07	.07 ± .05	1.10 ± .40
B. Cells with mitochondrial inclusions (2 muscles, 18 spectra)	.75 ± .15	1.03 ± .19	.79 ± .17	.67 ± .31	.62 ± .40	.12 ± .06	1.36 ± .47

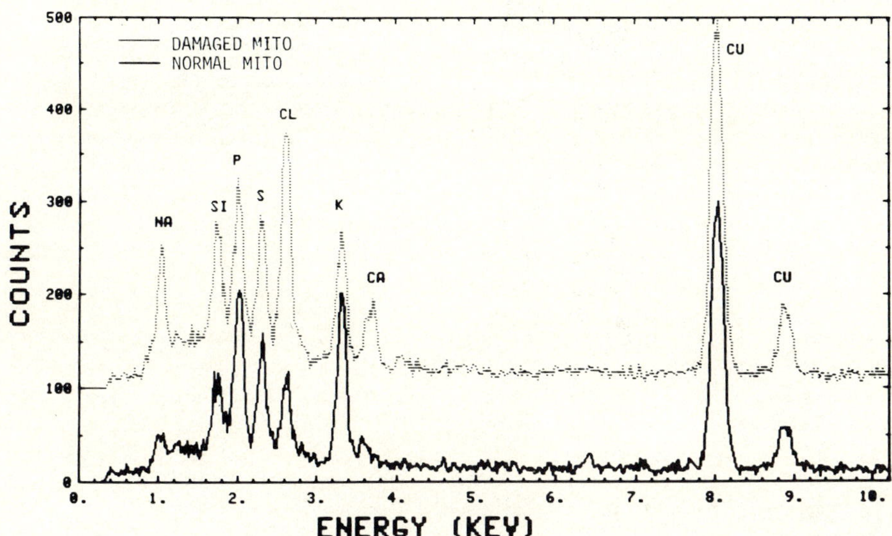

FIGURE 7. Typical x-ray spectra from normal (solid line) and damaged (dotted line) mitochondria from unfixed cryosections of rabbit myocardium.

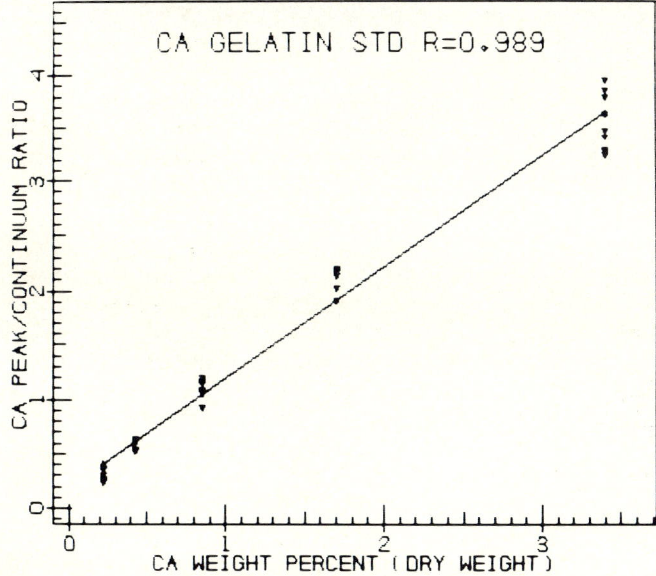

FIGURE 8. This graph shows data taken from the Ca standards in 20% gelatin. The Ca weight percent is in terms of dry weight. There is a very good linear relationship between peak-to-continuum ratio and Ca weight percent.

minutes of hypoxia, abnormal intracellular localization of lanthanum occurred in the form of electron-dense deposits on the myofibrils and in the mitochondria (Figure 9). The onset of abnormal intracellular lanthanum accumulation correlated with the transition from reversible to irreversible contractile depression in the isolated cat papillary muscles subjected to progressive intervals of hypoxia and reoxygenation (Burton et al., 1977). We have also shown that glucose deprivation during hypoxia hastens the onset of intracellular La accumulation and progression to irreversible injury (Burton et al., 1980b).

Recent studies in the isolated perfused septal preparation have shown a similar time course in the development of widespread abnormal intracellular La accumulation following ischemia and reperfusion (Burton et al., 1980a). These changes correlated with a progressive decline in the return of contractile function upon reperfusion. We have also found that pretreatment with the calcium-blocking agent, chlorpromazine, results in reduced intracellular La accumulation and better recovery of contractile function after 1 and 1/2 hours of ischemia with reflow (Burton et al., 1980a).

Significance of Analytical Electron Microscopic Studies

Electrolyte shifts and pathological calcium accumulation appear to represent consequences of altered membrane integrity induced by ischemic damage (Burton et al., 1977, 1980a,b; Jennings and Ganote, 1976; Trump et al., 1976, 1978). Our *in vitro* lanthanum probe studies have provided direct cytologic evidence of altered movement and intracellular accumulation of polyvalent ions at a transitional stage in the evolution of irreversible hypoxic and ischemic injury. Selective extracellular localization of lanthanum in unfixed freeze-dried cryosections indicates that fixation does not produce major alterations in lanthanum distribution and provides supportive evidence for the use of the lanthanum probe technique for the study of membrane pathophysiology.

The methods applied to the isolated perfused interventricular septal model permit the experimental study of intracellular electrolytes in normal and injured myocardium. The preliminary results presented here will serve as a foundation for the study of various forms of myocardial cell injury. The septal model is typically stable

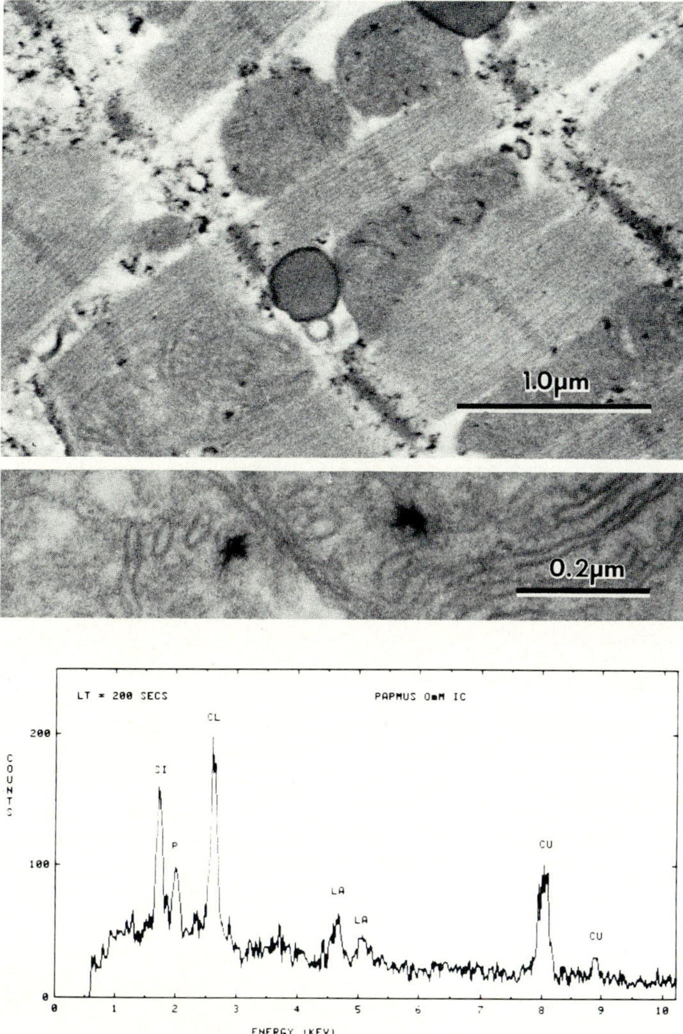

FIGURE 9. Abnormal intracellular accumulation of dense lanthanum deposits is present along the I bands of the myofibrils and within the mitochondria of a cat papillary muscle subjected to 75 minutes of hypoxia prior to exposure to 5 mM $LaCl_3$. Top panel. Tissue fixed in aldehyde and osmium. Unstained epoxy section. Transmission electron micrograph. Middle panel. Higher magnification transmission electron micrograph of La deposits in epoxy section stained with uranyl acetate and lead citrate. Bottom panel. Summed x-ray spectrum demonstrates lanthanum peaks in mitochondrial deposits from fixed, unstained epoxy sections.

for several hours. Performance based on physiological parameters of tension and rate of tension development are not changed significantly after 3 hours perfusion.

This model is well suited for the cryosectioning of heart tissue. The perfused papillary muscles provide a way to sample a preparation without having to obtain cryosections from a layer of cut and damaged tissue. The model reduces the alterations induced in electrolyte measurement due to other biopsy techniques and provides access to a piece of muscle which is large enough to work with easily and still has a very high freezing rate. When tissue sampling was optimal, electron microscopy of freeze-substituted tissue sections and cryosections revealed normal ultrastructure, and x-ray microanalysis showed similar elemental content of nonperfused and perfused control muscles. Our findings also indicate that less than optimal tissue preparation can result in cell damage characterized by formation of numerous small calcium deposits in the mitochondria and electrolyte shifts with decreased K and increased Na. Similar findings have been reported previously for other preparations (Barnard and Ruska, 1979; Séveus et al., 1978; Somlyo et al., 1975). These considerations emphasize the importance of performing repeated control studies during the course of evaluation of a particular experimental intervention. Our initial studies, however, have shown evidence of electrolyte shifts and mitochondrial calcium accumulation in hypoxic muscles compared to well-perfused control muscles. The combined application of the rabbit interventricular septal model with improved techniques for sampling, freezing, and cryoultramicrotomy holds promise for furthering our understanding of the normal regulation of intracellular electrolytes in myocardium and the pathophysiological role of electrolyte alterations in various types of myocardial cell injury.

ACKNOWLEDGMENTS

We acknowledge the excellent work of Ms. Linda Lopez and Dr. Mark Murphy in the preparation and microanalysis of tissues and standards; Mrs. Carol Greico, for cryoultra-microtomy; Mrs. Anna Siler, for transmission electron microscopy; Mrs. Rebecca Lunswick, for work with the isolated perfused septal preparation; and Ms. Kathy Handrick, for secretarial assistance.

This work was supported by NIH Ischemic Heart Disease SCOR Grant HL-17669 and by the Harry S. Moss Heart Center of The University of Texas Health Science Center at Dallas.

REFERENCES

Ashraf, M. (1979). Am. J. Pathol. 97:411-32.
Ashraf, M., and Bloor, C. M. (1976). Virchows Arch. B. Cell. Path. 22:287-97.
Ashraf, M., Sybers, H. D., and Bloor, C. M. (1976). Exptl. Molec. Path. 24:435-40.
Barnard, T., and Séveus, L. (1978). J. Microsc. 112:281-91.
Barnard, T., and Ruska, J. (1979). Exp. Cell Res. 124:339-47.
Bonucci, E., Derenzini, M., and Marinozzi, V. (1973). J. Cell Biol. 59:185-93.
Buja, L. M., Dees, J. H., Harling, D. E., and Willerson, J. T. (1976). J. Histochem. Cytochem. 24:508-16.
Buja, L. M., Tofe, A. J., Kulkarni, P. V., Mukherjee, A., Parkey, R. W., Francis, M. D., Bonte, F. J., and Willerson, J. T. (1977). J. Clin. Invest. 60:724-40.
Burton, K. P., Hagler, H. K., Templeton, G. H., Willerson, J. T., and Buja, L. M. (1977). J. Clin. Invest. 60:1289-1302.
Burton, K. P., Hagler, H. K., Greico, C. A., Willerson, J. T., and Buja, L. M. (1980a). (Abstract) Fed. Proc. 39:276.
Burton, K. P., Templeton, G. H., Hagler, H. K., Willerson, J. T., and Buja, L. M. (1980b). J. Molec. Cell. Cardiol. 12:109-33.
Burton, K. P., Hagler, H. K., Lopez, C. E., Greico, C. A., Lunswick, R. L., Wilderson, J. T. , and Buja, L. M. (1981). Abstract. Fed. Proc. 40:7-59.
Erkocak, A. (1977). Acta. Histochem. 58:360-63.
Gallop, P. M., Lian, J. B., and Hauschka, P. V. (1980). New Engl. J. Med. 302:1460-66.
Hagler, H. K., Burton, K. P., Browne, R. H., Reynolds, R. C., Templeton, G. H., Willerson, J. T., and Buja, L. M. (1977). In "Scanning Electron Microscopy/1977" (O. Johari, ed.), Vol. 2, pp. 145-52. IIT Research Institute, Chicago.
Hagler, H. K., Burton, K. P., Sherwin, L., Greico, C., Siler, A., Lopez, L., and Buja, L. M. (1979a). In "Scanning Electron Microscopy/1979" (O. Johari, ed.), Vol. 2, pp. 723-32. SEM, Inc., AMF O'Hare, Illinois.
Hagler, H. K., Sherwin, L., and Buja, L. M. (1979b). Lab. Invest. 40:529-44.

Hagler, H. K., Burton, K. P., Greico, C. A., Lopez, L. E., and Buja, L. M. (1980a). In "Scanning Electron Microscopy/1980" (O. Johari, ed.), Vol. 2, pp. 493-98 and 510. SEM, Inc., AMF O'Hare, Illinois.

Hagler, H. K., Lopez, L. E., Murphy, M. E., and Buja, L.M. (1980b). In "Proceedings of the Electron Microscopy Society of America," pp. 514-15.

Hagler, H. K., Lopez, L. E., Murphy, M. E., Burton, K. P., and Buja, L. M. (1980c). Lab. Invest. (Submitted).

Hall, T. A., and Gupta, B. L. (1974). J. Microsc. 100:177-90.

Hall, T. A., Clarke-Anderson, H., and Appleton, T. (1973). J. Microsc. 99:177-82.

Hoffstein, S., Gennaro, D. E., Fox, A. C., Hirsch, J., Streuli, F., and Weissmann, G. (1975). Am. J. Pathol. 79:207-28.

Jennings, R. B., and Ganote, C. E. (1976). Circ. Res. 38 (Suppl.I): I-80-I-91.

Jennings, R. B., Shen, A. C., Hill, M. L., Ganote, C. E., and Herdson, P. B. (1978). Exptl. Molec. Path. 29:55-65.

Johnson, G. G., Jr., and White, E. W. (1970). In "ASTM Data Series DS 46." American Society for Testing and Materials, Philadelphia.

Langer, G. A., and Brady, A. J. (1968). J. Gen. Physiol. 52:682-713.

Langer, G. A., and Frank, J. S. (1972). J. Cell Biol. 54: 441-55.

Landis, W. J., Paine, M. C., and Glimcher, M. J. (1977). J. Ultrastr. Res. 59:1-30.

Landis, W. J., and Glimcher, M. J. (1978). J. Ultrastr. Res. 63:188-223.

Lehninger, A. L., Reynafarje, B., Vercesi, A., and Tew, W. P. (1978). Ann. New York Acad. Sci. 307:160-76.

Levy, R. J., Zenker, J. A., and Lian, J. B. (1980). J. Clin. Invest. 65:563-66.

Martinez-Palomo, A., Benitez, D., and Alanis, J. (1973). J. Cell Biol. 58: 1-10.

Means, A. R., and Dedman, J. R. (1980). Nature 285:73-77.

Ornberg, R. L. (1979). National Institute of Neurological and Communicative Disorders and Stroke, National Institute of Health, Bethesda, Maryland. Personal communication.

Posner, A. (1969). Physiol. Rev. 49:760-92.

Roomans, G. M., and Seveus, L. A. (1977). J. Submicr. Cytol. 9:31-35.

Roomans, G. M. (1979). In "Scanning Electron Microscopy/1979" (O. Johari, ed.), Vol. 2, pp. 649-57. SEM Inc., AMF, O'Hare, Illinois.

Saetersdal, T. S., Myklebust, R., Berg Justesen, W. P., Engedal, H., and Olsen, W. C. (1977). Cell Tissue Res. 182:17-31.

Schamber, F. H. (1976). In "Workshop on X-Ray Fluorescence Analysis of Environmental Samples" (T. Dzubay, ed.), pp. 1-26. Ann Arbor Science Publications, Ann Arbor.

Séveus, L. (1978). J. Microsc. 112:269-79.

Séveus, L., Brdiczka, L., and Barnard, T. (1978). Cell Biol. Internatl. Rep. 2:155-62.

Shuman, H., Somlyo, A. V., and Somlyo, A. P. (1976). Ultramicrosc. 1:317-39.

Shuman, H., Somlyo, A. V., and Somlyo, A. P. (1977). In "Scanning Electron Microscopy/1977" (O. Johari, ed.), Vol. 1, pp. 683-72. IIT Research Institute Chicago.

Somlyo, A. V., Silcox, J., and Somlyo, A. P. (1975). In "Proceedings of the Thirty-Third Annual meeting of the Electron Microscopy Society of America" (G. W. Bailey and C. J. Arcenaux, eds.), pp. 532-33. Claitor's Publishing Division, Baton Rouge.

Sutfin, L. V., Holtrop, M. E., and Ogilvie, R. E. (1971). Science 174:947-49.

Trump, B. F., Mergner, W. J., Kahng, M. W., and Saladino, A. J. (1976). Circ. Res. 53 (Suppl. I): I-17-I-26.

Trump, B. F., Berezesky, I. K., Pendergass, R. E., Chang, S. H., Bulger, R. E., and Mergner, W. J. (1978). In "Scanning Electron Microscopy/1978" (R. P. Becker and O. Johari, eds.), Vol. 2, pp. 1027-39. SEM Inc., AMF O'Hare, Illinois.

DISCUSSION

SPEAKER: Herbert K. Hagler.

LECHENE: What was the precaution taken to avoid loss or redistribution of calcium in preparations fixed in osmium?

HAGLER: In our initial studies, we analyzed particulate mitochondrial inclusions which were at least partially preserved in aqueous fixatives. Different mitochondrial inclusions were clearly distinguishable by analytical electron microscopy. The amorphous matrix (flocculent) densities were devoid of calcium peaks. Discrete granular to spicular densities had calcium peaks of intermediate intensity. The confluent dense masses had large calcium peaks.

SOMLYO, Andrew: When you say peak-to-continuum, which region of the continuum is it?

HAGLER: Initially we used the region of 5-5.5 KeV. We are currently using the region 5-6 KeV. This seems to be a region that is reliably free of interfering peaks for our specimens.

SOMLYO, Andrew: Except from the grid. And also when you are comparing osmium-fixed material, for example, the continuum production is highly beam dependent. That is going to change the continuum.

HAGLER: The contribution from the grid is minimized by using 50-mesh grids and working well away from the grid bars. Even though osmium was present, clear differences were found in calcium content among the different inclusion types. Our more recent work has been performed with unosmicated tissue.

ORNBERG: Do you straighten your sections cut on glycerol when they are floating?

HAGLER: Yes.

ORNBERG: How do you straighten them?

HAGLER: With the glycerol sections we either use a small heat probe or hot wire, or the diamond knife can be removed from the microtome and gently heated on a hot plate until the sections relax.

SOMLYO, Andrew: The potassium-to-chloride ratio seems to be rather low doesn't it? Unless you have a very thick beryllium window, I think the potassium and chloride peaks should be similar.

BUJA: Our preliminary results using our peak-to-continuum ratios and gelatin standards for potassium and chloride have yielded reasonable values for these two elements.

RICK: Do these cells swell in ischemia?

HAGLER: Yes, to a variable degree. This is also true for hypoxia models we have studied.

RICK: Well, then I am a little bit worried about the fact that chloride does not increase. Because there would be two reasons: firstly, extracellular negative ions would enter the cells in order to account for other electrolyte shifts; and, secondly, you would have to divide the total electrolyte concentration by a lower parameter.

HAGLER: The exact relationships among cell swelling, sodium and potassium alterations, and chloride levels in our preparation will have to be determined in future studies.

SOMLYO, Andrew: I would just like to make a second point. I do not mean to detract from the value of calcium napthenate; that, indeed, is a very nice standard. When I tried out these standards, however, I could change the calcium concentration by a factor of two, depending on the dosage, because many of the plastics are heavily radiation sensitive. If you use epon/araldite, it may be less radiation sensitive. But if we want to use calcium napthenate as an absolute standard for calcium, we need to be concerned that we deal with it with precautions to avoid mass loss.

HAGLER: We have obtained very reproducible results using calcium napthenate in epon/araldite.

X-RAY MICROANALYSIS OF FREEZE-DRIED MUSCLE: TECHNIQUES AND PROBLEMS

Keith L. Monson
Thomas E. Hutchinson

Center for Bioengineering
University of Washington
Seattle, Washington

INTRODUCTION

Quantitative x-ray microanalysis offers a unique opportunity to determine the elemental composition of organelle regions to a resolution of at least 0.2 μm (Shuman et al., 1976), while they are being visualized with the transmission or scanning transmission electron microscope. This paper describes some experimental protocols developed in our laboratory for microanalysis of freeze-dried muscle sections, as well as approaches to problems encountered in the application of this method.

SPECIMEN FREEZING

The primary obstacle to tissue preparation for microanalysis is that it must be made vacuum compatible and electron transparent, yet the structural and compositional integrity must be preserved. Ultrastructural features must be recognizable in the electron microscope so that an electron beam can be focused upon them. If we are to obtain elemental analyses which are representative of an area when the cell is functioning normally, we must have no redistribution of elements during preparation and observation. The usual methods of fixation and dehydration are probably adequate for morphological study only, but the

ISBN 0-12-362880-6

chemical modifications and elemental losses which they induce make them unacceptable for microanalysis (Hayat, 1970; Sjöström and Thornell, 1975). Rapid freezing and frozen thin sectioning, followed by either frozen-hydrated electron microscopy or freeze-drying, satisfy the requirements.

Freezing rates have been measured by most researchers who do rapid freezing, but results vary according to the volume, geometry, and water content of the specimen on which the temperature is monitored; the velocity at which it is allowed to travel during immersion into the cryogen; and the temperature interval over which the rate is reported. Costello and Corless (1978) have published a very complete review of the freezing rate experiments reported over the last 30 years. We measured freezing rates to develop an experimental procedure and to help us choose from among the various cryogens.

To ensure that the specimen would always enter the coolant at the same rate, we constructed a simple appliance which utilizes acceleration by gravity. A 1 cm diameter stainless steel rod, 75 cm long, is locked above the coolant. A collet at the end grips a microtome chuck, to which the specimen is attached. When the lock is released, the rod travels downward a distance of 12.7 cm, whereupon it is brought to an abrupt halt by deformation of an O-ring. This arrangement provides a constant entry velocity of about 160 cm/sec. If the coolant is one of the liquid Freons, it is maintained at constant temperature by suspending the stainless steel beaker which contains it at the top of a dewar flask filled with liquid nitrogen, so that it is just in contact with the bottom of the stainless steel beaker. Uniform mixing is assured by an electric stirrer submerged in the Freon.

Aluminum screwdriver-shaped microtome chucks of the type described by Somlyo et al. (1977) are very convenient for freezing of small tissue pieces, as no trimming is necessary. To evaluate the approximate specimen freezing rate obtainable with this apparatus in various coolants, a thermocouple made from copper and constantan wires 0.002 inch in diameter was attached to one of these chucks in several configurations. A single sheet of lens tissue (Ross, 0.001 inch thick) was placed over the tip for insulation, and then the thermocouple leads draped over it and taped circumferentially. Another layer of lens tissue, if used, was placed over the thermocouple. Lens tissue, especially when moistened, was used to simulate the low thermal conductivity of biological tissue. The freezing rates obtained in four cryogenic fluids are tabulated in Table 1. The cooling rate of the uncovered miniature thermocouple represents the ultimate rate attainable in this apparatus. The ultimate freezing rate of Freon 21 is very high, and the rate for a wet specimen is only slightly less, so this coolant was chosen for the muscle studies.

FROZEN SECTIONING AND FREEZE-DRYING

200 nm frozen thin sections were cut on a dry knife at -80°C using a Sorvall MT-2B with FTS frozen thin sectioning attachment. Sections were deposited on a Formvar and carbon coated folding grid with a precooled eyelash probe. The folded grid was placed in a cold metal grid box, which serves as a module for transfer to the freeze-drying chamber. The small metal grid box was placed inside a massive brass cup precooled in liquid nitrogen, which was then placed on an insulating platform inside a vacuum evaporator. The chamber was evacuated to 10^{-6} torr, and the cup was allowed to slowly warm overnight, at which time freeze-drying was assumed to be complete.

Figure 1 illustrates a typical section of freeze-dried rat soleus muscle, unstained in any way. The characteristic banding pattern of striated muscle is apparent, with regions of thick and thin filaments clearly delineated. The paired mitochondria at the Z line are electron-dense, with no internal structure in well-frozen specimens. Nuclei with perinuclear mitochondria are present in this field, as well as collagen fibrils which constitute the endomysium.

A NEW COLD STAGE FOR THE JEM-100C

A reliable cold stage, optimized for biological microanalytical work, is essential to reduce or eliminate electron beam-induced loss of mass. The JEOL cold stage is a brass block which is tied via a strap of multiple copper foils to a solid copper conductor embedded in a tank of liquid nitrogen. The entire assembly is supported within a concentric brass cylinder at room temperature, isolated from it by stainless steel bellows, stainless steel knife edges, and glass beads. Insulation between the pieces (which sometimes are separated by only 0.003 inch) at temperature

TABLE 1 - Freezing rates measured in four cryogenic fluids.

Stirred coolant at temperature shown	Freezing Rate (degrees/second)		
	Uncovered 0.002" thermocouple	0.002" thermocouple sandwiched between single layers of <u>dry</u> lens tissue	0.002" Thermocouple sandwiched between single layers of <u>wet</u> lens tissue
Boiling liquid Nitrogen (-196°C)	5000 ($-57°C<T<9°C$)	1800 ($-62°C<T<5°C$)	12 ($94°C<T<-50°C$)
Freon 12 (-140°C to -150°C)	7800 ($-62°C<T<16°C$)	4400 ($-49°C<T<-5°C$)	1600 ($-5°C<T<20°C$) 900 ($-32°C<T<-50°C$)
Freon 22 (-155°C)	26,000 ($-94°C<T<10°C$)	5900 ($-58°C<T<-5°C$)	5300 ($58°C<T<-5°C$)
Freon 21 (-140°C)	23,000 ($-62°C<T<8°C$)	21,700 ($-94°C<T<20°C$)	20,000 ($-32°C<T<20°C$)

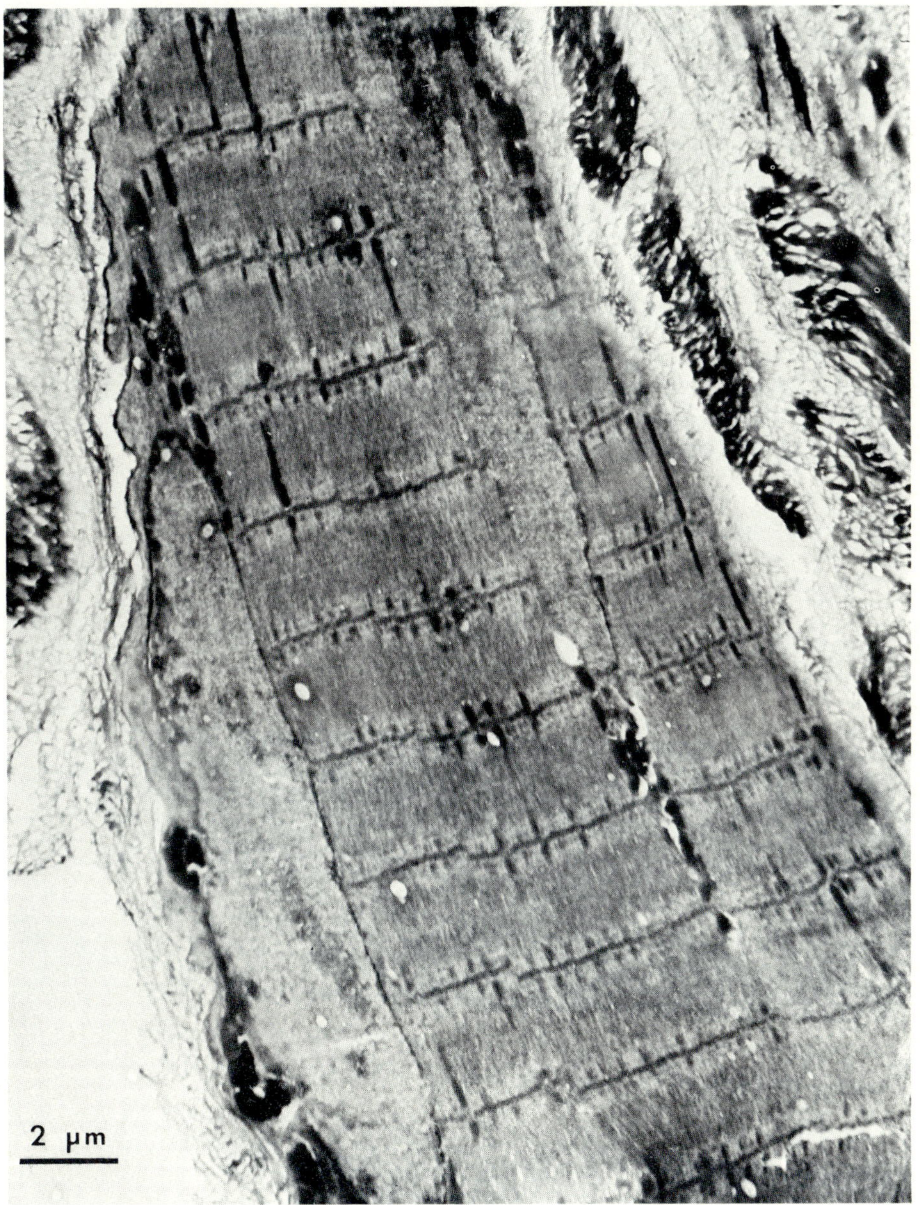

FIGURE 1. An unfixed, unstained, freeze-dried cryosection of rat soleus. Two nuclei are present in the field, with numerous perinuclear mitochondria. A, I, and M bands are visible, as well as a portion of the supporting network of collagen fibrils constituting the endomysium.

extremes is by the vacuum of the microscope, which is maintained by a sliding Viton O-ring on the outer, room temperature cylinder. Unfortunately, the specimen cooling holder as supplied by JEOL is inadequate for study of frozen hydrated specimens or for x-ray microanalysis. Loading of the specimen grid into the stage is tedious due to the many pieces involved, so that it is particularly unsuitable for cold transfer of a frozen specimen. In addition, the position of the grid deep within a brass block prevents any x-rays produced from reaching the detector.

A new stage insert (Figure 2) was designed to satisfy the conditions of operation below -175° C, efficient collection of x-rays generated within the specimen, minimal contribution of spurious background, wide field of view, and rapid specimen loading, even at low temperatures. The idea of a thin "flipper," which is pressed against the grid by a spring load, was adapted from the type used by JEOL in their high resolution side entry stage. The flipper was made thin, and the grid position high in the specimen rod, to allow closer approach by the x-ray detector. This distance

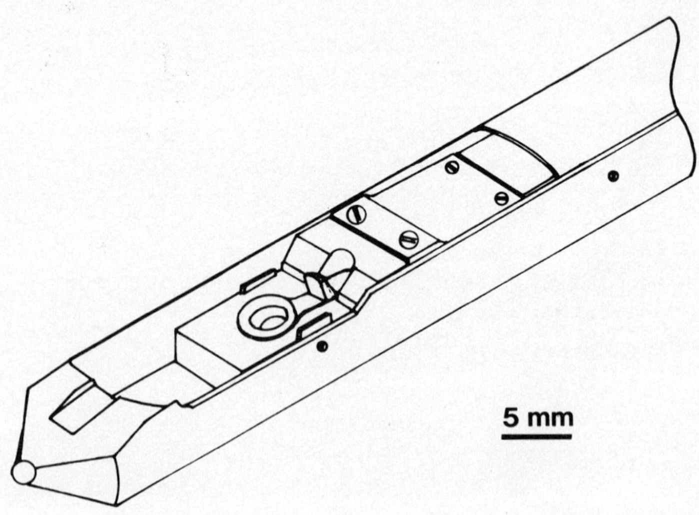

FIGURE 2. Detail of the modified cold stage insert, and modification to the supporting sheath.

was further decreased by fitting the detector snout with a conical collimator. The stage itself was fabricated from the lowest atomic number material which is convenient to machine, aluminum. The stage was made as large as possible to gain thermal stability, yet thin enough so that even for large tilt angles the field of view would be wide.

Numerous changes were made to the specimen rod. The near side of the sheath surrounding the stage was filed down to allow a direct line of sight from detector to specimen, and also to allow a shorter source-detector distance. The bottom cross-member of this sheath was also filed, to accommodate thermal contraction of over a millimeter, and the two glass beads between the top aspect of the sheath and the conducting rod were eliminated. The inner conducting rod had three knife-edged stainless steel beads at the level of the O-ring, and three more near the liquid nitrogen container; these were removed and replaced by Teflon rings, turned to a knife edge at the outer circumference.

EVALUATION OF TWO METHODS FOR CALCULATION OF MASS FRACTIONS

The definition of mass fraction of element x, C_x, is:

$$C_x = \frac{\text{mass of x/unit area}}{\text{total mass/unit area}} = \frac{M_x}{M} = \frac{AI_x}{\rho t} .$$

For "thin" specimens, the so-called Hall assumption holds that the mass of x, M_x, is proportional to the intensity of the characteristic x-ray line, I_x. A is a constant relating x-ray intensity to concentration. Total mass per unit area, or mass thickness, ρt, can be determined from the intensity of the Bremsstrahlung, or alternatively from the attenuation of the incident beam due to scattering by the specimen.

The Peak-to-Continuum Normalization Method

T. A. Hall developed this now well-known routine which has been described in numerous sources (e.g., Hall and Gupta, 1979; Shuman et al., 1976). For specimen and standards of similar average atomic number, the working equation derived by Hall is:

$$C_x = \frac{(I_x/I_w)_{sp}}{(I_x/I_w)_{std}} \; (C_x)_{std} \; .$$

Since the method depends on reference to the background created by the electron beam from the specimen alone (I_w), it may be necessary to apply a correction to backgrounds of both the specimen and standards to account for the portion of the background produced by matter which is not part of the specimen.

The causes and elimination of stray radiation in the electron microscope which give rise to spurious results in x-ray microanalysis have been studied by many authors. In a commercial STEM such as the JEM-100C, the strong prefield above the objective pole piece prevents backscattered electrons from reaching the detector. In these STEM/TEM units, the major contribution to spurious background is non-localized excitation by hard x-rays produced in the first and second condenser apertures (Headley and Hren, 1978). The extent of this contribution can be evaluated by directing the beam through a small hole in the specimen, and operating the microscope in standard conditions. After we installed a thick tantalum aperture just above the specimen, the background "hole count" was reduced by a factor of 8. The hard x-ray-preventing aperture is now used routinely for all analyses to effectively screen the specimen from the x-ray flux.

Even after the generation of x-rays from outside the area of interest by hard x-rays is effectively eliminated, excitation by backscattered electrons remains, and it is rather difficult to control, as it is specimen dependent. This contribution to the spectrum can be minimized by careful design of a detector collimator. The cylindrical collimator ordinarily supplied for this instrument was replaced by a conical type, which allowed reduction of the source-detector distance from 2.3 to 2.0 cm, increasing the solid angle of collection from 0.057 to 0.075 sr, improving the geometrical detection efficiency by one-third (Woldseth, 1973). The diameter of the collimator was chosen to allow use of the full detector area, but at the same time to shield it from x-rays originating outside of the specimen.

After these measures were taken which reduced excitation of the specimen and its surroundings by hard x-rays, and which minimized, by effective collimation, collection of x-rays produced outside of the analyzed area, the corrections to specimen and film background for bulk-generated continuum were small enough to be neglected. The low-energy region used to represent the Bremsstrahlung (1.1-2.9 kev) is fairly insensitive to the superposition of a contribution due to Cu, as the bulk Cu continuum intensity is low in this region. The Al peak due to the cold stage was in all cases the same order of magnitude as those of the light elements being analyzed, which are known to constitute only a few percent of the total mass. Thus, no corrections for bulk-generated continuum have been applied to the concentration values.

On theoretical grounds, a film correction must be applied to the Bremsstrahlung if one hopes to evaluate concentrations in the thin specimen alone. The most obvious method is to direct the beam on the film adjacent to the specimen, and then subtract its continuum from the total continuum. When this technique is applied, however, the results are not always predictable. To illustrate this, calculated concentration values for A band potassium and sulfur are presented in Figure 3. All analyses of standards and of various regions of freeze-dried cryosections were performed at 80 kv, using the cold stage at -180° C. It effectively eliminated mass loss for an exposure of 10 C/cm^2, over an analysis period of up to 500 s. Solutions of known concentrations of K, S, and Cl in epoxy resin (Spurr, 1975) were utilized as primary standards, and binary salts containing Na, P, or Ca as secondary standards (Shuman et al., 1976).

The film corrected results form an unusual distribution, which is not seen in the uncorrected data (Figure 3). In the corrected data, roughly two-thirds of the concentrations are of magnitude comparable to the uncorrected data. There are also, however, clusters of values which are several times greater than the expected values, and two values which are very large negatively. The large values do not occur randomly; rather, when this occurs all values collected from the same section under similar conditions are of this magnitude, and the concentrations of the other elements are similarly elevated.

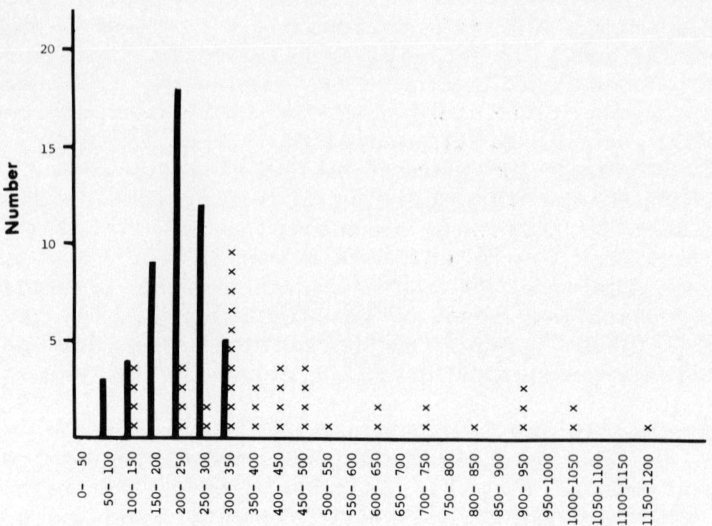

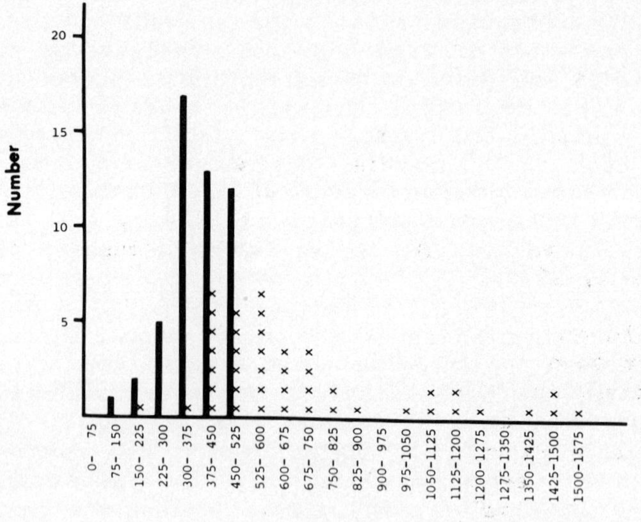

FIGURE 3. Concentrations of S and K in freeze-dried rat soleus were calculated by the continuum normalization method. The narrow distribution of the uncorrected values (solid bars) is broadened when specimen backgrounds are corrected by subtracting spectra from the supporting films (x x).

One explanation for this phenomenon is based on differences in x-ray collection efficiency from different areas of the grid. To test this, the electron beam was directed on different areas, relative to the grid bars and relative to the stage, of a Formvar and carbon coated grid. The background collected varied significantly. Thus, if the film correction to the total background is collected from an area of the grid from which x-rays are detected at higher or lower efficiency than the area containing the specimen, the correction will be too high or too low, resulting in calculated concentrations which are too low or too high. Occasionally, the film background exceeded the background from specimen and film, resulting in negative concentration values.

There is a second explanation for the introduction of large errors when the spectrum is corrected by subtracting a spectrum gathered from the supporting film. As reported by Jánossy and Neumann (1976), differences in lateral spreading of the beam in the supported specimen and in the film alone can result in large errors in this type of correction procedure, and this is especially problematic as the film thickness approaches specimen thickness. Also, the fact that we used a supporting film both above and below the specimen introduces the possibility of a gap between the specimen and one or both films, resulting in a larger excited volume.

Concentrations calculated from the raw data fall on a more narrow distribution. Two conclusions may be drawn from this fact. Collection efficiency continues to vary depending on the location of the specimen on the grid, but it affects the characteristic and continuum counts equally. Since only the ratio of these values is used to calculate concentrations, the effect is nullified. The second conclusion suggested from the narrow distribution of the uncorrected data is that the contribution of the supporting film to the observed spectrum must be relatively constant. Therefore, an average film correction could be inferred from the portion of the film corrected data which appears to have been properly corrected. This factor might then be applied to the mean concentrations calculated from the uncorrected data, for another estimation of actual concentrations in the tissue. For these data, the factor is about 1.6. When concentrations in the A band are calculated by this technique the mean values for potassium and sulfur are 600 and 350 mmol/kg dry mass, respectively. These values are in reasonable agreement with those from homogenized whole muscle. "Wet chemical" results, converted to dry weight by assuming 75 percent hydration, are for potassium, 410-506 mmol/kg dry weight (Spector, 1961). The elemental concentrations measured in this, as well as other regions of the muscle, will be discussed in detail in a subsequent paper.

The Mass Scattering Method

This method of quantification is a simplification of the technique described by Halloran et al. (1978). It should offer an independent check of the Hall technique and eliminate the need for correction of the Bremsstrahlung. An advantage incidental to the use of this method is the ease with which one may continuously monitor specimen mass thickness.

Halloran et al. (1978) assess ρt by using the STEM photomultiplier current to measure incident and transmitted beam current. Their method is based on the relation:

$$\frac{I}{I_o} = e^{-S_t \rho t},$$

where I_o and I_o are incident and transmitted beam current, respectively, and S_t is the total cross section for mass scattering. S_t is virtually independent of atomic number (Z), and is constant over a limited range of mass thicknesses for a given accelerating voltage and objective aperture semi-angle. Contrast, K, is defined as:

$$K = \ln\left(\frac{I_o}{I}\right) = S_t \rho t$$

Now, for both specimen and standard we can write:

$$C_{sp} = \frac{A(I_x)_{sp}}{K_{sp}} \text{ and } C_{std} = \frac{A(I_x)_{std.}}{K_{std}}$$

Taking ratios and rearranging we can eliminate the constant A, as well as the need for the absolute knowledge of ρt:

$$(C_x)_{sp} = \frac{(I_x)_{sp}}{K_{sp}} \cdot \frac{K_{std}}{(I_x)_{std}} \cdot (C_x)_{std.}$$

The practical application of this equation was investigated, along with the next logical extension of the technique, viz., to monitor transmitted intensity as analysis proceeds, providing a simultaneous record of mass loss. The mass scattering method is complementary to the Hall method, as it uses the same x-ray intensity data and concentration standards, along with a few more parameters. In addition to characteristic x-ray intensities and operating voltage, the mass scattering technique requires knowledge of the operating beam current (or something proportional to it) before and after specimen transmission. This measurement can be performed using the STEM photomultiplier.

For a fixed photomultiplier gain, photomultiplier output current (or voltage) was determined to increase linearly with beam current up to 6 nA, as measured with a Faraday cup. The standard plastic scintillator displayed a slow decrease in light output with time, however, due to radiation damage. It was replaced with a CaF_2 (Eu) detector (Harshaw Chemical Co.), which was much more radiation resistant.

As mentioned earlier, the transmitted current measured with the STEM photomultiplier can be sampled repeatedly while x-ray analysis is in progress, giving an indication of change in mass thickness. An increase in transmitted intensity due to specimen thinning is interpreted as loss of mass, but the results will be convoluted with vacuum contamination if it is present, which tends to increase specimen thickness. The absence of contamination must be demonstrated by other means before the results can be entirely unambiguous. The electrometer which is used to measure photomultiplier voltage produces an amplified analog output signal, and it is convenient to digitize this information and store it in the Kevex 7000 spectrometer, which can be operated in an XES-SEQ mode (i.e., pulse height analysis in memory group one, with multi-channel scaling in group two).

Our application of the mass scattering method failed to give dependable concentration values, although it was very useful as a sensitive monitor of mass thickness during an analysis. The fundamental limitation is due to differences in x-ray detection efficiency for different areas of the grid. As explained in the previous section, this effect cancels when concentrations are calculated using the continuum normalization method, as it affects characteristic peaks and continuum equally. It became a problem when a correction for the supporting film was attempted. In the case of the mass scattering method, collection efficiency of characteristic x-rays will continue to vary with position of the specimen relative to the grid bars and to the stage, while the transmitted intensity will remain constant with position. The result is a wide range of concentration values.

In addition, the mass scattering method is subject to failure if specimens become sufficiently thick to make plural scattering a probable event. This will occur long before x-ray absorption and secondary fluorescence cause breakdown of the continuum normalization method. Determination of the onset of plural scattering is equivalent to determination of the mass thickness (ρt) at which the mass scattering cross section (S_t) ceases to be constant. When we reach this condition, the measured transmitted intensity becomes less sensitive to increases in mass thickness. Using data from this range with the cross section for single scattering would lead to underestimation of mass thickness, and concentration values which are too large.

Unfortunately, the mass thickness at which deviation from linearity occurs is quite imprecisely known. Cosslett (1965) gives a range of values from theory and experiment for S_t for carbon using an objective aperture semi-angle of about 5 milliradians and 80 kV, typical for the work described herein. The transition point is around $\rho t = 30\,\mu g/cm^2$, which for carbon corresponds to a thickness of about 150 nm, and which the freeze-dried sections, plus their supporting films, may approach. According to the curves presented by Zietler and Bahr (1965), increasing the aperture angle from 1×10^{-3} to 2×10^{-2} would extend the value of ρt at which linearity with contrast ceases by 10 $\mu g/cm^2$, which for carbon would correspond to 500 nm. The rate of change of contrast with mass thickness would be reduced by about two-thirds, however, which would increase errors in measurement.

CONCLUSIONS

Mammalian muscle frozen in Freon 21, and frozen thin sectioned and freeze-dried by the methods described, exhibits excellent ultrastructural morphology, and intracellular ionic gradients are preserved, within the spatial resolution limits of x-ray microanalysis. The cold stage developed allows efficient collection of the x-ray signal and sufficiently depresses specimen temperature so that beam-induced loss of mass is effectively eliminated in these specimens.

One must be cognizant of possible differences in collection efficiency between specimen and its supporting film. Specimens mounted on folding grids are particularly troublesome in this respect. The use of an empirical film correction factor which may be applied in such cases to uncorrected concentration data is proposed.

A second method for quantifying x-ray data was evaluated. It estimates total mass of the analyzed area by the differences in contrast, i.e., the differences in amount of electron scattering, between the specimen and its supporting film. Several problems remain to be resolved in the application of this quantitative technique. It is immediately useful as a means to assess total mass loss during analysis.

ACKNOWLEDGMENTS

We thank Dr. Dale Johnson for critically reading the manuscript. We are grateful for the excellent technical assistance of Eunice Wang and acknowledge the help of Dr. Walter Sembrowitz in the animal surgery.

REFERENCES

Cosslett, V. E. (1965). Lab. Invest. 14:24.

Costello, M. J., and Corless, J. M. (1978). J. Microsc. 112:17.

Hall, T. A., and Gupta, B. L. (1979). In "Introduction to Analytical Electron Microscopy" (J. J. Hren, J. I. Goldstein, and D. C. Joy, eds.), p. 169. Plenum Press, New York.

Halloran, B. C., Kirk, R. G., and Spurr, A. R. (1978). Ultramicroscopy 3:175.

Hayat, M. A. (1970). "Principles and Techniques of Electron Microscopy," Vol. 1. Van Nostrand Reinhold, New York.

Headley, T. J., and Hren, J. J. (1978). Proc. 9th Intl. Conf. Electron Microscopy, Vol. 1, p. 504.

Janossy, A. G. S., and Neumann, D. (1976). Micron. 7:225.

Shuman, H., Somlyo, A. V., and Somlyo, A. P. (1976). Ultramicroscopy 1:317.

Sjöström, M., and Thornell, L. E. (1975). J. Microsc. 103:101.

Somlyo, A. V., Shuman, H., and Somlyo, A. P. (1977). J. Cell. Biol. 74:828.

Spector, W. S. (1961). "Handbook of Biological Data," p. 72. Saunders, Philadelphia.

Spurr, A. R. (1975). J. Microscopie Biol. Cell. 22:287.

Woldseth, R. (1973). "X-Ray Energy Spectroscopy." Kevex Corp., Burlingame.

Zeitler, E., and Bahr, G. F. (1965). Lab. Invest. 14:208.

DISCUSSION

Speaker: Keith L. Monson

ORNBERG: What is the chemical composition of Freon 21?

MONSON: Freon 21 is dichloromonofluoromethane.

LECHENE: How is the cold stage cooled?

MONSON: Unfortunately, I did not include a slide that showed the entire stage, but it is by conduction through a metallic rod.

LECHENE: And by conduction you get -180°C?

MONSON: Yes. I did a lot of other things with the stage: improving the insulation, and reducing the number of standoffs, and things like that, but this is basically just a copper rod insulated by the microscope vacuum.

FERRIER: The temperature is measured by thermocouple right at the tip to get that sort of reading, isn't it?

MONSON: Yes, there is a thermocouple that is located at the back of the stage insert.

FERRIER: I think you may very well find a clear temperature gradient up there to the specimen. We mounted one like it on the specimen and you don't get -180°.

MONSON: That certainly may be the case.

SOMLYO: What is the error of your measurement in the film? I mean, unless you count for a relatively long time, you will get a relatively small number of counts from your support film; therefore, could it be that perhaps they are Poisson-statistic limited? That would introduce an error in the correction if there are a small number of counts. Secondly, we have noticed Formvar to stretch and move under the beam, and could this stretching and thinning be different when there is tissue between the two sides of the Formvar or when there is none? So I am just wondering whether there may be additional reasons, in addition to the ones that you are mentioning, that make the correction less than desirable.

MONSON: As far as your first point goes, I analyzed the film for a much longer time period than the sections, so that the counting statistics of the white region in the two analyses were rather comparable. Certainly your second comment is very possible and is consistent with the data.

CANTINO: Wouldn't you expect that to give you an error in both directions also, whereas this shows an error in one direction?

MONSON: I would expect the thinning of the film in the beam would be greater in the center than toward the grid bars, but not all film analyses are performed in the very center of the grid square, so that I think you are right.

HALL: What fraction of the total count is the film count?

MONSON: I would say it is about 10% or less.

HALL: Have you subtracted the contribution of electrons scattering off the grid bars to the white count?

MONSON: Yes, we did that at first, but we found that if we carefully align our system, and we have a hard x-ray shielding aperture above the specimen, this contribution is small, although it is still present, and there is of course scattering within the specimen.

FERRIER: I think Ted is right; I think you can get beam smearing effects which would explain those observations. In the mass scattering method,

I_o is through the support only and I is support film plus section?

MONSON: That is right. If there were no film, I_o would be simply the incident current.

FERRIER: Would you like to say a few words about the effects of plural scattering?

MONSON: If we visualize a specimen with an incident beam, in regions of thickness where single scattering prevails, a scattered electron will deflect sufficiently so that it will be caught on the objective aperture and disappear from the transmitted signal. However, if we have a second scattering event, it is possible that the electron could rescatter so that the signal could appear in the transmitted intensity. As Dr. Ottensmeyer pointed out, this is an unlikely process, especially in an amorphous sample, but the point I wish to make is that we are starting from a zero signal from a singly-scattered electron. A few of the secondary scattering events could add to the transmitted intensity, while most will only deflect further toward the apertures. Thus, for a given ρt, I will be larger than it would be if we had only single scattering, causing underestimation of thickness and overestimation of concentration.

RICK: What do these I-values directly reflect?

MONSON: Before I began, I varied beam currents which I measured by a Faraday cup, and then I monitored photomultiplier current and voltage, and it turned out I could use either one to represent actual beam current.

RICK: In those instances where you said that I is even larger than I_o did you collect the electrons or did you use a scintillation device and photomultiplier?

MONSON: It is a scintillation device.

RICK: We have all sometimes seen a sort of reverse contrast when we use plastic scintillators in order to detect a transmitted electron. The center spot, after a while can have a very poor efficiency so when you have a little bit of a scatter, you get out of that area. This is a very simple explanation for having a lower I_o than I.

MONSON: This is exactly what did happen with the standard plastic scintillator, and why I went to calcium fluoride. It is very easy to determine if this is indeed a problem by focusing the beam directly on the scintillator and scanning across it. Even after many months of use our Ca F_2(Eu) scintillator shows no local deterioration in efficiency.

FERRIER: But, if I_o is with zero film, you cannot possibly get an intensity greater than when passing through free space, surely.

HUTCHINSON: We agree with that, Robert!

ANDREWS: I wanted to comment a little bit on the difficulties of your support film correction because our system operates quite a bit differently than yours, and also because I think it is relevant to this question on whether or not you can correct for the extraneous scattering that occurs in the tissue. Professor Ferrier has expressed this reservation during my talk and during yours as well. We find that if you put the beam on a carbon film which looks to be uniform in thickness and measure the actual energy spectrum around the square of a 100 or 200 mesh grid, that the raw data definitely shows less Bremsstrahlung in the area in the middle than it does near the end. If we run it through the ML routine that we described, that came from Henry Shuman, after the correction for extra scattering which is greater near the lower grid than it is on the upper one which is tipped, and when you plot that as a function of distance, you don't find any great difference. I think this is reasonably good evidence that scaling the extraneous continuum to the Cu L peak works pretty well, and you want it with a curve which is deviant only very near the edges of the copper grid. Now when we do this kind of correction, we have to be very careful about this because our films are more than 10% and yours are only 10. We do a number of them at various regions around the section and take the average of those and use the area of the film only if the film is consistent. What normally happens is the standard deviation on the corrected continuum is not any worse than it is on the continuum from the tissue, so when you carry the variance through, it just doesn't make that big a difference on the final data, and you don't see any really great spread in the distribution of your data. The uncertainty in the potassium values, for example, might go from say 5% to 7% or something of this sort. That 2% is the price we pay for having the film underneath, which given biological variability, I do not think is that bad, so I think that under those circumstances you can do that subtraction reasonably well. I am not sure what is going on in your system, but it seems to behave quite a bit differently.

MONSON: Basically what you do is average a large number of film counts?

ANDREWS: Yes. Four or five. They ought to be the same when you do them in different regions around the grid square. You should not find great variation on the part of the film. If there were, I wouldn't know which one would be the correct region.

PREPARATION AND USE OF ERYTHROCYTE SECTIONS AS METHODOLOGY FOR VALIDATING PROCEDURES FOR X-RAY MICROANALYSIS OF ELECTROLYTES

John McD. Tormey

Department of Physiology
School of Medicine
University of California
Los Angeles, California

INTRODUCTION

The 1973 Battelle Conference on biological microprobe analysis had a large impact on many of us who were about to enter the field. Reading the book which resulted from that conference (Hall et al., 1974), one was impressed by some encouraging results but even more by the difficulty of obtaining reliable quantitation of electrolytes in tissues. A sea of poorly charted artifacts waited to be navigated. In other words, biological microanalysis in the mid-1970s was in an uncertain state, similar to that of nonbiological microanalysis more than a decade earlier.

Faced with this situation, it was apparent that a biological test system was needed. The accuracy and reliability of microanalysis could be validated, if one had a standard biological preparation consisting of small compartments whose electrolyte composition was accurately known. Suspensions of erythrocytes were (and are) the best biological preparation to meet this need. The cells are small, their subcellular structure is homogeneous, and their intracellular composition is more accurately measurable by wet chemical methods than that of any other tissue.

This paper will describe a method of obtaining erythrocyte preparations in the form of sections that are

ISBN 0-12-362880-6

suitable for validating biological microprobe methods. It will present, in narrative fashion, results we have obtained using electron probe x-ray analysis to quantitate their electrolyte composition. Some of these results have been published previously (Tormey, 1978; Tormey, 1979; Tormey and Platz, 1979), while others are described here for the first time.

The electrolyte preparation has proven extremely useful for "tuning up" our methods. We have come in several years from poor to excellent results and along the way have encountered and overcome several pitfalls.

METHODS

Blood was freshly drawn from adult animals, citrated, and immediately washed three times in cold isotonic saline. A Ringer's solution containing 143 mM NaCl, 6mM KCl, 2.125 mM Na_2HPO_4, 0.375 mM NaH_2PO_4, and 10.1 mM glucose per liter H_2O was prepared; 20 g of Sigma Type 60C dextran (average MW 83,000) was added per 100 ml of this solution to make 20% dextran Ringer. Washed pellets of erythrocytes were resuspended in this solution and incubated with constant agitation for two hours at 37°C .

Millipore (Type MF, ∿ 150 μm thick) filters that had been precut into narrow wedges and presoaked in dextran Ringers were dipped into the suspension and then slowly withdrawn vertically, so that thin films of suspension were formed on either side. Each filter was immediately plunged into vigorously stirred liquid propane at -175° C, taking care to avoid any drying of the adhering films.

At the same time that samples were frozen, aliquots of suspension were taken for chemical analyses and immediately centrifuged at 10,000 G. H^3-mannitol was included as a marker for extracellular fluid. Concentrations ($mM/10^3cm^3$) were determined in both supernatant and erythrocyte pellet, the latter being corrected for trapped supernatant. Na and K concentrations were determined by flame photometry, Cl by potentiometric titration with $AgNO_3$, and Fe by a standard method for hemoglobin Fe. Concentrations/kg dry mass were estimated after weighing thoroughly dried, precisely measured volumes of cell and supernatant.

Frozen sections were cut at thicknesses (between 0.2 and 0.5 μm) using a Sorvall FTC cryostat attached to a Porter-Blum MT-2 microtome, usually at -80°C. Sectioning the filters was carried out easily and was facilitated by mounting them in a specially designed vise chuck so that just the tip of a filter protruded from its jaws. Sections were usually mounted on folding copper grids covered by either Formvar or pure carbon films. They were then transferred to a freeze-dryer for two days at -75°C. After gradual warming to room temperature they were handled in a dry box and kept dry until use. (Sections left 24 hours in a covered Petri dish in room air developed extremely fine salt crystals that quickly vanished under probing electron beams, leaving the sample bereft of measureable salt.)

Specimens were probed in either a ETEC Autoscan SEM (operated in STEM mode) or a JEOL 100C(X) STEM at various accelerating voltages. Both were equipped with energy dispersive x-ray spectrometers. Spectra were usually analyzed with a Northern 880 system, but similar results were obtained on both microscopes with a Kevex 7000 system. Various quantitation methods were used, as described below.

In order to understand the basis of discrepancies between chemical analyses and microprobe results, it has been extremely useful to analyze two types of erythrocyte of widely varying electrolyte compositions within the same section. We sometimes used mixtures of sheep HK (high potassium) and LK (low potassium) erythrocytes (Tosteson and Hoffman, 1960); however, since sheep with HK blood proved difficult to find, mixtures of human (high potassium) and sheep LK cells were more often used. (Working with mixed suspensions, it was of course necessary to do chemical analyses on separate parallel suspensions of the two cell types.)

RESULTS AND DISCUSSION

Initial Results

Although the sections consist mostly of Millipore filter, they are easy to cut. Figure 1 is a STEM micrograph of part of a typical section. The Ringer's solution is present as a uniform extracellular matrix, by virtue of its dextran content. Both matrix and cells are usually virtually free of ice crystals.

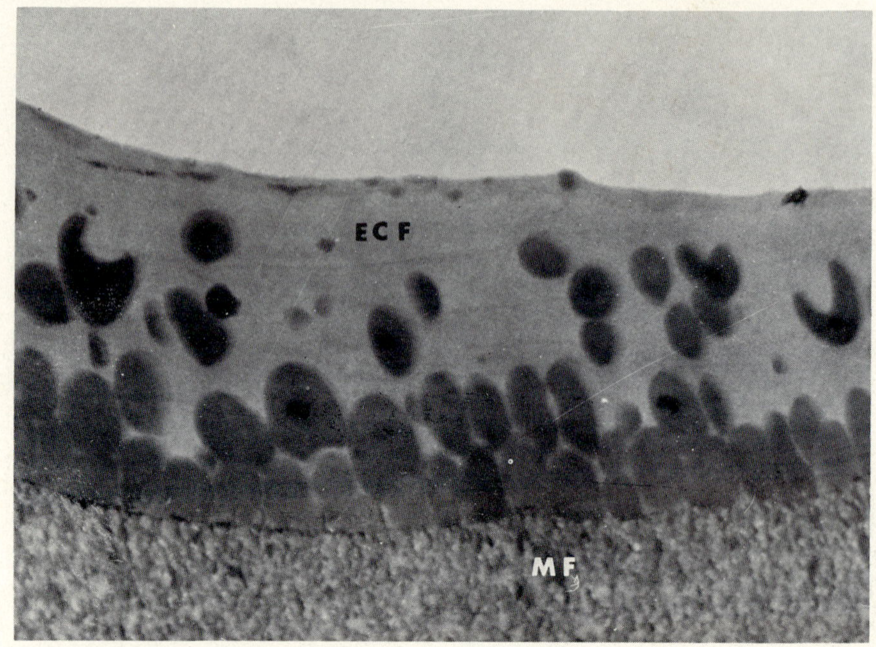

FIGURE 1. STEM micrograph of freeze-dried frozen section of erythrocyte standard. A layer of extracellular fluid (ECF) containing numerous erythrocytes rests on a Millipore filter (MF). Magnification 8,000X.

We have thus prepared a three-phase test system: two types of erythrocyte of different electrolyte composition, plus a well-preserved extracellular phase. Figure 2 illustrates typical compositions of these phases.

As a first approach to quantitation of microprobe data, the intracellular concentration of each element, X, was estimated from x-ray intensity ratios by

$$[X_{cell}] = \frac{I_{x\ cell}}{I_{x\ ecf}} [X_{ecf}]_{chem} \tag{1}$$

where $[X_{ecf}]_{chem}$ is the concentration of X measured in the extracellular fluid by wet chemical analysis, and I_{xcell}/I_{xecf} is the ratio of the net peak intensities of X measured over a cell and over nearby extracellular matrix. Assuming that a cell and the extracellular material within 5 μm of it have

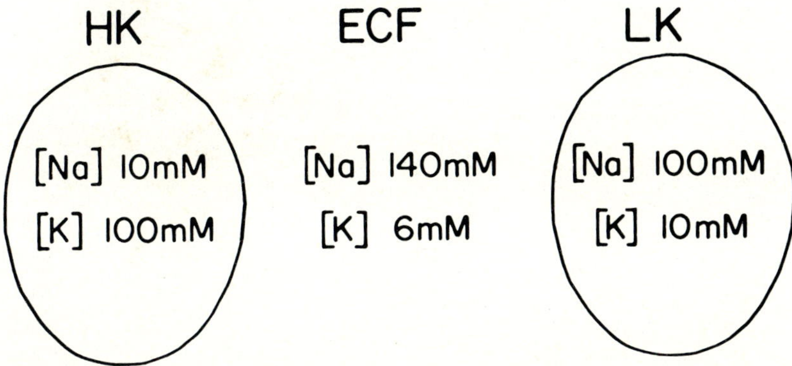

FIGURE 2. Diagrammatic representations of typical Na and K concentrations in the three phases of sections containing both high potassium (HK) and low potassium (LK) erythrocytes.

virtually the same thickness, $[X_{cell}]$ should have the same value as the intracellular concentration measured chemically.

Table 1 shows some early results using this approach with a SEM at 20 kV. The errors for Na and Cl are very large. (It was subsequently learned that a large part of the Cl error could be eliminated by cooling the specimen to about -150°C and surrounding it with a cold trap. However, even when this was done, a large Cl error persisted.)

Nystatin Experiments

In an effort to understand the origin of such results, an erythrocyte preparation with zero aqueous concentration gradients was studied. Cells were treated with the macrocyclic antibiotic Nystatin in order to increase the cationic permeabilities of their plasma membranes, and they were titrated with CO_2 to a pH equal to the isoelectric point of hemoglobin. They were then incubated in a medium containing equal concentrations of NaCl and KCl. As a result of this treatment we obtained suspensions in which there were no aqueous concentration gradients of Na, K, or Cl between phases (Cass and Dalmark, 1973). This minimized the likelihood of ions moving along concentration gradients during specimen preparation and also provided an opportunity to test whether specific ions might migrate under the electron beam.

TABLE 1. Preliminary Microprobe Measurements of Intracellular Electrolyte Concentrations in HK and LK Erythrocytes

Cell Type-Element	$[X_{cell}]$	$[X_{cell}]_{chem}$
HK-Na	47.3	20.9
K	111.1	102.0
Cl	325.1	97
LK-Na	141.8	97.3
K	22.3	17.1
Cl	350.7	99

$[X_{cell}]$ calculated from X-ray data using equation (1) (external ratio method.)
Data obtained at ambient temperature in SEM operated at 20 kV.

Table 2 shows that all the values of $I_{x\ cell}/I_{x\ ecf}$ measured by microprobe analysis were too high by about 40%. This rules out the possibility of movements or losses of specific ions, either during specimen preparation or under the electron beam.

TABLE 2. Results with Erythrocytes Treated with Nystatin to Eliminate Aqueous Concentration Gradients[a]

Element	$I_{x\ cell}/I_{x\ ecf}$	$[X_{cell}]_{chem}/[X_{ecf}]_{chem}$
Na	1.22 ± .05 (s.e.)	.85
K	1.09 ± .04	.84
Cl	1.24 ± .05	.88

[a]Data obtained with -160° C cold stage in SEM operated at 20 kV.

Mathematical Modeling

In order to better understand our results, the following equation was fitted to data from normal erythrocyte preparations:

$$[X_{cell}] = A[X_{cell}]_{chem} + B[X_{ecf}]_{chem} \qquad (2)$$

where [X] is the concentration of each electrolyte species measured by microprobe or chemical methods, in cellular or extracellular phases, as indicated by subscripts. Ideally, the microprobe would measure intracellular concentrations equal to those measured chemically, i.e., A = 1, B = 0. Values for $[X_{cell}]$ were first estimated from microprobe data using equation (1). Equation (2) was then fitted by multiple linear regression analysis to a large set of data consisting of values for Na, K, and Cl from both HK and LK cells. It was found that A = 1.116 and B = 0.309 fitted the data best. Since the correlation coefficient for the fit was 0.992, equation (2) is a good mathematical model (Tormey, 1978).

In order to grasp the physical basis for the discrepancy between microprobe and chemical measurements, the physical significance of the coefficients A and B must be understood. A scales the chemically measured intracellular concentrations for each element, and A = 1.116 could be due to an 11.6% increase in the density of the cells relative to the density of extracellular fluid matrix. This would occur, for instance, if the cells shrank more than the extracellular fluid during freeze-drying. B greater than 0 indicates that intracellular measurements are being contaminated by x-ray signals originating from the extracellular compartment. This could occur if a fraction of the extracellular material were picked up by the microtome knife and smeared over the cells, or if stray radiation were exciting the extracellular matrix when the cells were being probed.

Comparison of Quantitation Methods

Results obtained with our original method of quantitation were compared with those obtained from the same x-ray data by three additional methods. The methods may be summarized as follows:

(a) External Peak Ratios Method - As already described and defined by equation (1), this method uses the extracellular fluid as a standard and compares the intracellular peak intensity for each element with the peak intensity of that element outside of the cell (hence, "external"). Intracellular concentrations are in units of moles/volume, since extracellular concentrations are in these units. An advantage of this method over the peak-to-continuum method (b) is its insensitivity to changes in mass under the probing electron beam. However, it requires negligible changes in the relative densities of the cellular and extracellular phases during specimen preparation and analysis. It also requires preservation of the extracellular fluid, and short-range uniformity of section thickness.

(b) Peak-to-Continuum Ratio Method - This approach, introduced by Hall (1971), utilizes the relationship

$$[X_{cell}] = \frac{I_{x\ cell}}{I_{b\ cell}} W_x \qquad (3)$$

where $I_{x\ cell}/I_{b\ cell}$ is the ratio between an intracellular characteristic peak and a simultaneously measured region of interest of the background spectrum, and W_x is a proportionality constant determined empirically from standards. This method has the advantages of not requiring the retention of extracellular fluid close to the cell being analyzed, and of being insensitive to changes in cell density and section thickness. Since the background (continuum) radiation is used as an index of excited mass, the resulting concentrations are in units of moles/unit dry mass. It thus requires negligible mass change during irradiation under the probing beam and careful corrections for stray continuum radiation, as well as accurate determination of W_x values.

(c) Internal Peak Ratios Method - This method requires the presence of an element that is homogeneously distributed within a cell and that is exclusively present in a form which is irreversibly bound to a nondiffusible substance. Hemoglobin Fe in erythrocytes meets these criteria. When $[Fe_{cell}]$ has been determined by chemical methods, intracellular electrolytes can be estimated from

$$[X_{cell}] = \frac{I_{x\ cell}}{I_{Fe\ cell}} k\ [Fe_{cell}]_{chem} \qquad (4)$$

where k is a proportionality constant which can be determined empirically from standards, but which can also be estimated with good accuracy from published values of ionization cross section and fluorescence yield, together with a knowledge of the attenuation characteristics of the x-ray detector's beryllium window.

This method has the advantages over methods (a) and (b) of being insensitive to both absolute and relative changes in section density or mass, and of requiring no correction for continuum radiation from outside the tissue. However, its applicability appears limited to erythrocytes.

The three approaches to quantitation gave very similar results when applied to the same x-ray data, as shown in Table 3.

In order to achieve agreement between the external ratios method and the other two methods, it was necessary first to determine the coefficient A by fitting all the data to equation (2). In this case, the empirical value of A was 1.126; therefore, the external ratio data presented in Table 3 was corrected for a presumed 12.6% relative shrinkage of cells with respect to extracellular matrix. The excellent agreement then obtained allows the conclusion that the mathematical model used to fit the external ratios data

TABLE 3. Intracellular Concentrations of Erythrocytes Estimated by Three Methods of Quantitating Microprobe Data[a]

Cell Type-Element	External Ratio[b]	Peak-Continuum	Internal Ratio[c]
HK-Na	70.6 ± 5.5 (s.e.)	73.4 ± 5.5	70.4 ± 5.5
K	110.3 ± 6.0	111.2 ± 9.6	106.1 ± 2.7
LK-Na	154.9 ± 11.1	145.8 ± 3.9	159.9 ±11.7
K	10.6 ± 2.2	10.5 ± 2.9	11.3 ± 1.9

[a]Data obtained at ambient temperature in SEM operated at 20 kV.

[b]Corrected for 12.6% relative increase in density, as determined by multiple linear regression analysis using equation (2).

[c]Calculated by multiplying moles/kg dry mass by independently measured dry mass/wet mass ratio.

provides a fit that corresponds to a physical reality. The fact that the coefficient A comes into play only when the extracellular fluid is used as a standard shows that A is actually related to a change in relative densities of the cellular and extracellular phases. Therefore, approaches that use the extracellular fluid as a standard in the way we have used it are liable to introduce a substantial quantitative error, which may, however, be acceptable for some purposes. Over a twelve-month period, A was observed to have a mean value of 1.21 $\pm$ 0.21. The reason for this large standard deviation is unknown. When 20% albumin was used in place of 20% dextran, the value of A became even larger.

The internal ratio, though not a generally applicable technique, has the advantage of being extremely straightforward to apply in cases where the concentration of cellular Fe is easily measured by accurate chemical methods. It does not involve continuum radiation, yet its results can readily be converted to mM/kg dry weight when the erythrocyte preparation is used as we have described. Therefore, this approach should be quite useful when setting up and debugging a peak-to-background quantitation method, both for checking out W_x values and for assessing the accuracy of corrections for background originating from other than the section itself.

Parenthetically, we originally avoided using the peak-to-continuum method because we observed rapid mass loss, especially from the dextran matrix, even when specimen chamber vacuum was 2×10^{-6} Torr and we were using a -160° C cold stage (Tormey, 1979). Subsequently it was found that with similar specimens in a better vacuum environment no mass change was often detectable at all, even with the specimen at ambient temperature and total radiation dosages of 5×10^4 coulombs/cm^2. This suggests that much better detection limits can be achieved in small areas than were originally proposed, e.g., by Shuman et al. (1977) based on their experience with a particular instrument. It also raises the question of how much so-called radiation damage is intrinsic and inevitable, and how much is due to the interplay between the electron beam and a particular combination of specimen material, temperature, and vacuum environment.

Source of Quantitation Artifact

All of the data for Na and Cl presented thus far show microprobe measurements of concentrations that are much too large, regardless of the method of quantitation used. This is shown again in Table 4, which compares Na and K results obtained at 20 kV with the internal ratio method. There is only a small excess K concentration, but a large excess Na concentration, relative to that predicted by chemical analysis. It is noteworthy that the excesses for each element are proportional to the concentration of each in the extracellular fluid and not to its concentration in the cell. They are what would be expected if the intracellular signal were augmented by stray radiation producing a signal at 30% of the normal level from the extracellular matrix (corresponding to a value of about 0.3 for the coefficient B in equation (2)).

A number of different observations have led to a tentative explanation of this type of result. For instance: (1) When the specimen was replaced by a 5 μm diameter aperture, the "hole count" observed was two orders of magnitude too low to explain the effect (Tormey, 1978). This rules out the possibility that it might be due to electron beam tailing or to hard x-rays originating from electron column apertures, etc. (2) When section thickness was doubled, the excess Na signal doubled (Tormey and Platz, 1979). (3) When accelerating voltage was reduced from 20 to 10 kV, the excess Na signal more than doubled (Tormey and Platz, 1979). Both this and the previous observation show that the excess intracellular signal is related to the probability of electron scattering by the specimen. (4) When sections which showed a large error in the SEM at 20 kV

TABLE 4. Excess Na and K Concentrations Measured in Erythrocytes by Microprobe Analysis[a]

Cell Type-Element	$[X_{cell}]$	$[X_{cell}] - [X_{cell}]_{chem}$
HK-Na	64.5 ± 4.8 (s.e.)	47.0 ± 4.8
K	89.8 ± 2.2	1.1 ± 2.2
LK-Na	133.2 ± 7.5	36.2 ± 7.5
K	12.1 ± 1.0	1.7 ± 1.0

[a] $[X_{cell}]$ calculated from x-ray data using equation (4) (internal ratio method). Data obtained at ambient temperature in SEM operated at 20kV.

were examined in a (S)TEM at 80 kV, the error disappeared (see Table 5). This clearly shows that the effect is not due to an artifact of specimen preparation but depends on microprobe operating parameters.

Although the dependence on accelerating voltage and section thickness strongly suggests that excess signals are related to electron scattering by the section, the question remains, how could enough electrons be scattered into the extracellular matrix to explain the size of the excess? Geometrical considerations, as well as the relatively small values of backscatter coefficients, make it highly unlikely that sufficient excitation could result from electrons scattered back to the specimen from either above or below it. Figure 3 illustrates an experiment to test this conclusion. A specimen grid is shown supported at a 45 angle in a substage configuration that we have frequently used for microprobe analysis in the SEM. It is held on a thin sheet of carbon which is supported by an aluminum mount with 4 mm diameter holes for transmitting electrons. Both the aluminum support and the face of the x-ray detector are heavily coated by carbon (Dag 154). This configuration combines good x-ray detection efficiency with a low probability that transmitted electrons will scatter back to the grid. After making measurements on this stage, we inserted the aluminum tubes A and B, which were configured to

TABLE 5. Effect of Type of Microprobe and Accelerating Voltage on Excess Na[a]

Instrument, Cell Type	$[Na_{cell}]$	$[Na_{cell}]-[Na_{cell}]_{chem}$
TEM, 80 kV	10.1 ± 1.19 (s.e.)	0.9 ± 1.9
HK	91.0 ± 8.3	-11.2 ± 8.3
LK		
SEM, 20 kV		
HK	70.5 ± 5.6	61.3 ± 5.6
LK	152.5 ± 5.9	50.3 ± 5.9

[a]Data obtained on the same cells in both instruments. Quantitation using internal ratio method.

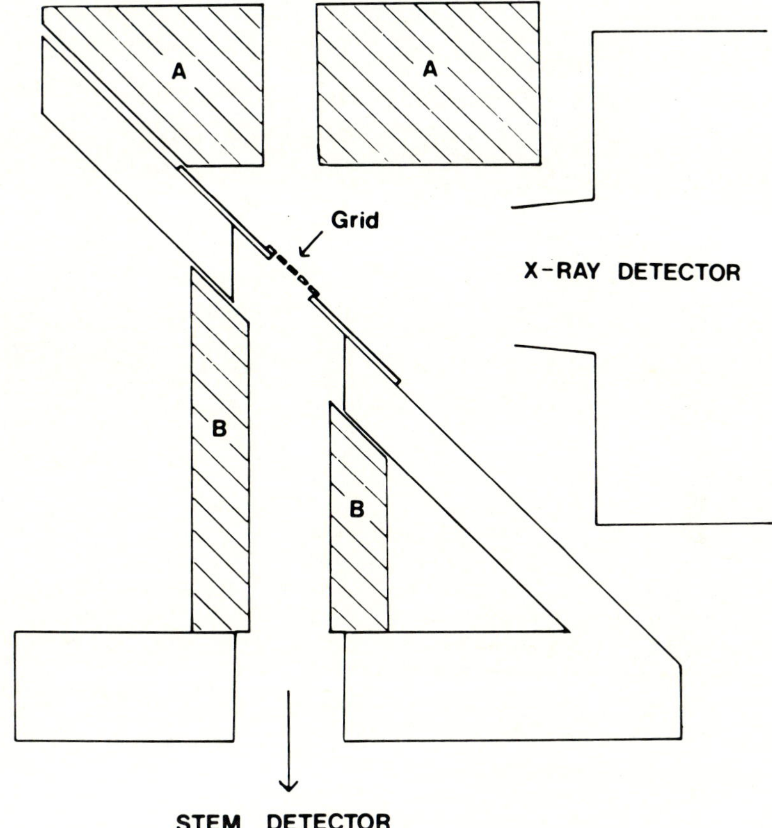

FIGURE 3. Scale drawing of specimen substage used to study the contribution to SEM results of stray radiation from above and below the sample. The addition of removable tubes (A and B) increases the probability that backscattered electrons and fluoresced x-rays will excite the sample (grid).

increase the probability that electrons scattered by the specimen would be scattered back to it. When the cells were remeasured, a large aluminum peak was detected, but there was no increase in the excess Na signal. This rules out the possibility that the excess Na was generated by either backscattered electrons or by secondary fluorescence originating from x-rays fluoresced from the specimen stage.

Scattering within the plane of the section is therefore the most likely explanation. However, application of a simple, single elastic scattering model developed by Goldstein et al. (1977) predicts the scattering of an insufficient number of electrons to explain the effect quantitatively. Monte Carlo calculations are thus in order but have not been pursued.

The improvement using (S)TEM at 80kV is not simply the result of a higher accelerating voltage, since, although some excess Na was measured on the same instrument at 20 kV, it was much less than at 20 kV in a SEM. This surprising result might possibly be related to the fact that in the former type of instrument the specimen is immersed in a very strong magnetic field, which alters the trajectories of scattered electrons--perhaps in a manner favorable to improved quantitation.

CONCLUSION

We have developed a simple method for preparing samples that can readily be used to validate both preparative and instrumental methods of microanalysis of biological samples. The results obtained at 80 kV clearly demonstrate that these samples are properly prepared.

In our hands, this approach has been essential for arriving finally at the point where we know that we can achieve valid results. Hopefully, such preparations will prove useful to others who seek to apply microprobe analysis with confidence to samples of uncertain composition.

The primary purpose of this paper has not been to describe a particular quantitation artifact and its probable origin. The point has been to show the usefulness of this preparation, first for determining the accuracy of microprobe results, and then for tracking down sources of error. For instance, fitting a mathematical model both quantitates the errors and suggests possible physical explanations. Such fitting would not have been possible without data from two types of well-characterized cells. Also, the internal ratio method affords the most straightforward approach to reliable quantitation, and therefore is a good point of departure for testing and debugging other, more demanding methods of quantitation.

Our experience suggests that microanalysis of biological sections is best carried out at the highest feasible accelerating voltage and with specimens of the lowest feasible mass thickness. In addition, there are probably advantages to a TEM configuration.

Successful biological microanalysis has been carried out, of course, in SEMs at relatively low accelerating voltages. Some of the difference between our results and those of others might lie in differences in sample geometry. Ours consists of quite small cells surrounded by a large extracellular phase that represents 75% of the volume of the sample. Stray extracellular signals should be less of a problem when one is working with, say, a multilayered epithelium with a relatively small amount of extracellular fluid. Also, they should prove no problem when probing areas where an element is accumulated at much higher concentrations than in its surroundings.

We have arrived by a rather involved series of observations at the same point reached by others at this Conference: namely, the conviction that present day methods of freezing, frozen sections, and freeze-drying are capable of yielding accurate quantitative localization of electrolytes in biological systems. Clearly, this field has become much more firmly established in the years since it was last the subject of a Battelle Conference.

ACKNOWLEDGMENTS

This work has been supported by grants from the National Institutes of Health (Program Project #HL 11351), the American Heart Association, Greater Los Angeles Affiliate (Research Award #510), and the Muscular Dystrophy Association. Thanks to Mr. Robert M. Platz for his excellent technical assistance.

REFERENCES

Cass, A., and Dalmark, M. (1973). Nature (New Biology) 244:47-49.

Goldstein, J. I., Costley, J. L., Lorimer, G. V., and Reed, S. J. B. (1977). In "Scanning Electron Microscopy/1977," Vol. 1, pp. 315-24. IIT Res. Inst., Chicago.

Hall, T. A. (1971). In "Physical Techniques in Biological Research," 1A, 2nd Ed., pp. 157-275. Academic Press, New York.

Hall, T., Echlin, P., and Kaufmann, R., eds. (1974). In "Microprobe Analysis as Applied to Cells and Tissues." Academic Press, New York.

Shuman, H., Somlyo, A. V., and Somlyo, A. P. (1977). In "Scanning Electron Microscopy/1977," Vol. 1, pp. 663-72. IIT Res. Inst., Chicago.

Tormey, J. M. (1978). In "Scanning Electron Microscopy/1978," Vol. 2, pp. 259-66. SEM, Inc., Chicago.

Tormey, J. M. (1979). In "Microbeam Analysis in Biology" (C. P. Lechene and R. R. Warner, eds.), pp. 615-33. Academic Press, New York.

Tormey, J. M., and Platz, R. M. (1979). "Scanning Electron Microscopy/1979," Vol. 2, pp. 627-34. SEM, Inc., Chicago.

Tosteson, D.C., and Hoffman, J.F. (1960). J. Gen. Physiol. 44:169-94.

DISCUSSION

SPEAKER: John McD. Tormey.

LECHENE: Did you use the TEM at 20 kV?

TORMEY: We did and found that there was some Na excess, but much less than in the SEM.

HUTCHINSON: John, was that the data you got using our TEM with the hard x-ray aperture in place?

TORMEY: Our first data at 80 kV was obtained on your instrument with the hard x-ray aperture. However, we found that the hard x-ray aperture had no effect at all on the excess sodium, but only on copper x-rays from the grid.

HALL: I don't understand your explanation for the difference between your results and those of other people. You say that with an epithelial tissue you wouldn't expect such adverse effects, because these cells are not surrounded by extracellular space. But if one goes right to the edge of a cell with an internal standard right next to it, one has a situation at least 50% as bad as you have with your samples. So why don't too many people see Na right at the edge of the cell?

TORMEY: I am not so sure that the situation you describe would be half as bad as mine. Assume that electrons are scattered uniformly in all directions within the plane of section and travel many µm. Since the adhering layer of extracellular material would be thin, I would expect much more of the excitable volume within a 10-20 µm radius of the probe to be intracellular than extracellular.

RICK: I would like to add a comment. When we studied toad urinary bladder a couple of years ago, there were many areas where the epithelium was no thicker than 2-3 µm, so that we had a virtually infinite extracellular space on both sides of the epithelium over an angle of about 300 . Yet the sodium concentration was as low as in other epithelial cells. I want to ask you what does the intensity profile of an intracellular electrolyte look like when you go through one of your cells from one end to the other? Did you get a plateau-shaped distribution?

TORMEY: We have only made x-ray intensity profiles at 80 kV. There the intensity profile for K across a HK cell was plateau-shaped.

RICK: This observation is very difficult to bring in line with a direct short-range scattering in the specimen itself. I would think it more likely that stray excitation was caused by stage or some other structure within the specimen chamber. Can you exclude the possibility that secondary fluorescence of the specimen from electrons striking your specimen stage might have caused your results?

TORMEY: First, if there were secondary fluorescence, it is unlikely that Na, K, and Cl would all be affected to the same degree. However, we did worry about this possibility, which is the reason we altered specimen stage geometry so as to increase its likelihood (cf. Figure 3). We found no increase in excess intracellular signals under conditions in which increased amounts of backscattered electrons and fluoresced x-rays should have impinged on the specimen.

LECHENE: John, if you have increased sodium contamination when you are at 20 kV in the SEM mode, this still points to an instrumental factor, rather than to something in the preparation.

TORMEY: Well, I think that the good results at 80 kV make clear that there is no problem with our preparation.

LECHENE: No, No. By preparation, I mean scattering. . .

RICK: Within the specimen itself.

LECHENE: I don't see why you should have different results at 20 kV between a SEM and a TEM.

TORMEY: I don't have a totally satisfying answer for the difference. However, we made every effort to rule out the possibility that stray radiation from outside the sample has contributed to our SEM results. The possibility remains that immersion of the sample in a strong magnetic field in the TEM instrument might be a cause of the improved results at 20 kV.

HALL: Do you have a copper K signal?

TORMEY: Our earlier SEM measurements were made with samples mounted on carbon tubes, in which case there was no Cu signal. With more recent measurements on Cu grids, we do, of course, have a Cu signal.

HALL: Then how much copper L do you have? How do you distinguish between Cu L and Na?

TORMEY: We have substantial Cu L at 20 kV. To deconvolute the overlap between Cu L and Na we have used multiple least squares fitting routines as well as the simplex routine from

Kevex. We have gotten the same results from these routines with samples on Cu grids, as well as with samples on carbon tubes that produced no Cu L at all. It is clear that problems with deconvoluting this part of the x-ray spectrum have made no contribution to our results.

RICK: Do you have any idea of the relative sizes of the Cu L line and the Na line in normal intracellular measurements?

TORMEY: With specimens on carbon tubes there is no Cu L line. Depending on the cell type studied, the Cu-to-Na ratio varies from about 1:1 to about 20:1 on Cu grids.

THE CONTINUUM-FLUORESCENCE CORRECTION IN BIOLOGICAL TISSUE

R. R. Warner
D. A. Taylor

The Proctor & Gamble Company
Cincinnati, Ohio

INTRODUCTION

Continuum fluorescence results from a multistep process that produces characteristic x-rays in addition to those produced by the primary electron beam. In this multistep process, electron de-acceleration in the sample due to inelastic interactions with a nucleus results in the emission of x-rays having a continuum of energies up to the accelerating potential. These continuum (bremsstrahlung) x-rays can penetrate far into the sample or standard, where some are absorbed and result in secondary ionizations with emission of characteristic x-rays (fluorescence). Some of the characteristic x-rays fluoresced by the continuum are collected by the detector, thus augmenting the characteristic x-ray signal produced by the primary electron beam. A model of this process is shown in Figure 1. If this effect is significantly different in the sample versus standard, then quantitative results can be in error.

Quantitative microprobe analyses did not address fluorescence effects due to continuum radiation until just prior to the last decade. An accurate expression for the continuum fluorescence was developed (Hénoc et al., 1973) and incorporated into COR2, a computer program for data reduction (Hénoc et al., 1973). Investigation of the magnitude of the continuum-fluorescence correction using this program (Myklebust et al., 1970), as well as with the less rigorous development by Springer (1967), showed that for the analysis of minerals and metals the continuum fluorescence accounted for only a small fraction of the emerging x-ray intensity.

ISBN 0-12-362880-6

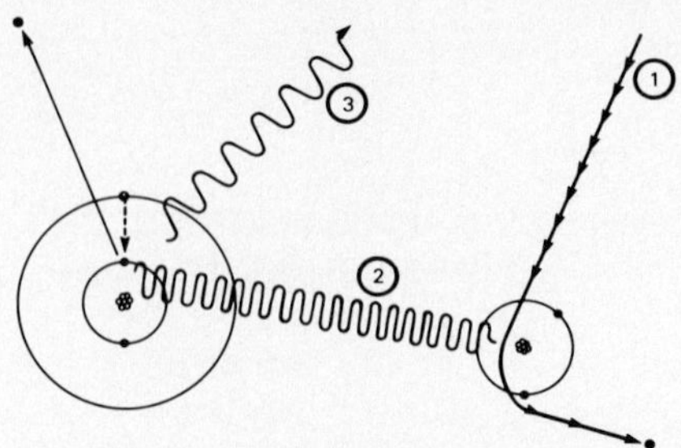

FIGURE 1. Continuum-fluorescence excitation. (1) Electron de-acceleration due to inelastic interactions with a nucleus. (2) Emission of continuum x-rays. (3) Fluorescence of characteristic x-rays by the continuum.

This fraction was approximately equal for sample and standard, and consequently for the sample/standard ratio the continuum correction was nearly always negligible. This conclusion supported the common practice of neglecting the continuum correction and is seen today in the omission of this rather lengthy correction from data reduction procedures of manufacturer-supplied software. Although the continuum-fluorescence correction can generally be neglected in the analysis of minerals and metals, this may not be true for biological tissue. Myklebust et al. (1970) noted from investigations with COR2 that analysis of hard x-ray lines in a low atomic number matrix (which is typical of a biological analysis) always required a continuum fluorescence correction. Hall, in the only previous paper to address the continuum fluorescence from a biological perspective, approximated the continuum-induced fluorescence and calculated upper limits for its magnitude. Under certain conditions the continuum-induced fluorescence was exceptionally large (Hall, 1971).

The continuum-fluorescence correction may not be a factor in the analysis of sectioned material using thin standards (Hall, 1971; Tixier, 1979). However, for other sample types such as frozen-hydrated bulk tissue, the continuum-fluorescence correction could be important. In addition, the appropriate choice of standards for bulk frozen-hydrated tissue has not been ascertained and could profit from a theoretical analysis; for instance, the use of mineralogical standards could require a large continuum-fluorescence correction and, thus, not be the standards of choice. The object of this investigation of the continuum-fluorescence effect in thick biological tissue was to delineate conditions that would give accurate analyses, and to spotlight conditions that would lead to grave errors if the continuum-fluorescence effect were neglected.

METHODS

Our procedure was to model the composition of a cell or extracellular space, choose analysis conditions, and then run the computer program COR2 backwards to calculate the sample/standard x-ray intensity ratio (K-value) that would be expected in the presence of continuum fluorescence. The COR2 output was modified to print out the magnitude of the continuum-fluorescence correction; from this value one can directly calculate the K-value that would be obtained if the continuum fluorescence were omitted. Results are presented as the error in the K-value that would occur due to the omission of the continuum fluorescence (CF), or:

$$\frac{K\ (CF) - K\ (\text{no CF})}{K\ (CF)}$$

Had we expressed the results as the error encountered in a concentration determination if the continuum fluorescence were omitted, the reported errors would be larger and in some cases considerably larger. Extensive modifications were made in the output of COR2 to print out various correction factors and mass absorption coefficients. Three problems were found with COR2: (1) there were many omissions in the permanent data file that prevented calculation of mass absorption coefficients for the organic elements; (2) the computer followed a code to select the analyzed line, but for the organic elements the code did not agree with the permanent data file; and (3) values for the intensity ratios were incorrectly computed unless pure element standards were used. These minor errors were corrected, but changes were not made in the fundamental correction procedures of COR2.

We have modeled the composition of a frozen-hydrated cell, frozen-hydrated extracellular space, plastic-embedded cell, and plastic-embedded extracellular space. In addition, we modeled elements in a pure carbon or boron matrix. The composition of a frozen-hydrated cell was modeled by assuming typical(weight percent) values for the major molecular components of a cell[1], approximating the contributions of their constituent elements, and obtaining an element distribution as shown in the first column of Table 1. The composition of plastic was modeled after Spurr's low-viscosity resin. As seen from Table 1, the major constituents of the tissue compartments are carbon and oxygen, and the extremes are represented by the frozen-hydrated and plastic-embedded extracellular space.

TABLE 1. Element weight percent (wt %) distribution modeled for frozen-hydrated or plastic-embedded cellular or extracellular tissue compartments.

	Frozen-Hydrated Cell (Wt.)	Frozen-Hydrated Extracellular Space (Wt.%)	Plastic Embedded Cell (Wt.%)	Plastic Embedded Extracellular Space (Wt.%)
H	9.6	10.9	8.1	8.8
C	12.6	.03	63.9	68.6
O	72.8	88.1	23.0	21.6
N	3.3		3.3	
P	.51		.5	
S	.21		.2	
Na	.05			
Mg	.02			
Cl	.27			
K	.63			
Ca	.01			
Ions	(itemized above)	1.0	1.0	1.0

RESULTS

We investigated the error that would be encountered in the analysis of our modeled frozen-hydrated cell of Table 1 if the continuum-fluorescence correction were omitted. Analysis was modeled at 5 KV, the voltage one would like to use for bulk tissue in order to optimize spatial resolution. The analysis was based on the use of mineralogical metallurgical standards. As shown in Table 2, the error due to omitting the continuum fluorescence increased with the

[1]The values we have chosen are H_2O, 74%; protein, 15%; RNA, 4%; carbohydrate, 3%; lipid, 2%; DNA, 1%; and inorganic ions, 1% (Lehninger, 1970; Giese, 1964; Rick et al., 1978; Mountcastle, 1974).

TABLE 2. Error in K-value due to omission of the continuum-fluorescence correction for the constituent elements of a frozen-hydrated cell (Table 1). Analysis is modeled at 5KV. The standards modeled are indicated. "A" is atomic number. Take-off angle for this and succeeding analyses is 40°.

Z	Sample Element	Standard	Cont. Flr. Error (%)
6	C	C	0
7	N	BN	.01
8	O	MgO	−.01
11	Na	NaF	.01
12	Mg	Mg	.01
15	P	$CaHPO_4$	.05
16	S	$BaSO_4$	.03
17	Cl	KCl	.37
19	K	KI	.50
20	Ca	CaF	1.42

atomic number (Z). However, the largest error (for calcium) was less than 2%, and for the major physiological elements the continuum-fluorescence correction can be ignored.

The conclusion that the continuum-fluorescence correction can be ignored for the major physiological elements is also valid at higher accelerating potentials. Figure 2 shows the error due to omitting the continuum-fluorescence correction as a function of accelerating potential (KV) for Ca $K\alpha$ or Zn $K\alpha$ lines in a pure carbon matrix. The error is larger in a pure carbon matrix than the predominantly oxygen matrix of a frozen cell, and the error in the calcium determination is nearly 4% at 5 KV. However, as accelerating potential increases, the error decreases by a substantial fractional amount. For the major physiological elements ($Z \leq 20$), the continuum-fluorescence error will always be small, at most a few percent.

From Figure 2 it is apparent that for elements above calcium, the continuum-fluorescence correction cannot always be ignored; for Zn $K\alpha$ lines the error can exceed 35%. We have investigated this atomic number dependence of the continuum-fluorescence correction as shown in Figures 3, 4, and 5.

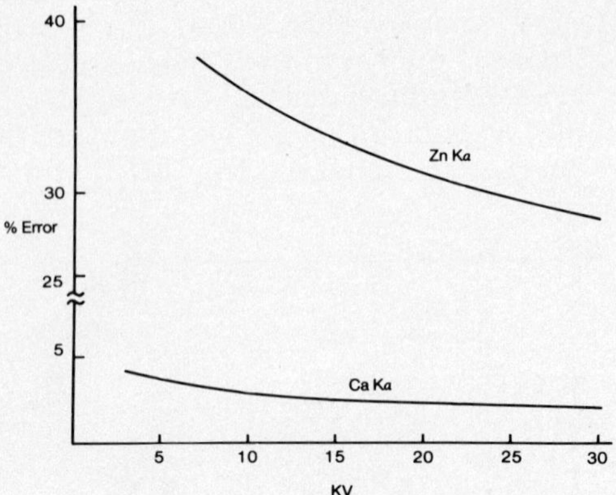

FIGURE 2. The effect of accelerating potential (KV) on the error in K-value due to omission of the continuum-fluorescence excitation correction for CaKα and ZnKα lines Pure carbon matrix, pure element standards. The concentration of Ca or Zn is 0.1%.

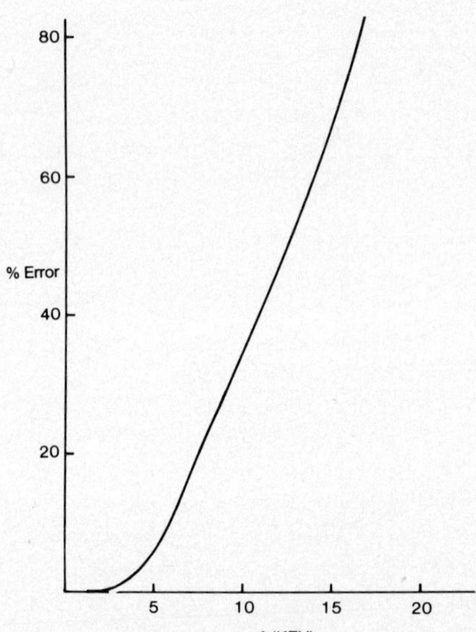

FIGURE 3. Percent error in K-value due to omission of the continuum-fluorescence correction as a function of x-ray line energy (λ) in KeV. Carbon matrix, 30 KV, pure element standards, analyzed element concentration 0.1%.

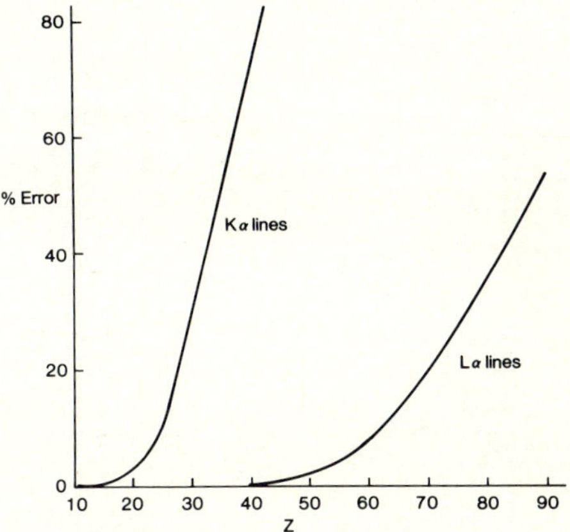

FIGURE 4. Atomic number (Z) dependence of the error in K-value due to omission of the continuum-fluorescence correction. The error is shown as a function of Kα and Lα lines. Analysis conditions are identical to those of Figure 3.

For Figures 3 and 4 we have assumed a 0.1% concentration of the element of interest in a pure carbon matrix. Pure element standards were assumed. In order to excite a broad range of elements, the analysis was modeled at 30KV. As seen from Figure 2, this KV will minimize reported errors. As shown in Figure 3, the error due to omitting the continuum-fluorescence correction is a continuously increasing function of the x-ray line energy. The error is less than 5% for x-ray lines less than 5 KeV, but thereafter the increase in error with x-ray line energy is rather rapid. If the data of Figure 3 is replotted with atomic number along with the abscissa rather than x-ray line energy, as shown in Figure 4, separate curves are obtained for each x-ray line. The advantage of using Lα lines whenever possible is clearly illustrated in this figure; for instance, the error in the K-value for Zr (Z = 40) due to omitting the continuum-fluorescence correction is 73% for analysis by the Kα line, whereas the error is only 0.2% for analysis by the Lα line. Again, one can see that for the major physiological elements the error is negligible.

The effect of matrix on the continuum-fluorescence correction is shown in Figure 5 for the Kα lines. The matrix for plastic-embedded ECF (extracellular fluid) and frozen-hydrated ECF are reported in Table 1. These matrices and that of boron are modeled to contain 1% of the element of

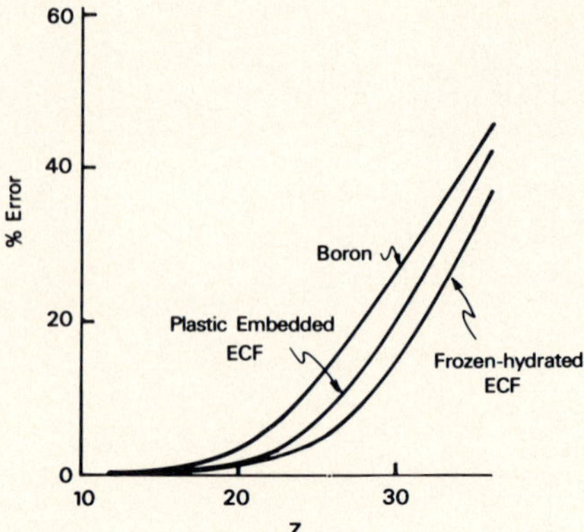

FIGURE 5. Atomic number dependence of the error in K-value due to omission of the continuum-fluorescence correction for Kα lines in a matrix of pure carbon, plastic-embedded extracellular fluid (ECF), or frozen-hydrated extracellular fluid. Analyzed element concentrations for the two latter materials are 1%, 30KV, pure element standards.

interest; analysis conditions are otherwise identical to those of Figure 4. For the boron matrix the average atomic number[2] ($\overline{Z}$) is 5; for plastic-embedded ECF the $\overline{Z}$ is 6; for the frozen-hydrated ECF the $\overline{Z}$ is 7.2. As seen in Figure 5, the error due to omitting the continuum-fluorescence correction falls with increasing average atomic number of the matrix, although other factors such as absorption and overvoltage play a role.

The effect of concentration of the analyzed element on the error in K-value due to omission of the continuum-fluorescence correction is shown in Figure 6 for ZnK and ZrK lines in a plastic-embedded ECF matrix. The analysis conditions are the same as in Figure 5 with the exception that the concentration of Zn or Zr was varied from 0.01% to 10% at the expense of carbon. As seen in Figure 6, there is a large effect of concentration on the continuum fluorescence error.

---

[2]Z is the weight-averaged atomic number defined as Z = C_i Z_i where C_i is the weight percent concentration of element i.

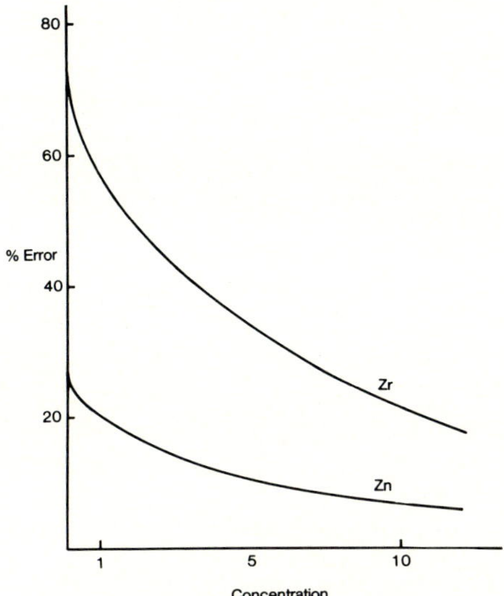

FIGURE 6. Effect of concentration of the analyzed element on the error in K-value for Zn and ZrKα lines due to omission of the continuum-fluorescence correction. Analysis conditions are 30KV, pure element standards, plastic-embedded ECF matrix.

We wish to stress that we have reported the % errors in K-value determinations. The relative error that would be encountered in running COR2 in the opposite direction (from an experimentally measured K-value to a concentration determination) can be much larger as shown in Figure 7. This data is for an element at 0.1% concentration in a carbon matrix with the same analysis conditions as before. The line marked "K-value" is a reproduction of the Kα line for carbon in Figure 4. The 73% error in K-value for Zr corresponds to a 270% relative error in the concentration determination; for the Ag Kα line the error in concentration is over 700%.

The reported errors are partly due to the unequal contribution of the continuum fluorescence in the sample vs. the standard. These errors could be reduced, and a more accurate analysis obtained, if a standard was chosen that was more representative of the organic sample. Figure 8 shows the error in K-value due to omitting the continuum-fluorescence correction for a variety of cations at 1% concentration in a plastic-embedded ECF matrix. Results for each cation obtained usinq a high $\overline{Z}$ (iodine) salt for a

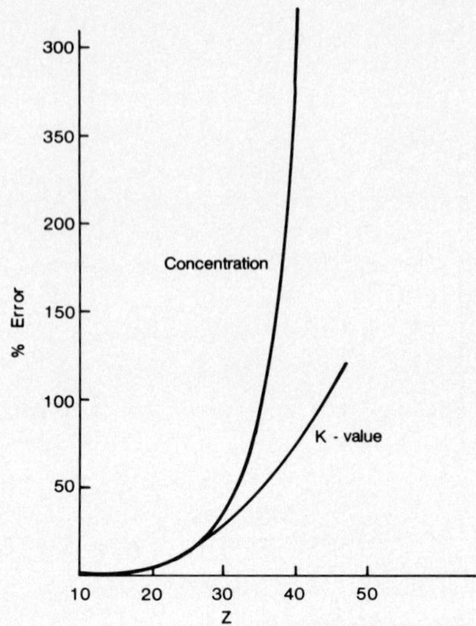

FIGURE 7. Comparison of the relative error in K-value vs. the relative error in element concentration due to omission of the continuum-fluorescence correction as a function of atomic number for Kα lines. Analysis conditions are identical to those of Figure 4.

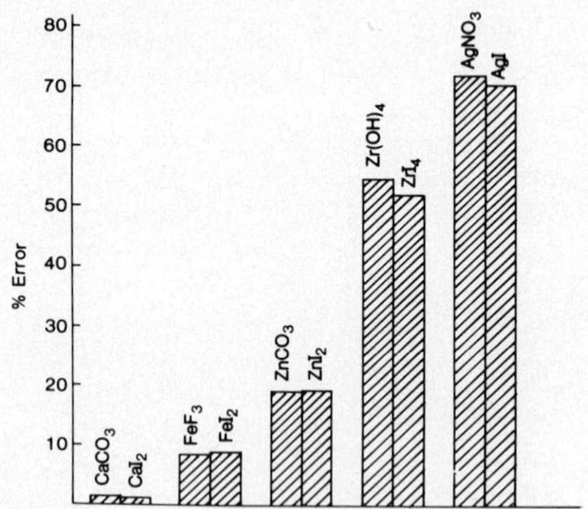

FIGURE 8. Effect of high $\overline{Z}$ (iodine) standards vs low $\overline{Z}$ standards on the error in K-value due to omission of the continuum-fluorescence correction for Ca, Fe, Zn, Zr, and Ag. Analysis conditions are Kα lines, 30KV, and 1% concentration of the analyzed element in a plastic-embedded ECF matrix.

standard are compared with those using a lower $\overline{Z}$ standard such as a carbonate or hydroxide. From Figure 8 there is little improvement in the % error with the lower $\overline{Z}$ standards, and in some cases the errors are actually worse, probably due to absorption effects. This lack of an improvement was investigated in more detail for the Zn salts.

Shown in Table 3 are the results of a modeled 30 KV analysis of frozen-hydrated ECF containing 1% [Zn] using a variety of standards. The standards are ranked in Table 3 in the order of decreasing $\overline{Z}$. There is essentially no change in

TABLE 3. Effect of Zn standards on the error in K-value due to omission of the continuum-fluorescence correction. The sample modeled is a 1% concentration of Zn in frozen-hydrated ECF. $\overline{Z}$ is the average atomic number of the standard. [Zn] is the fractional Zn concentration by weight in the standard. Analysis is modeled at 30KV.

Compound	$\overline{Z}$	[Zn]	C.F. % Error
ZnI_2	48.3	.205	13.77
Zn_3Sb_2	41.6	.446	13.89
$ZnBr_2$	33.6	.29	13.10
Zn	30.0	1.00	14.49
ZnO	25.7	.813	14.18
$ZnCl_2$	23.2	.48	14.25
ZnF_2	22.3	.632	13.93
$ZnCO_3$	19.3	.521	13.53
$Zn(C_{12}H_{23}O_2)_2$	11.8	.231	11.74
$Zn(C_{18}H_{35}O_2)_2$	8.1	.103	8.05

the continuum-fluorescence error as $\overline{Z}$ decreases from 50 to 20. Only when $\overline{Z}$ is below 10, and the Zn concentration approaches 10%, is a significant improvement in the continuum-fluorescence error obtained.

CONCLUSION

The magnitude of the continuum-fluorescence correction in thick biological tissue is a function of the sample matrix composition, the atomic number or line energy of the analyzed element, the analyzed element concentration, the accelerating voltage, and the composition of the standard. This correction is shown to be of negligible importance for the analysis of the major physiological elements (Z $\leq$ 20) in an organic matrix, amounting at most to errors of only a few percent. However, for elements above atomic number 20, the continuum fluorescence can lead to substantial concentration overestimations. This effect is greatest with very low concentrations of the analyzed element. These overestimations can be minimized by analyzing for L-lines as much as possible, as has been reported by others (Myklebust et al., 1970; Hall, 1971). Increasing the accelerating voltage will decrease the concentration overestimation, but at the expense of penetration depth and x-ray spatial resolution for bulk specimens. The continuum-fluorescence correction can be minimized by the use of standards with a relatively low concentration of the analyzed element and a mean atomic number as close to that of the sample as possible. For instance, the analysis of frozen-hydrated samples, the use of frozen salt-solutions as standards may be ideal.

The conclusions of this paper are valid only for bulk biological tissue analyzed with bulk standards. It is unlikely that the continuum-fluorescence correction will be important for thin tissue analyzed with thin standards (Hall, 1971; Tixier, 1979). However, for the analysis of thin unsupported tissue using bulk standards, it should be recognized that the omission of the continuum-fluorescence correction can lead to concentration underestimations as has been recently reported (Kuptsis et al., 1980).

REFERENCES

Giese, A. C. (1964). "Cell Physiology," 2nd Ed., p. 592. W. B. Saunders Co., Philadelphia.

Hall, T. A. (1971). In "Physical Techniques in Biological Research" (G. Oster, ed.), 2nd Ed., Vol. 1A, pp. 157-275. Academic Press, New York.

Hénoc, J. (1968). In "Quantitative Electron Probe Microanalysis," Nat. Bur. Stand. Spec. Publ. 298 (K. F. J. Heinrich, ed.), pp. 197-214. U.S. Dept. of Commerce.

Hénoc, J., Heinrich, K. F. J., and Myklebust, R. L. (1973). N.B.S. Technical Note 769. U.S. Dept. of Commerce.

Kuptsis, J. D., Cardone, F., and Savoy, R. J. (1980). In "Microbeam Analysis-1980" (D.B. Witty, ed.), pp. 51-52. San Francisco Press, San Francisco.

Lehninger, A. L. (1970). "Biochemistry." Worth Publishers, Inc., New York.

Mountcastle, V. B. (1974). "Medical Physiology," 13th Ed., Vol. 1. C. V. Mosby Co., St. Louis.

Myklebust, R. L., Yakowitz, H., and Heinrich, K. F. J. (1970). Proc. 5th Nat'l Conf. Electron Probe Analysis, New York.

Rick, R., Dorge, A., Macknight, A. D. C., Leaf, A., and Thurau, K. (1978). J. Membr. Biol. 39:257-71.

Springer, G. (1967). N. Jb. Miner. Abh. 106:241-56.

Tixier, R. (1979). In "Microbeam Analysis in Biology" (C. Lechene and R. Warner, eds.), pp.209-24. Academic Press, New York.

DISCUSSION

SPEAKER: R. R. Warner

HALL: Could you please explain in more detail what your "K-value" is? Are you using total x-ray counts, just characteristic intensity, or what?

WARNER: What we call the "K-value" is perhaps more frequently referred to as the "K-ratio." The K-value or K-ratio is usually obtained by taking the experimentally measured x-ray intensity for an element in the sample and dividing this measurement by the intensity obtained from the standard for that element. The ratio is formed using only characteristic intensities, which are obtained by correcting the total measured x-ray counts for dead-time losses and subtracting the background counts. These characteristic intensities thus include contributions from continuum excitation in addition to the primary electron-beam excitation. The intensity ratio is usually the starting point from which a computer program calculates the sample concentrations; we run the computer backwards from known concentrations to calculate an expected intensity ratio or "K-value." We calculate a K-value both considering and ignoring the contribution of continuum x-ray fluorescence; the relative difference in these two numbers is what we graph as "% Error."

HECKER: Why did the error blow up at a particular value?

WARNER: The rapid increase in error as seen in Figure 3 is primarily due to a difference in the ionization cross sections between the matrix elements and the analyzed elements. For x-rays, the ionization cross section varies as the binding energy or critical excitation potential raised to some exponent greater than +1, and consequently is considerably larger for heavy elements than for light elements. The continuum x-radiation penetrating into the specimen is preferentially absorbed by the heavy element in much greater proportion than its relative atomic concentration due to the disparity in ionization cross sections. This process is in contrast to the production of characteristic radiation by the primary electron beam which is directly proportional to the element concentration. For a heavy element at low concentration, where characteristic x-ray production by the primary beam is small, the relative enhancement of the characteristic x-ray signal by continuum fluorescence can be considerable as we have shown. The contribution of continuum fluorescence to the characteristic x-ray signal is usually less in the standard since the element of interest is usually present in much higher concentrations and there is frequently less of a discrepancy in the ionization cross sections among the remaining elements.

There are additional factors such as absorption that influence the continuum contribution. Since x-rays penetrate matter much better than electrons, continuum-induced characteristic x-rays can be generated deep within the specimen. However, the characteristic (high energy) radiation of the heavier elements is poorly absorbed by a low Z matrix, and most of this continuum-generated radiation escapes the sample where it can be detected. Depending on the composition of the standard, the continuum-generated characteristic radiation in the standard can be significantly attenuated, thus decreasing its contribution to the primary-beam-generated characteristic radiation.

HECKER: Is the point at which the error blows up indicative of the ionization cross section or x-ray generation depth?

WARNER: Definitely ionization cross section. For instance, a plot of the critical excitation potential (which is proportional to the ionization cross section) vs. atomic number gives a curve with a shape very similar to that of Figure 4.

QUICK FREEZING AND FREEZE SUBSTITUTION FOR X-RAY MICROANALYSIS OF CALCIUM

R. Ornberg
T. Reese

Laboratory of Neuroanatomical and Neuropathological Sciences
National Institute of Neurological Communicative Disorders and Stroke
National Institutes of Health
Bethesda, Maryland

INTRODUCTION

A wide variety of processes within cells is regulated by calcium ions. These include muscle contraction, secretion, cytoskeletal movements, and membrane permeability to mention a few. For calcium to independently regulate a number of these functions within the same cell, the internal calcium concentration and distribution must be carefully controlled as well. Thus, knowledge of calcium distribution in cells afforded by x-ray microanalysis would greatly add to our understanding of each of the functions. Based on a variety of different measurement methods (Caswell, 1979), calcium movements in cells are believed to be transient and fast due to short-lived influxes, rapid internal buffering, and extrusion mechanisms. For example, in squid nerve terminals, the calcium influx which supports neurotransmitter secretion following an action potential occurs on a time scale of hundreds of microseconds (Llinas and Steinberg, 1977). Consequently, preparative procedures for analytical x-ray microanalytical techniques must be fast enough to capture these events in addition to minimizing artifactual redistribution during preparation.

Quick-freezing as a preparative technique has proven useful for studying the morphological events of secretion

ISBN 0-12-362880-6

(Heuser et al., 1979; Ornberg and Reese, 1980) even at the millisecond time scale. In addition to the short time resolution, other advantages such as the non-selective nature of freeze-fixation and the avoidance of noxious effects of chemical fixatives make quick-freezing ideal for preserving natural distributions of soluble molecules in cells. Recently, we described a method using freeze-substitution to localize calcium in quick-frozen tissue for x-ray microanalysis (Ornberg and Reese, 1980). Despite the apparent success of this method, however, it is limited to ions which are insoluble in acetone and can only be regarded as qualitative until it is shown that no calcium movements occur during any step of the processing. Some ion movements might be expected during substitution in a polar solvent like acetone which could be minimized by substituting tissue at the lowest temperature and for the shortest time allowable followed by a transfer to a less polar solvent, at low temperature, for subsequent warm-up and processing. For these experiments and future developments of freeze-substitution methods, we have measured substitution rates for acetone by measuring the efflux of tritiated water from frozen ferritin solutions. In addition, we describe here the use of acrolein as an alternate fixative for osmium tetroxide for preparations where osmium interferes with images of elements and elemental analysis. Finally, we have developed a new x-ray standard for calcium in epoxy sections for quantitative x-ray microanalysis.

MATERIALS AND METHODS

Sea urchin eggs from Strongylocentrotus purpuratus were prepared by Laurinda Jaffe according to procedures which will be described in a later paper.

Amebocyte blood cells from horseshoe crabs, Limulus polypheums, were obtained as previously described (Ornberg and Reese, 1980b). Ferritin solution containing tritiated water as prepared by diluting stock ferritin solution Sigma to a final concentration of 33 mf per milliliter in frog saline solution (Ornberg and Reese, 1980a) that contained 0.1 mCi/ml ^{3}H-water. Calcium naphthenate in mineral spirits containing 5% calcium by weight was a generous gift from Witco Chemicals, Chicago. It was diluted without purification into Araldite epoxy resin (CY212, Polysciences) on a wet weight/weight basis to make a series of calcium concentrations from 500 mmole/kg to 2 mmole/kg. The resin-naphthenate mixtures were cured at 60° to 65°C for 24 to 36 hours.

Suspensions of amebocytes and sea urchin eggs and aliquots of ferritin solution (10 µl droplets) were quick-frozen by pressing them against a cold copper block cooled with liquid helium (Heuser et al., 1979). Frozen amebocyte preparations were freeze-substituted in acetone containing 4% OsO_4 or 10% acrolein (volume/volume) as previously described (Ornberg and Reese, 1980, 1980a). Frozen ferritin solutions were substituted in pure acetone at -83°C by first placing them in a polyethylene scintillation vial containing 10 ml of frozen acetone and roughly 10 ml of liquid nitrogen. Immediately after the vial had warmed and the acetone had melted (this was measured with a thermocouple), the vials were transferred to a freezer for the remainder of the substitution at 83°C. After the acetone had melted, and at various times thereafter, the ferritin sample was removed from the substituting acetone, placed in one ml of fresh acetone at room temperature, and then counted in a liquid scintillation counter along with the substitution acetone in a triton-toluene based liquid scintillation cocktail. The tritium counts were corrected for quenching and the activity in the substitution acetone was expressed as a percentage of the total activity, acetone plus sample.

Thin sections, 100 nm thick, of the calcium naphthenate standards were analyzed in the scanning transmission mode from reduced areas at 5000X with a JEOL 100CX equipped with a KEVEX energy dispersive spectrometer. The analytic conditions were as follows: specimen current 4 nA, accelerating voltage 80 KV, analysis time 200 seconds, and beam diameter 50 nm. X-ray counts from 110 eV wide windows placed in the spectra at 3.1, 3.25, 3.45, 4.12, 4.34, 4.55, and 4.70 KeV served as background windows for calculating the background under the calcium K peak at 3.60 - 3.80 KeV.

RESULTS

Calcium Naphthenate as an X-Ray Standard

X-ray spectral data from thin sections of calcium naphthenate-Araldite standards are graphed in Figure 1.

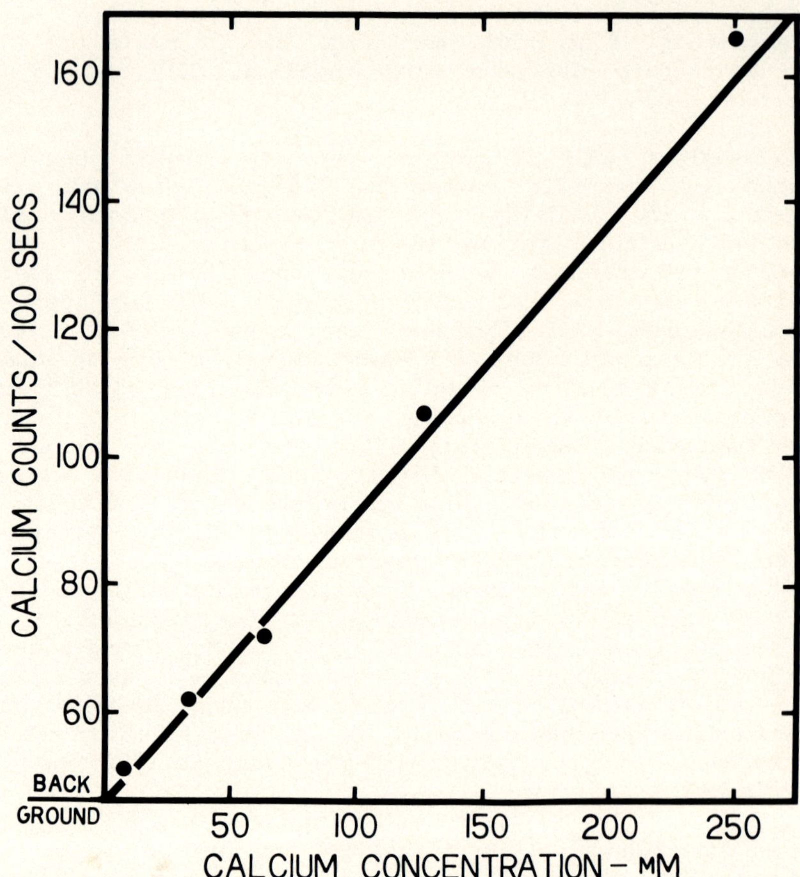

FIGURE 1. Calcium Kα counts for 100 nm thick sections graphed according to calcium concentration using calcium naphthenate-Araldite epoxy standard

Calcium concentrations up to 500 mmoles per kg were prepared without noticeably affecting the cutting properties of the plastic. Sections cut and floated on either water or glycerol gave statistically indistinguishable peak-to-background ratios; however, any calcium solubilized from the surface of the section during cutting would be too small to be measured by x-ray analysis. Repeated analysis over the same area resulted in higher peak-to-background ratios, probably as a result of radiation damage since this was greatly reduced after the sections were coated with a light layer of carbon. Dr. A. P. Somlyo (personal communication) has experienced similar effects which he also attributes to radiation damage.

With low concentrations of calcium, we were able to determine our present minimum detectable limit to be somewhere between 5 to 10 mmole per kg.

Freeze-Substitution Rates

The amount of water substituted by acetone at -83°C is graphed as a function of time in Figure 2. Clearly, the quick-frozen ferritin samples substituted much more rapidly, being 50% complete within 4 to 5 hours, than did those quench-frozen in Freon 22. However, the quick-frozen samples had smaller ice crystals that extended much deeper into the preparation (Figure 3a) than quench-frozen samples (Figure 3b). Thus, the disparity in substitution rates might be expected if the substitution rate depends on the surface-to-volume ratio of the ice crystals.

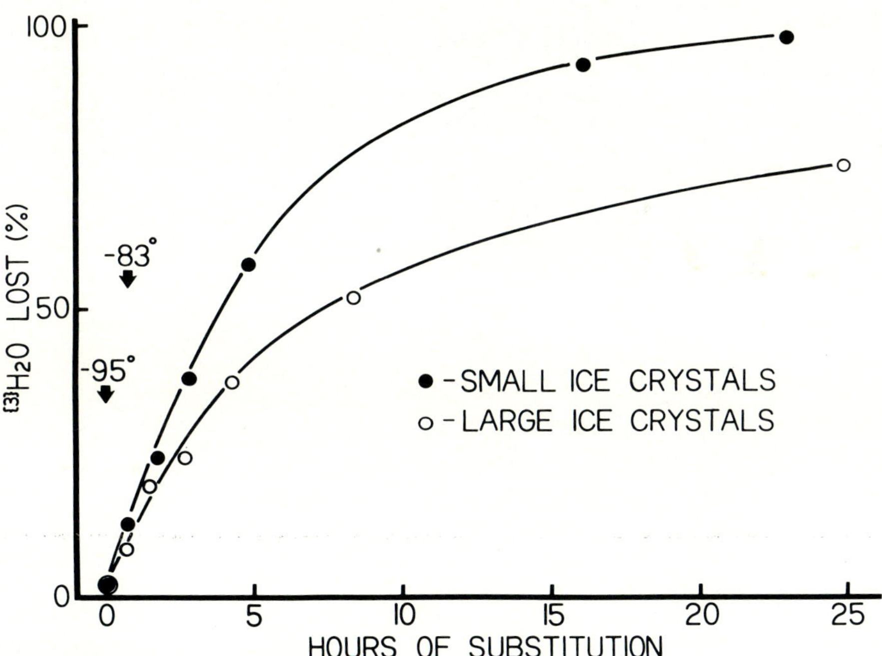

FIGURE 2. Efflux of tritiated water from quick-frozen (small ice crystals 0) and quench-frozen (large ice crystals 0) ferritin solutions. The corresponding ice crystal patterns are illustrated in Figure 3.

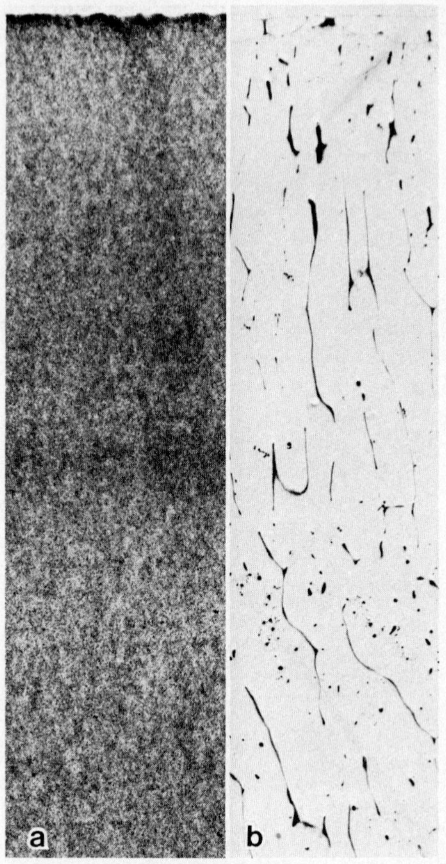

FIGURE 3. Electron micrographs of quick-frozen ferritin solution (a) and ferritin quench-frozen in Freon 22 (b). The top of each figure is the other surface from which the freezing process began. The minute ice crystals (< 10 nm) in the quick-frozen specimen are not resolvable at this magnification (2750X).

Acrolein as a Freeze-Substitution Fixative

Preliminary results describing the redistribution of calcium in sea urchin eggs during the fertilization rate and in amebocytes dramatically illustrate a possible difficulty with the use of osmium tetroxide as a fixative during freeze-substitution. The cortical granules in Strongylcentrotus purpuratus sea urchin eggs contain a characteristic crescent-shaped core which is disposed to one side of the granule (Figure 4). X-ray analysis of the granule produces a strong calcium emission which we originally believed originated from the dense matrix. However, separate analysis of the dense matrix and the clear region within the same granule indicate that calcium is present in both regions while the increased electron density is due to osmium and

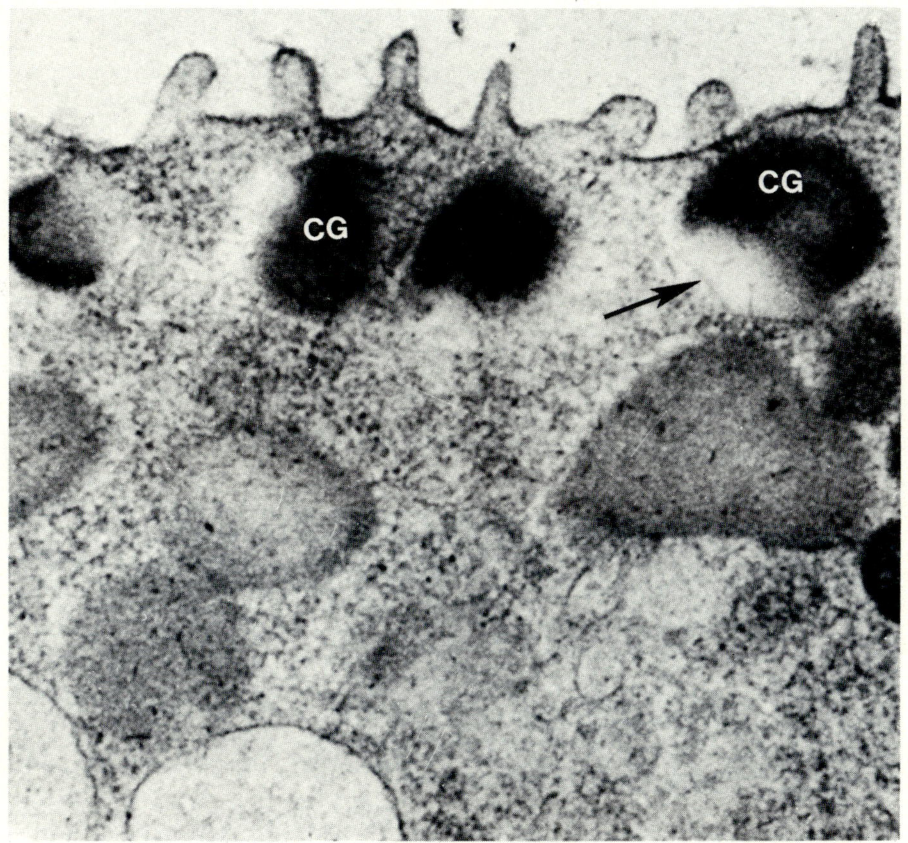

FIGURE 4. Scanning transmission electronmicrograph of quick-frozen, freeze-substituted sea urchin egg. The cortical granules (CG) usually appear with two distinct regions, one being the crescent-shaped, electron-dense matrix and the other being the electron-translucent space (arrow). Both regions contain roughly equal amounts of calcium (45,000x).

sulfur. Thus, it became clear that calcium is not always visible in osmicated tissue. Omission of the osmium tetroxide and substitution in pure acetone eliminates this problem, but the tissue has a shrunken appearance that is obvious in <u>Limulus</u> amebocytes from light microscopic observations (not shown). Acrolein added to the substitution acetone minimizes this shrinkage and allows the natural electron-dense structures to be visualized. One such structure is a dense calcium deposit located in secretory granules of <u>Limulus</u> amebocytes (Figures 5a,b). Like the sea urchin egg cortical granules, the amebocyte granules contain large amounts of sulfur and are the osmiophilic to the extent that these dense bodies are unnoticed in osmicated preparations. Unfortunately, acrolein/acetone-substituted tissue has low contrast and provides less structural detail for high-resolution microscopic observations.

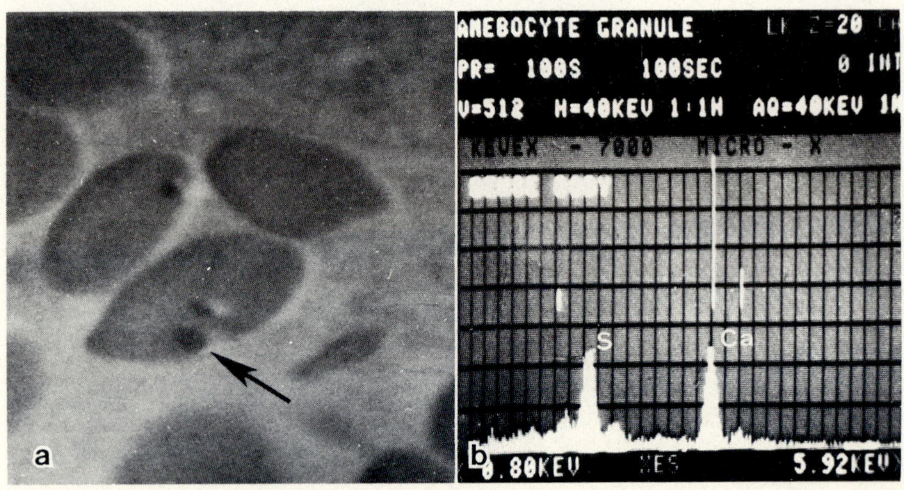

FIGURE 5. (a) Scanning transmission electron micrograph of a <u>Limulus</u> amebocyte that was quick-frozen and freeze-substituted in acetone containing 10% acrolein and 20 mM oxalic acid. Electron-dense deposits of calcium which are usually not seen in osmicated preparations can now be seen in association with secretory granules(arrow). (b) x-ray spectra obtained from probing the dense deposit reveals that it is almost completely comprised of calcium and sulfur (24,000x).

CONCLUDING DISCUSSION

The use of some methods of freezing followed by frozen thin sectioning and x-ray analysis on either the frozen-hydrated or freeze-dried section has been widely used for microanalysis. However, some very basic problems with these methods have caused us to use freeze-substitution and ultramicrotomy of plastic section. Conventional plastic sectioning, even when the sections are cut and floated on glycerol, is less difficult than ultracryomicrotomy. Frozen sections have inherently poor contrast which renders small well-defined structures in cells invisible for subsequent microanalysis. Although freeze-dried sections have improved contrast, there is no guarantee that ions stay in their natural positions during the drying procedure for high-resolution analysis. While our freeze-substitution method provides images adequate for such detailed analysis and our results with frog skeletal muscle (Ornberg and Reese, 1980) are encouraging, freeze-substitution for microanalysis is still tentative. Clearly any ion which is "free" in the cytosol will undoubtedly move and "fix" itself to a neutralizing charge when its hydration shell is removed and substituted by a molecule like acetone. However, those ions such as calcium, which are bound to intracellular stores and thus have a low activity, may withstand substitution and thus be fixed more readily than more active ions such as sodium and potassium. This issue will only be resolved by careful comparisons of microanalytical data from cryosections and freeze-substituted plastic sections from identical preparation.

Our freezing method and those which produce similar freezing may be the only methods with which freeze-substitution for microanalysis will work. As evidenced in Figure 2, quick-freezing with a liquid helium-cooled copper surface is clearly superior to quench-freezing in liquid nitrogen-cooled Freon 22. Since ice crystals grow by exclusion of ions and other solutes, quick-freezing immediately minimizes ion redistributions that may result from inferior freezing methods. The fact that quick-frozen preparations with smaller ice crystals freeze-substitute faster at low temperature than quench-frozen preparations strongly suggests that the substitution process is a function of the ice crystal size or, more directly, on the surface-to-volume ratio of the crystals. Just how the substitution rate enters into the fixation of ions for microanalysis is unclear. However, the rapid removal of tissue water at low temperature may result in less bulk movement of ions in the water acetone phase that comprises the freeze-substitution front.

Precise knowledge of the freeze-substitution process, based on the movements of tritiated water, may prove helpful for future development of substitution methods. Solvents other than acetone, such as diethyl ether have been successfully used for retaining sodium, calcium, and chloride (Spurr, 1972; Harvey et al. 1976). Diethyl ether is, however, a poor substitution fluid, since it substitutes very slowly and may not remove all of the tissue water for subsequent embedding procedures. Such incomplete dehydration may explain the embedding problems reported by Harvey et al. (1976) for diethyl ether-substituted tissue. Presumably, frozen tissue could be substituted in acetone and then be transferred at low temperature to a less polar solvent such as hexane or diethyl ether for subsequent warm-up and processing. Such a method may retain tissue sodium and potassium as well as calcium. These experiments are currently underway in our laboratory.

The replacement of osmium tetroxide by acrolein as a substitution fixative is useful not only for the visualization of the natural elements but also for the analysis of biologically important ions, such as phosphorus and strontium with which the osmium $M\alpha,\beta$ emissions interfere. Although the omission of osmium takes away the advantage of improved contrast provided by our freeze-substitution method, certain structures such as secretory granules and mitochondria are readily recognizable. With improved electron-imaging techniques in scanning tranmission electron microscopes, such as zero-loss and dark field images and digital image processing, the problem of low contrast may be tolerable.

The calcium naphthenate standard should prove useful for quantitative analysis since few calcium standards exist. The calcium crown-ether complexes mentioned by Roomans (1979) have a low solubility limit in epoxy resin. Calcium naphthenate is readily soluble and very easy to use since it comes in an organic solvent. Calcium naphthenate is used industrially to augment the drying of paints and as such might be expected to alter the polymerization and hardness of epoxy resins. We have not noticed any such effects with Araldite resin at concentrations of calcium up to 500 mmoles/kg or 2.0%. Naphthenate salts of a variety of metals including cobalt, iron manganese, zinc, and lead are also available. These naphthenate salts, particularly zinc and cobalt naphthenate, may also be included in the embedding resin for tissue for subsequent section thickness measurements based on their characteristic peak-to-background ratio.

REFERENCES

Caswell, A. (1979). International Review of Cytology 56:145-81.

Harvey, D.M., Hall, J.L., and Flowers, T.J. (1976). Journal of Microscopy 107:189-98.

Heuser, J.E., Reese, T.S., Dennis, M.J., Jan, Y., Jan, L., and Evans, L. (1979). J. Cell Biol. 81:275-300.

Llinas, R.R., and Steinberg, I. (1977). Neurosciences Res. Prog. Bull. 15:565-74.

Ornberg, R.L., and Reese, T.S. (1980). Fed. Proc. 39:90-95.

Ornberg, R. L., and Reese, T.L. (1980a). J. Cell Biol. Submitted.

Roomans, G.M. (1979). Scanning Electron Microscopy, Vol. 2, pp. 649-57.

Spurr, A. R. (1972). Bot. Gaz. 133:263-72.

DISCUSSION

SPEAKER: R. L. Ornberg

HUTCHINSON: If you use nitrogen as a coolant, instead of helium, do you obtain data that would fit those curves you showed previously for He?

ORNBERG: I do not know. We are just starting to work with that with some ferritin standards and freeze-substitution rates.

MACKENZIE: I might just mention that the freeze-substitution rates I measured in a setup in a light microscope showed, to a first approximation, that the substitution rate was independent of gross ice crystal diameter, and I put that down to the fact that the rate of limiting step was the diffusion of water through solvent and not the rate of dissolution of ice or surface area.

Secondly, to explain the difference, I would suggest that the lower freezing rate gives more time during the cooling of any individual volume for a recrystallization or grain growth which is a sealing-off process and the branching ice necks off and becomes separate whereas the faster freezing leaves you with a continuous skeleton through which spaces your substitution proceeds.

ORNBERG: That is certainly true. Certainly another valid explanation.

SAUBERMANN: How much drying do you think occurs in these very small samples prior to freezing as they are dropped? You have a sample with very large surface-to-volume ratio and I am wondering whether or not that might contribute some drying.

ORNBERG: That is a very important point and it is something we have worried about. In particular with muscle, when we actually put the muscle onto the stage, at the very last instant take a piece of filter paper and wick off that excess water lying on the surface; then freeze it. We really have not thought of a good way to measure that yet. We are aware of the problem. We fussed around with different types of freezing machines in effect to eliminate that possibility and to see if there is any way we could measure anything, but those freezing machines did not work.

SAUBERMANN: I think it is very significant because the method that you described really does an outstanding job, if in fact that is what is happening, and in the fact that you are freezing tissue that has not been drying, then you are freezing tissue plus water.

SOMLYO, AVRIL: Did you cut it into cryosections with this type of freezing?

ORNBERG: No, we haven't yet. John Tormey and I hope to do this soon. We have to make a careful comparison between frozen sections and the freeze-substituted material. The main reason for going to the embedding is that for the work on synapses and terminal cisternae, we need a magnification of 60,000 x to find the areas of interest and we want to be able to probe areas of about 100 nm. We must be able to see these small structures very, very well. We are morphologists and are not willing to give up anything in our image to the analysis if we absolutely do not have to.

SOMLYO: Since we have also taken up freeze-substitution for similar reasons, we would like to see where we are when it works, it will be easier, I think one of the things that is worth reemphasizing is what you have already alluded to and I would like to reemphasize it here because contrary claims have been made in that when you do freeze-substitution, the diffusible monovalent ions diffuse; they go. In other words, they are just not there, but some physical chemists tell me that there is some spectroscopy work that shows that if you have an apolar solvent with a few molecules of water, any ion that is there will seek out those very few molecules out of billions and combine with them so as the water goes out . . .

ORNBERG: That is, if you strip the hydration shells of those ions. Now it is possible that there may be substituting agents which do not do that, but substitute them in the bulk water.

SOMLYO: But the ions go. Your experience is that potassium chlorides are gone. The second question is that while freeze-substitution with this method preserves deposits of calcium, for example, mitochondrial rocks and such, on occasion then, we have quantitated for example, calcified mitochondria in atherosclerotic vascular smooth muscle, that we were interested, to our horror, the "cytoplasmic concentration," which is really plastic at this point and which are also around 10-15 millimoles per kg dry weight . . . Now we know. Have you seen, for example, when you have sea urchin rocks and you go nearby and probe the area of cytoplasm and see what you get?

ORNBERG: I haven't seen anything.

TORMEY: However, we have in skeletal muscle.

ORNBERG: John Tormey has mentioned to me that he has an observation that there is a cytoplasmic calcium concentration which is much too high and we are going to do the frozen section work and compare. There are two possible steps in the process where this may be occurring: during the freeze-substitution step, which is why we are careful in checking substitution rates and determining when all the water is gone, so that maybe we can put it into a different solvent, and during sectioning.

BUJA: What size samples of muscle have you used and still had success with oxylate method? We have had some difficulty using the freezing substitution with oxylate in preserving the Ca and P which we think should be present in blocks of cardiac muscle.

ORNBERG: On the subcutaneous pectoris of the frog.

BUJA: How thick is that?

ORNBERG: 120 μms.

SOMLYO: I think it depends; on the other hand, I think I should also point out that when in collaboration with Tony Scarpa we measured in situ and isolated platelet-dense bodies where the ATP with firefly luminescence, magnesium with atomic absorption spectroscopy and we measured phosphorus divalent cation ratios with the probe, we got exactly and embarrassingly to the first decimal correspondence of the two sets of data, so that there are circumstances when you have these circumscribed deposits. Would that really stay under the freeze- substitution conditions stoichemically? There is no way in freeze-substitution that one can keep anything except when it is in the form of a precipitate.

LECHENE: You are doing this very sophisticated freezing technique to see movement which can occur in 1 millisecond which have movement basically of free diffusible calcium, so how could you say that during the freezing substitution step you will not have redistribution of calcium which will go naturally to try to escape through some of the protein?

ORNBERG: We cannot, we can simply look and see. With the synapse experiments. . .

LECHENE: Did you look at calcium in the synapse?

ORNBERG: Yes, we did. We looked at it after repetitive stimulation and essentially the micrograph I skipped is identical to the one that Avril Somlyo and Kathy McGraw showed. It shows the Ca in the mitochondria, calcium in the large circular cisternae and also calcium in the other cisternae. But now the question is: Is this the experiment which we wish to do? What we are talking about is calcium that is flowing in during 200 microsecs and in that time if we can freeze, it is only diffused from the membrane at the point of entry about 1500Å and there is probably nothing around to hold it.

I think that for some of the problems that we are looking at we may be aided by binding proteins which are effectively taking the calcium out right away and that is what we are looking for, but nothing works very well for Sr where for most of these systems you can remove the external Ca, add Sr and effectively look at the Sr that is coming in during the activity. Sr is much easier to see since it has a high Z number.

SOMLYO: You can even do that with Os fixed material.

ORNBERG: Yes, very heavily fixed Os material. I think that the real key to the method for freeze-substitution will be with the fluorinated material and energy-loss analysis where we have to cut a very thin section and where we have to be very careful about mass thickness changes and that is why I do not think we will ever completely abandon it. Particularly if we can get enough F into some of these compartments to do some interesting analyses so it will be useful, but how useful for Ca is yet to be seen.

Part III

ENERGY-LOSS SPECTROSCOPY IN THE ELECTRON MICROSCOPE: THEORETICAL CONSIDERATIONS

R. P. Ferrier

Department of Natural Philosophy
University of Glasgow
Glasgow, Scotland

INTRODUCTION

In recent years, the power of the transmission electron microscope as a structural tool has been greatly extended with the development of techniques which allow the chemical nature of microareas of the specimen to be explored. The first of these which should be mentioned, and the one most fully developed, is x-ray microanalysis, which is discussed in detail in other chapters in this volume. The information gained from the study of such an x-ray spectrum can be complemented (and indeed supplemented) if an electron spectrometer is attached to the microscope and the spectrum of inelastic electrons transmitted by the specimen is recorded. An analysis of such a spectrum can, in principle, lead to information on the electron states in the sample, on the chemical composition and on its bonding. However, before quantitative information can be extracted from electron energy-loss spectroscopy (EELS), it is necessary that the basic physics of the various energy loss processes is understood and that accurate cross sections differential in energy and scattering angle are available. In this paper, progress towards this goal will be reviewed with particular reference to two energy regions currently of interest, viz. (a) the low-loss regions 5 $\sim$ 50eV and (b) the high-loss region 50 $\sim$ 2000 eV.

ISBN 0-12-362880-6

THE ELECTRON ENERGY-LOSS SPECTRUM

In the electron microscope, a beam of fast electrons of energy (E_o), velocity (v_o), and rest mass (m_o) is incident on a specimen of thickness (t). In discussing the inelastic interaction of this beam with the specimen, we are interested in two effects, firstly the energy transfer distribution which describes the probability of an electron suffering a particular energy loss ΔE and secondly the momentum transfer which describes the probability that the electron is scattered into an elementary solid angle $d\Omega$ about a particular scattering angle θ. These two effects are generally expressed in combination in the form of the double differential scattering cross section $d^2\sigma/d\Delta E d\Omega$. In the electron microscope the normal mode of recording a spectrum is to insert on axis a small aperture to define the maximum scattering angle from the specimen θ_o which contributes to the spectrum. Thus, in such a recorded spectrum the intensity at any particular energy $I(\Delta E)$ will be of the form:

$$\int_o^{\theta_o} \frac{d^2\sigma}{d\Delta E d\Omega} 2\pi \sin\theta d\theta$$

the inelastic scattering being assumed axially symmetric. Two examples of energy-loss spectra are shown in Figure 1. In each case the elastic (zero-loss) peak has been omitted for clarity. For very thin specimens, the elastic peak can dominate the scattered intensity distribution; however, as the specimen thickness increases, the scattering is progressively redistributed into inelastic events. In setting up the formal mathematical description of the various inelastic processes we consider the spectra in two stages.

The Low-Loss Region

In this region, extending roughly from 5 - 50 eV, the energy is transferred to the conduction or valence electrons of the solid through individual or collective excitation processes. In many specimens (including most metals and semiconductors and some insulators) the dominant feature at low ΔE is a single peak or a series of equally spaced peaks which can vary considerably in "sharpness," depending on the particular material examined. These are the volume plasmons arising from collective oscillation of the quasi-free electrons in the specimen. The plasmon frequency ω_p in the classical long wavelength limit is given by:

$$\omega_{op} = \sqrt{\frac{4\pi Ne^2}{m}} \qquad (1)$$

where N is the number of free electrons (of mass m) per unit

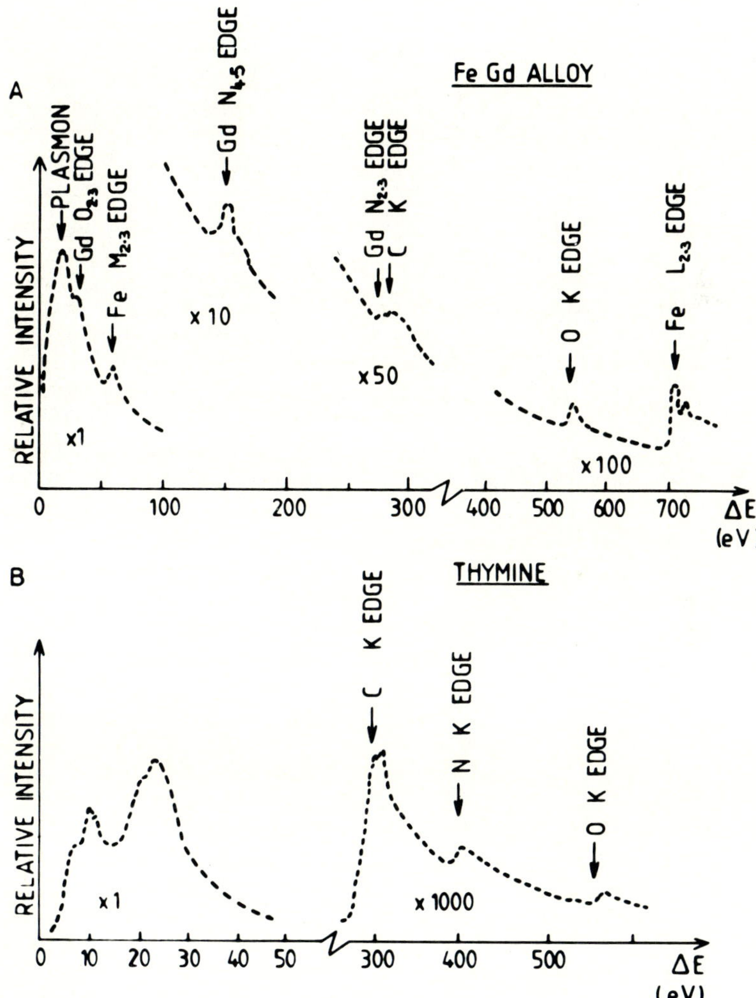

FIGURE 1. Energy-loss spectra obtained over a wide energy range. (a) Iron-gadolinium alloy from Colliex and Trebbia and (b) Thymine ($C_5N_2O_2H_6$) from Isaacson and Johnson.

volume. The peak energies for plasmons $\Delta E_p = \hbar\omega_p$ lie typically in the range 15 - 30 eV.

A quantum mechanical calculation shows that the dispersion relation $\omega_p(\kappa)$, where $\kappa = 2\pi/\lambda$ is the wave vector of the scattered electrons, is of the form:

$$\omega_p^{\ 2} = \omega_{op}^{\ 2} = \alpha\kappa^2 \tag{2}$$

where is a constant and typically ~ 0.4. This relation holds for wave-vector κ up to a critical value κ_c at which the collective mode breaks down. This occurs for a wavelength $\lambda_c = 2\pi/\kappa_c$ which is of the order of the average separation of the electrons in the plasma. For shorter wavelengths than λ_c, the energy transfer is principally to single electron excitations which will be discussed in the next section.

To set up a complete theoretical model for the excitation of plasmons, we must develop a formulation describing the screening of a longitudinal field varying in space and time. Such information is contained in the complex dielectric constant $\varepsilon(\Delta E,\theta)$ of the material and, in particular, the imaginary part of that function. The formula for the differential cross section (Raether, 1968) is of the form:

$$\frac{d\sigma_p}{d\Delta E} = \frac{1}{\pi a_o N T_o} \; \frac{\theta d\theta}{\theta^2 + \theta_E^{\ 2}} \; \mathrm{Im}\left[-\frac{1}{\varepsilon(\Delta E,\theta)}\right] \tag{3}$$

when $a_o = \dfrac{\hbar^2}{me^2} = 0.529 \times 10^{-8}$cm is the Bohr radius,

$t_o = \dfrac{1}{2} m_o v_o^{\ 2}$, and

$\theta_E = \dfrac{\Delta E}{2T_o}$

is the characteristic scattering angle for the plasmon event (typically $\approx 10^{-4}$ radian). The cross-section decreases rapidly with angle so that the spectrum is dominated by the small angle values.

The strength of a particular inelastic event can be characterized by its mean free path, and a knowledge of the value of this parameter is essential in assessing how much plural scattering or multiple loss events can occur for a given specimen thickness. We can estimate the mean free paths of plasmons in a particularly simple way by the following approximate calculation. The variation of ω_p with K is slow and hence we can rewrite equation (3) in the form:

$$d\sigma \approx \frac{\hbar\omega_p}{\pi a_o N T_o} \frac{\theta d\theta}{\theta^2+\theta_E^2} \operatorname{Im}\left(-\frac{1}{\varepsilon}\right) \tag{4}$$

To integrate this we should know the exact form of $\operatorname{Im}(-\frac{1}{\varepsilon})$ and to avoid this we make the sweeping assumption that for all values of θ up to θ_c, $\operatorname{Im}(-\frac{1}{\varepsilon})$ (the angle corresponding to θ_c) takes the value unity and is zero for $\theta>\theta_c$. We then have an upper limit to the cross section:

$$\sigma_p \simeq \frac{\hbar\omega_p}{\pi a_o N T_o} \ell n\left(\frac{\theta_c}{\theta_E}\right) \frac{\theta_E \ \ell n\left(\frac{\theta_c}{\theta_E}\right)}{N a_o} \tag{5}$$

The mean free path Λ_p is then given by:

$$\Lambda_p = \frac{1}{N\sigma_p} \simeq \frac{0.5}{\theta_E} \ell n \frac{\theta_c}{\theta_E} \ \text{Å} \tag{6}$$

Values obtained from equation (6) are typically in the range 50-100 nm for 100 KeV incident energy and agree with experimental values to within an accuracy of $\sim$10%.

Since the energy loss to the plasmon is proportional to $N^{\frac{1}{2}}$ we can probe the local free electron density by studying the electron energy-loss spectrum. However, because the plasmon peaks are generally broad and if N varies, it does so only slowly, the use of plasmons for microanalysis purposes is likely to remain confined to a few special cases such as the study of segregation in Al-Mg alloys (Doig et al., 1975). The function $\operatorname{Im}(-\frac{1}{\varepsilon})$ does also contain information on band structure of materials, and this is reflected in the energy-loss spectrum of several solid materials (transition metals, wide-gap insulators, etc.) which display strong interband transitions. The analysis of such spectra to recover the band structure information is likely in the future to prove the most useful aspect of this range of energy loss.

The Moderate Energy-Loss Region

This region of the energy-loss spectrum, extending roughly over the loss range 50 - 2000 eV, is characterized by contributions of various kinds which give rise to a continuously decreasing background superimposed on which are characteristic "peaks" (Figure 1) arising from ionization processes. As will become clear, these features can provide information on: (a) the chemical composition of the specimen from measurements of the number of electrons contributing to the peaks; (b) the chemical state of the constituents from analysis of the fine structure close to the onset of the edge; and (c) information on the nature and the bond lengths of neighboring atoms, i.e., the average environment surrounding a particular species of atom, from a study of the fine structure extending several hundred eV beyond the ionization edge. The last two of these are essentially solid-state effects and in the first instance, it is instructive to discuss the basic physics of the microanalytical capability in terms of ionization as an atomic phenomenon.

In the ionization process, an inner-shell electron of the atom is promoted to a state in the continuum through interaction with an incident electron. The most convenient formulation for such an event written in the first Born approximation and for small scattering angles is of the form per electron (Inokouti, 1971):

$$\frac{d^2\sigma}{d\Delta E d\Omega} = \frac{2e^4}{T_o \Delta E} \cdot \frac{1}{\theta^2 + \theta_E^2} \cdot \frac{df(\kappa, \Delta E)}{d\Delta E} \qquad (7)$$

where the momentum transfer variable κ, the scattering angle θ and the incident electron wave-vector k are related by:

$$\kappa^2 = k^2 \, (\theta^2 + \theta_E{}^2) \qquad (8)$$

$\theta_E = \Delta E/2T_o$ is again the characteristic scattering angle for the event and $df(\kappa, \Delta E)/d\Delta E$ for any particular inner shell excitation, (labeled K, L_1, L_{23}, M_{45}, etc. by analogy with x-ray usage) is a property of the atom known as the Generalized Oscillator Strength (GOS) by analogy with the optical oscillator strength for $\kappa = 0$. It should be noted in passing that equation (7) is also suitable to first order for the relativistic regime (up to $\sim$0.5 MeV) of incident energies provided T_o is retained and is not replaced by the relativistically corrected kinetic energy.

Before proceeding to a detailed discussion of the GOS, it is instructive to consider the form it would take for the interaction between a free stationary electron and a fast incident electron. Equation (8) can be rewritten:

$$K^2 = k^2\theta^2 + \frac{\Delta E^2 k^2}{4T_o^2} \quad (9)$$

The energy gained by the stationary electron will be just that lost by the incident electron and, therefore:

$$K = \sqrt{\frac{2m\Delta E}{\hbar}} \quad (10)$$

Hence $$\frac{2m\Delta E}{\hbar^2 k^2} = \frac{\Delta E}{T_o} = \theta^2 + \frac{\Delta E}{4T_o^2} \quad (11)$$

For $\Delta K << T_o$ and $\theta \sim \left(\frac{\Delta E}{T_o}\right)^{\frac{1}{2}}$ (12)

We might expect a similar result for bound electrons if the energy loss is very much greater than the binding energy, although in this case a range of scattering angles will occur rather than the single angle of the classical calculation. Thus, if we consider an energy loss $\Delta E = 1$ keV and $T_o = 100$ keV while $\theta_E \simeq 5$ mradian, df/dE may peak around $(1/100)^{\frac{1}{2}} \simeq 100$ mradian and hence the need to allow for large scattering angles into the electron spectrometer is established, albeit in a crude manner.

The GOS may be calculated from the square of the matrix element between the initial core state and the final continuum state (Inokouti, 1971; Madison, 1972).

$$\frac{df\ (K,\Delta E)}{d\Delta E}\ \alpha\ K^{-2}\left|\langle\psi_f\right|\exp\ (i\ \underline{K}\cdot\underline{r})\left|\psi_i\rangle\right|^2 \quad (13)$$

Where ψ is the initial state wave function and ψ_f, the final state wave function, is normalized per unit energy range in the continuum. Recent computations by Leapman et al. (1978)

make use of a Hartree-Slater central field model. The initial state wave function was a one-electron, Herman and Skillman (1963) wave function, which is a solution of the radial Schrödinger equation. The final state wave function was found by solving the Schrödinger equation with the same central field for the continuum energies. The authors have performed calculations for a reasonably extensive range of elements and excitations. The result of their calculation for the K shell GOS of boron is illustrated in Figure 2a. This shows the function to be forward peaked at the edge, but for higher losses it begins to peak at non-zero momentum transfer $\kappa_B \approx k(\Delta E/E_o)^{\frac{1}{2}}$; this is called the Bethe ridge and arises from the predominance of optically forbidden transitions. The energy spectrum calculated from the GOS is shown in Figure 2b and exhibits the expected hydrogenic shape. For other losses, e.g. L_{23} and M_{45}, there can be considerable departure from the simple hydrogenic shape and the model predicts this well. Two examples (Rez and Leapman, to be published) of calculated ionization-loss peaks for magnesium L_{23} and molybdenum M_{45} are shown in Figure 3. In each case the computed GOS on which they are based

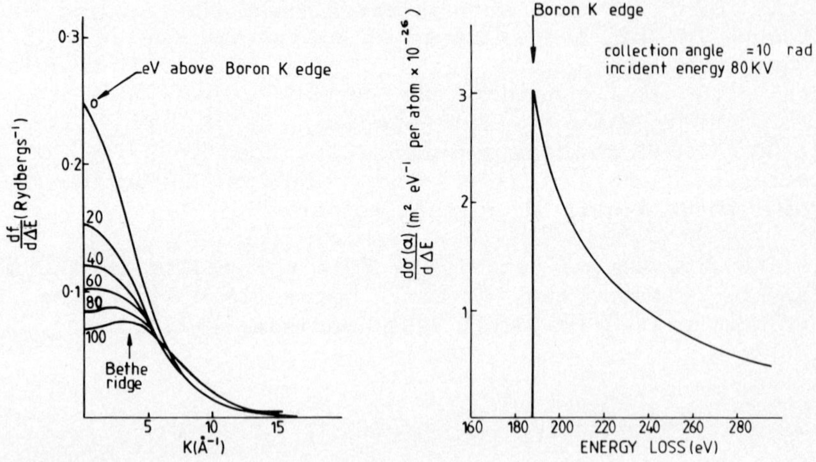

FIGURE 2. (a) The computed generalized oscillator strength df/dΔE for the boron K edge as a function of the momentum transfer variable K. (b) The energy-loss spectrum of boron for a collection angle of 10^{-2} radian calculated from (a) (From Leapman et al., 1978).

exhibits a maximum value at non-zero ΔE; this arises from a centrifugal barrier (Fano and Cooper, 1968) in the atomic potential, which reduces the matrix elements in equation (13) for several eV above the edge when the final state wave function was high-angular momentum.

In the methods developed for quantitative microanalysis (Egerton et al., 1976; Leapman, 1979), which are discussed elsewhere in the volume, partial or total cross-sections are required and these, in principle, can be computed by integration of equation (7) over the range of scattering angle and energy loss appropriate to the given experimental conditions. The calculations of Leapman et al. (1978) are not, however, sufficiently complete to allow such integration in a completely general case and it is, therefore, of interest to discuss a relatively simple procedure due to Egerton (1979) which allows such cross sections for K-shell excitations to be computed rapidly. Earlier we said little about the range of applicability of the calculations which were discussed, and it is, therefore, worthwhile to list the basic assumptions of the Egerton model, which are as follows:

(1) The Born approximation is applicable. This requires that

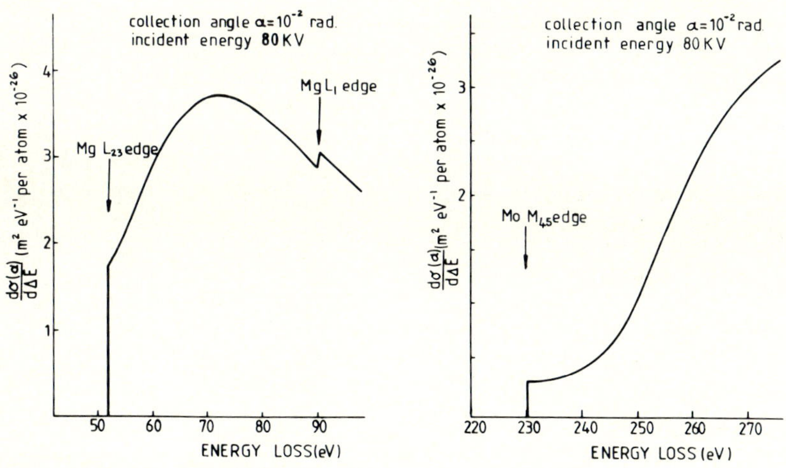

FIGURE 3. (a) Computed spectrum for the magnesium L_2 and L_{23}. (b) Computed spectrum from the molybdenum M_{45} edge (Rez and Leapman, to be published).

the energy of the incident electron is much greater than the ionization energy and that the average scattering angle and momentum transfer are also small. These restrictions hold for elements in the first and second rows of the periodic table for $T_o > 30$ KeV.

(2) Relativistic effects within the atom are ignored. Those associated with the incident electron are included by using $T_o = \frac{1}{2}m_o v_o^2$, as discussed previously.

(3) Exchange effects are ignored.

(4) An atomic model is assumed which neglects all solid-state effects. If the cross section is integrated over an energy range > 50 eV, however, the results should be independent of the physical and chemical environment.

(5) Wave functions for the initial and final states of the 1s electrons are solutions of the Schrodinger equation for hydrogen suitably scaled to take account of the effective nuclear potential. This is a good approximation for the K shell since we have an interaction which takes place predominately close to the atom center where outer shells do not distort appreciably the inner shell wave functions. The approximation is less good for shells other than the K.

(6) The screening effects of the outer shells are accounted for to first order by adding a nuclear potential E_s assumed constant over the K-shell region. The screening of the nuclear field by the second K-shell electrons (if present) is included by using an effective nuclear charge $Z_s = Z - 0.3$ where the numerical factor is due to Zener (or Slater) and is the most appropriate value for low Z elements.

Under these conditions, Egerton finds an expression for the GOS in the form:

For $k_H > o$

$$\frac{df_K}{d\Delta E} = \frac{2^7 N_K \Delta E \, (Q + \frac{1}{3} k_H + \frac{1}{3}) \exp (2_\beta / k_H)}{Z_S^4 R^2 \, [(Q - k_H^2 + 1)^2 + 4k_H^2]^3 [1 - \exp(-2\pi / k_H)]} \tag{14}$$

where $k_H = \dfrac{\Delta E}{Z_S^2 R} - 1$

$Q = |\kappa a_o|^2 / Z_S^2$

$N_K = 2$, denoting the result is per atom

$\beta = \tan^{-1}[2 k_H / (Q - k_H^2 + 1)]$ which lies in the range 0 to π

For

$k_H \leqslant 0$

$$\frac{df_K}{d\Delta E} = \frac{2^7 N_K \Delta E \, (Q + \frac{1}{2} k_H^2 + \frac{1}{3}) \exp (y)}{Z_S^4 R^2 [(Q - k_H^2 + 1) + 4k_H^2]^3} \tag{15}$$

where $y = -(-k_H^2)^{-\frac{1}{2}} \ln \left[\dfrac{Q + 1 - k_H^2 + 2 (-k_H^2)^{\frac{1}{2}}}{Q + 1 - k_H^2 - 2 (-k_H^2)^{\frac{1}{2}}} \right]$

These are ugly, but perfectly usable, formulae and the evaluation of the GOS as a function of κ and ΔE is readily achieved. An example for carbon is shown in Figure 4.

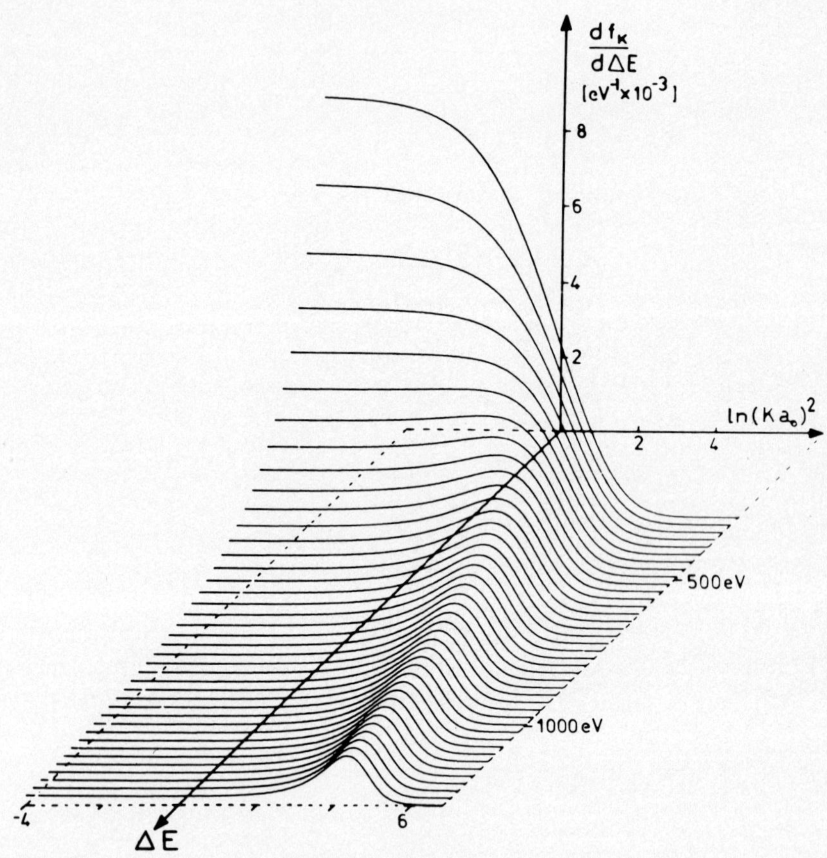

FIGURE 4. The Bethe surface for K-shell ionization in carbon, calculated from the hydrogenic model (Egerton, 1979).

In EELS we have a finite known entrance angle (α) into the spectrometer, and if we wish to calculate the form of the energy loss spectrum, we require the partial cross section for K ionization which can be written as:

$$\frac{d\sigma_K(\alpha)}{d\Delta E} = \int_{\kappa\,min}^{\kappa\,max} \frac{8\pi a_o{}^2 R^2}{T_o \Delta E} \frac{df_K(\kappa,\Delta E)}{d\Delta E} \frac{d\kappa}{\kappa}$$

$$\int_{(\kappa a_o)_{min}}^{(\kappa a_o)_{max}} \frac{4\pi a_o{}^2 R^2}{T_o\ \Delta E} \frac{df_K(\kappa,\Delta E)}{d\Delta E}\ d\left[\ln\ (\kappa a_o)^2\right] \tag{16}$$

Where R = 13.6eV is the Rydberg constant.

The expression for $d\sigma_K(\alpha)/d\Delta E$ is written in the latter form because integration over a log grid is most convenient for numerical evaluation. The value of this partial cross section is proportional to the area under a section through the Bethe surface (Figure 4) at fixed ΔE between $(\kappa a_o)_{min}$ and $(\kappa a_o)_{max}$. $\hbar\kappa_{min}$ is the minimum momentum transfer to a K-shell electron, i.e., at zero scattering angle, while $\hbar\kappa_{max}$ is the maximum that may be transferred at scattering angle α. A more general form of equation (8) relating κ to θ is of the form:

$$(\kappa a_o)^2 = \frac{2T_o}{R} - \frac{\Delta E}{R} - \frac{2T_o}{R}\left(1 - \frac{\Delta E}{T_o}\right)^{\frac{1}{2}} \cos\theta \tag{17}$$

The values of $(\kappa a_o)_{max}$ and $(\kappa a_o)_{min}$ can be found from Kincaid et al. (1978) by inserting $\theta = \alpha$ and $\theta = 0$, respectively, giving (in a form suitable for evaluation by computer):

$$(\kappa a_o)^2{}_{min} \approx \frac{\Delta E^2}{4RT_o} + \frac{\Delta E^3}{8RT_o{}^2} \tag{18}$$

$$(\kappa a_o)_{max} \approx (\kappa a_o)^2_{min} + \left(\alpha^2 - \frac{\alpha^4}{12} + \frac{\alpha^6}{360}\right)\left(\frac{T}{R} - \frac{\Delta E}{2R} - \frac{E^2}{8RT_o}\right)$$

Egerton (1979) has evaluated equation (16) numerically using equation (18), and his results for carbon for energy losses beyond the ionization edge are shown in Figure 5. Similar results should hold for all low z elements and indicate that, to a good approximation, the intensity fall-off beyond an edge is such that:

$$\frac{d\sigma_K}{d\Delta E} \quad \alpha \ \Delta E^{-s} \tag{19}$$

in agreement with experimental observations (Colliex et al., 1971; Egerton, 1974). A careful analysis of Figure 5 shows that: (1) s is constant over a limited range of ΔE but depends critically on α ; (2) for large such that most of the scattered electrons reach the spectrometer s varies from a value ≈ 3 for the low ΔE to value ≈ 2 for $\approx 10E_K$ as expected theoretically; and (3) for low values of α, s increases with energy loss; the largest value of s ≈ 6 occurs for large ΔE and small α.

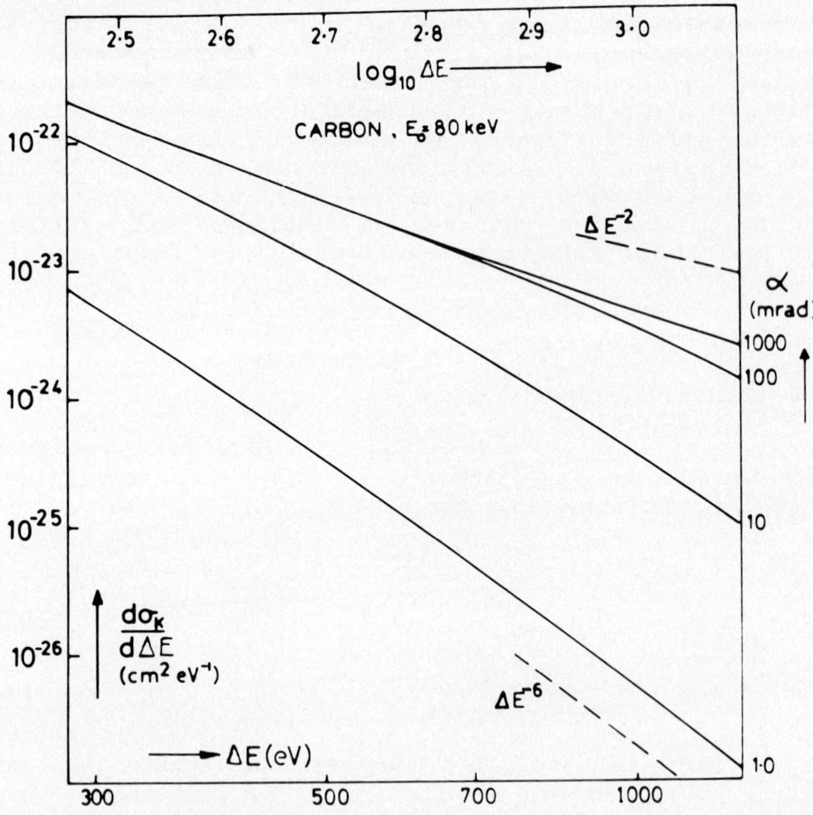

FIGURE 5. The K-shell, energy-loss spectrum of carbon for 80 KeV incident electrons and various collection semi-angles α (From Egerton, 1979).

So far we have talked of $d\sigma/d\Delta E$ for a particular ionization loss, and we have noted that the signal extends over a considerable energy range, falling off as ΔE^{-s}. Similar theory applies to the excitation of all inner shell electrons and even to valence electron excitations. Hence the inner shell loss "peak" must sit on the "tail" of lower energy losses. If the sample contains more than one element, then the inner shell loss tail of one may well constitute the major contribution to the background of another with a characteristic edge at higher energy. The importance of the background as far as quantitative microanalysis is concerned cannot be over emphasized. The accuracy with which it can be subtracted to give the intensity in the loss peak over some range of ΔE is a limiting feature, although the process is simplified because of the background behavior summarized by equation (19).

Provided we continue to restrict our discussion to single scattering processes, the background intensity should be calculable from the GOS which we have discussed previously, provided we know the initial and final state wave functions sufficiently accurately. This, however, is not simple and the hydrogenic description is likely to be less accurate. We would, however, expect a similar general form to that of the Bethe surface and to be such that $d\sigma/d\Delta E \propto \Delta E^{-s}$. We would also expect the background intensity to increase as the atomic number increases since there are more electrons in outer shells to be ionized into the continuum. Provided these expectations are valid, we can deduce how the signal to background will vary with collection angle. Figure 6 (Doig et al., 1975) illustrates this.

$\Delta E/R = 1$ for hydrogen is fairly typical in form to what we see close to the threshold for ionization and with a small spectrometer acceptance angle we will record virtually all of the signal. The background, however, arises from the tail of some loss with a threshold energy very much less than the characteristic loss of interest. If the energy loss $\gg$ the background, the background GOS is likely to be much more like $\Delta E/R > 10$. Here we are on the Bethe ridge and so the contribution to background, with a small spectrometer acceptance angle, will be very small. However, it is not signal/background which is important in microanalysis but signal/noise and this means that an optimum spectrometer acceptance angle exists for any particular energy-loss situation. Provided we have some idea of the quantitative GOS for the specimen, this optimum angle should be calculable.

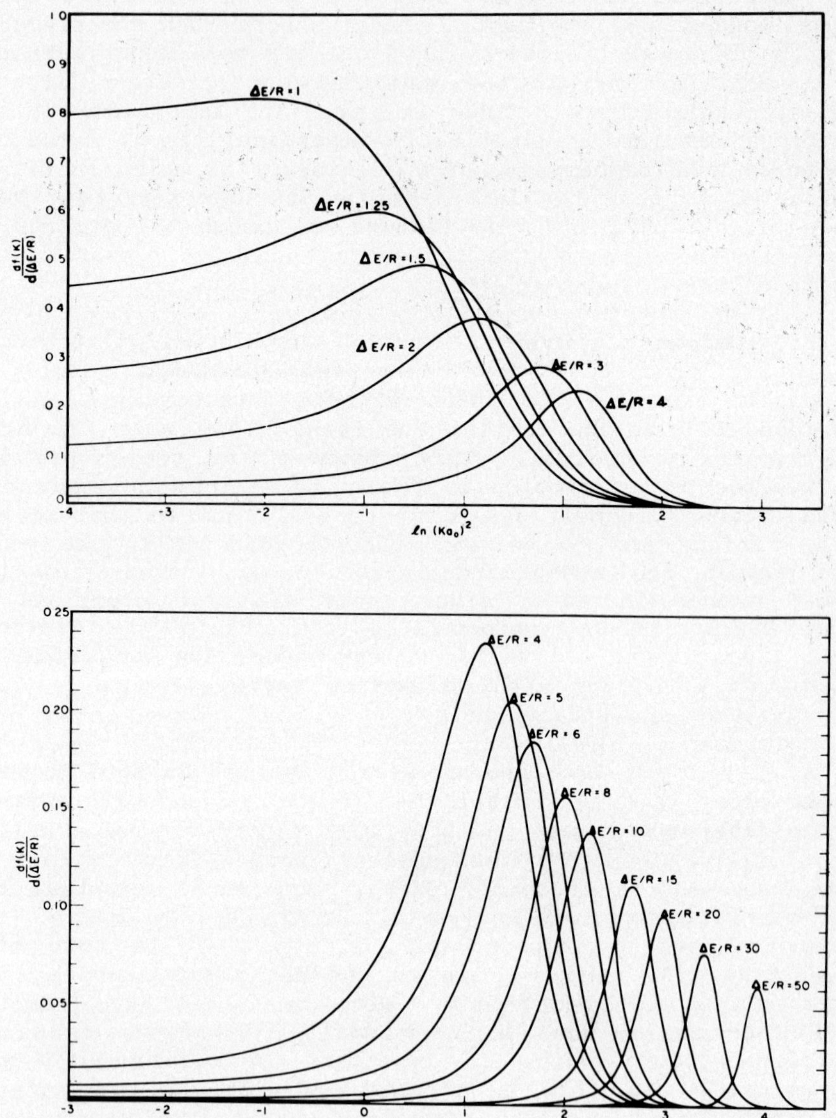

FIGURE 6. Density of the GOS per unit range of excitation energy $1 < \Delta E/R < 50$ for hydrogen (Inokouti, 1971).

Before we conclude our discussion of background, the influence of specimen thickness must also be noted. As the specimen thickness increases, the probability of multiple losses increases, and background distribution will change resulting in poorer visibility of the ionization-loss edges since the pattern in the range $0<\Delta E<50$eV will be replicated at the ionization edges. If we are only concerned about areas under the loss peaks over a range of $\Delta E>50$ eV, this may not cause a major problem since the area will remain unchanged at least to first order. However, if it is the fine structure of the spectrum that is of interest, then the confusion caused by the replication may require the application of deconvolution techniques (Leapman, 1979).

It is now timely to discuss the influence of the solid-state nature of the specimen on the shape of the ionization-loss peaks, which until now have been assumed purely atomic in character. In the neighborhood of the ionization edge, the inner-shell electrons are excited to unoccupied states above the Fermi level, the density of which constitutes the band structure of the solid (Egerton, 1979; Colliex et al., 1972). For small momentum transfer the dipole selection rules are obeyed ($\Delta\ell\pm1$), and the final states must have the correct symmetry if they are to be available to the excited electron. Thus the fine structure close to a K edge (ℓ = 0) will be different from that close to an L edge (ℓ = 1). An example (Leapman, 1979) of fine structure at the boron K edge for a sample of boron nitride is shown in Figure 7. The atomic potential may also be altered by bonding of the valence electrons. This can give rise to what is known as a chemical shift of the energy at which the edge occurs and the effect can have a magnitude of up to $\pm$ 5 eV. An analysis of such shifts can give chemical information on the structure (Isaacson, 1972).

At energies above about 20 eV from the ionization edge extended fine structure (EXELFS) (Kincaid et al., 1978; Batson and Craven, 1979; Leapman and Cosslett, 1976) similar to the fine structure (EXAFS) (Stern, 1974)) appearing in x-ray absorption spectra can occur. An example is shown in Figure 8 (Leapman and Cosslett, 1976). This fine structure arises from the back scattering of the outgoing photoelectron by the surrounding atoms. For any given energy of the photoelectron, the amplitude and phase of the backscattered wave (which can then interfere with the outgoing photoelectron wave) will depend on the distance to the neighboring atom (the bond length) and on the types of atoms present. Thus, in systems with more than one element

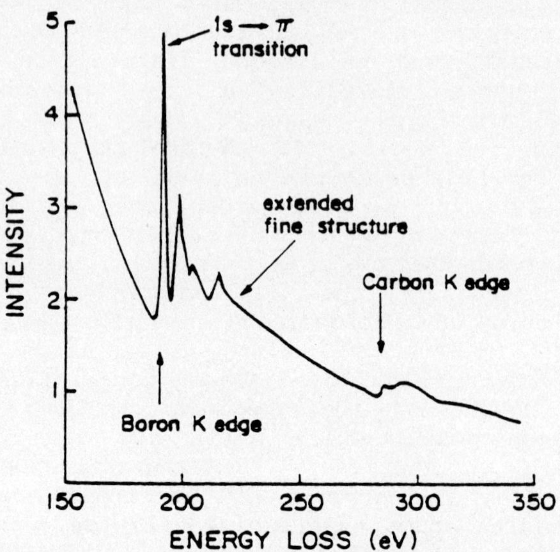

FIGURE 7. The boron K edge from hexagonal boron nitride (Leapman, 1979).

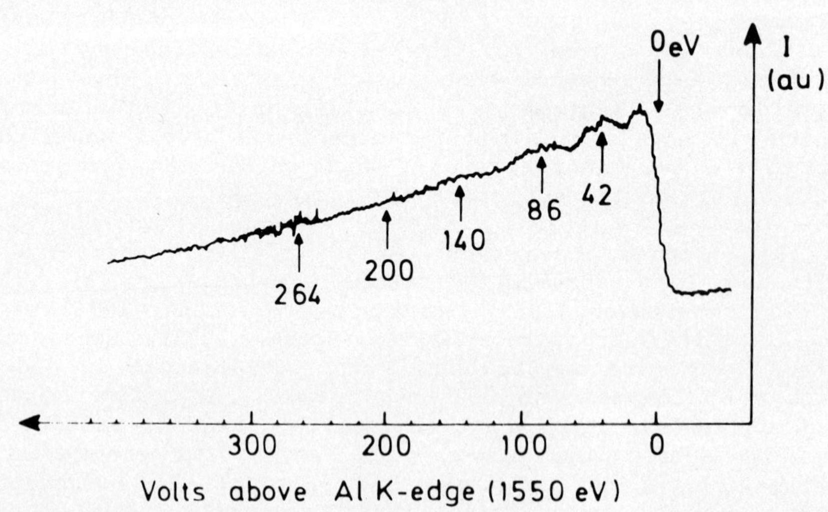

FIGURE 8. Extended energy loss fine structure (EXELFS) above the K-excitation edge in aluminum. (Leapman and Cosslett (1976).

present, it is possible, in principle, to deduce the average environment around each type by analyzing the fine structure beyond the ionization edges of each species.

CONCLUSIONS

In this paper we have seen that the basic physics of the electron-specimen interaction, which gives rise to the observed form of the electron energy loss spectrum, is complex and our knowledge of it is certainly not as yet complete. However, the wealth of information which will be available to us, if we can improve our understanding, should prove a sufficient incentive to encourage further studies both theoretical and experimental.

ACKNOWLEDGMENTS

The author is grateful to, among others, Drs. J. N. Chapman, V. E. Coslett, R. F. Egerton, R. D. Leapman, and D. B. Wittry for many helpful discussions over a period of years.

REFERENCES

Batson, P. E., and Craven, A. J. (1979). Phys. Rev. Lett. 42:893.

Colliex, C., Cosslett, V. E., Leapman, R. D., and Trebbia, P. (1976). Ultramicroscopy 6:301.

Colliex, C., and Jouffrey, B. (1972). Phil. Mag. 25:491.

Colliex, C., and Trebbia, P. (1976). In "Developments in Electron Microscopy and Analysis" (J. A. Venables, ed.), p. 123.

Doig, P., Edington, J. W., and Jacobs, M. H. (1975). Phil. Mag. 31:285.

Egerton, R. F. (1974). Phil. Mag. 30:739.

Egerton, R. F. (1974). Proc. 8th International Conference on Electron Microscopy, Vol. 1, p. 280. Canberra.

Egerton, R. F. (1979), Ultramicroscopy 4:169.

Egerton, R. F., Rossouw, C. J., and Whelan, M. J. (1976). In "Developments in Electron Microscopy and Analysis," (J. A. Venables, ed.), p. 129.

Fano, U., and Cooper, J. W. (1968). Rev. Mod. Phys. 40:441.

Herman, F., and Skillman, S. (1963). "Atomic Structure Calculations." Prentice Hall, Englewood Cliffs.
Inokouti, M. (1971). Rev. Mod. Phys. 43:297.
Isaacson, M. S. (1972). Chem. Phys. 56:1813.
Isaacson, M. S., and Johnson, D. (1975). Ultramicroscopy 1:33.
Kincaid, B. M., Meixner, A. E., and Platzman, P. M. (1978). Phys. Rev. Lett. 40:1296.
Leapman, R. D. (1979). Ultramicroscopy 3:413.
Leapman, R. D., and Cosslett, V. E. (1976). J. Phys. D9:L29
Leapman, R. D., Rez, P., and Mayers, D. F. (1978). Proc. 9th International Conference on Electron Microscopy, p. 526. Toronto.
Manson, S. T. (1972). Phys. Rev. A6:1013.
Raether, H. (1968). Springer Tracts in Modern Physics 38:84.
Rez, P., and Leapman, R. D. (To be published).
Stern, E. A. (1974). Phys. Rev. B10:3027

DISCUSSION

SPEAKER: R. P. Ferrier.

SOMLYO: I gather from Colliex that there was some debate as to what the calculated shape of the calcium L-edge should take. Later on, Henry Shuman's presentation will show some experimentally determined L edges, so there is an ulterior motive to the question.

FERRIER: If you have sufficiently good wave function for the element that you are interested in, then you should be able to place a fair amount of reliance at least in the shape. What is more in doubt is what is the absolute cross section. I think Dennis Maher will probably talk about this later on this morning. There is some question as to the absolute values in the oscillator strength and hence in the cross sections, although that is going to be needed before one really has a good deal of confidence in quantitation schemes which are put forward.

QUANTITATIVE LOCAL MICROANALYSIS WITH EELS

C. Colliex
C. Jeanguillaume
P. Trebbia

Laboratoire de Physique des Solides
Universite de Paris-Sud
Orsay, France

INTRODUCTION

Depending on the problem under investigation (localization of a given metallic or halogen species within a thin section of tissue or study of structural and molecular bonding of homogeneous specimen), the EELS technique offers different possibilities which are very sensitive to the choice of several critical parameters.

Our contribution will be focused on two main topics which are presently under investigation in our group:

A) Use of inner-shell signals for a quantitative local microanalysis technique: after a brief summary of the main experimental parameters and proposed formula for achieving microanalysis without standards, we will mainly deal with the problems of background stripping under the characteristic signal, of the validity and accuracy of theoretical models, and of the confidence interval that we can determine for the estimation of the apparent ratio between two elements.

B) Theoretical study of the energy resolution and collection efficiency of a magnetic spectrometer: the characteristics and properties of a spectrometer have been sketched out when it is coupled with a microscope in different ways (CTEM, STEM with/without post-specimen

ISBN 0-12-362880-6

lenses). This preliminary study has been made for the design of systems which will be adapted to our microscopes Philips EM 300 (CTEM) and VG HB 5 (STEM).

ABOUT THE ACCURACY OF CORE LOSS SIGNALS MEASUREMENTS

1. General Considerations

As previously mentioned (Ferrier, 1980), the existence in the specimen of a given atomic species can be related in an electron energy-loss spectrum by the occurrence of a jump in the recorded intensity of transmitted electrons. The onset of this jump is located at an energy loss ($\Delta E = E_C$ corresponding to the difference between the excited electronic level and the Fermi level. For energy losses greater than this threshold, the spectrum presents in some cases one or several maxima and decreases slowly towards high energy losses.

The characteristic spectrum $I_C^A(\Delta E)$, associated to the presence in the analyzed volume of the sample of an atomic species A, is defined as the difference in the energy-loss region of interest, between the actually recorded intensity and the intensity that would have been recorded if the species A was not present (background). If we make the assumption (Trebbia, 1979) that for energy losses smaller than E_C, the spectrum is independent on the presence of atoms A, the background can be defined (Figure 1) as the extrapolation for $\Delta E > E_C$ of the recorded spectrum for $\Delta E < E_C$.

The characteristic spectrum $I_C^A(\Delta E)$ is then integrated over an energy window Δ and the resultant signal,

$$S_C(\alpha,\Delta) = \int_{E_C}^{E_C+\Delta} [I(\Delta E) - B(\Delta E)]\, d(\Delta E) = I(\alpha,\Delta) - B(\alpha,\Delta)$$

α : angle of collection of the scattered electrons

$$I_C^A(\Delta E) = I(\Delta E) - B(\Delta E)$$

can be related to the number N_A of atoms A detected per unit volume by the formula (Egerton, 1978; Joy and Maher, 1979; Trebbia, 1979; Colliex and Trebbia, 1979):

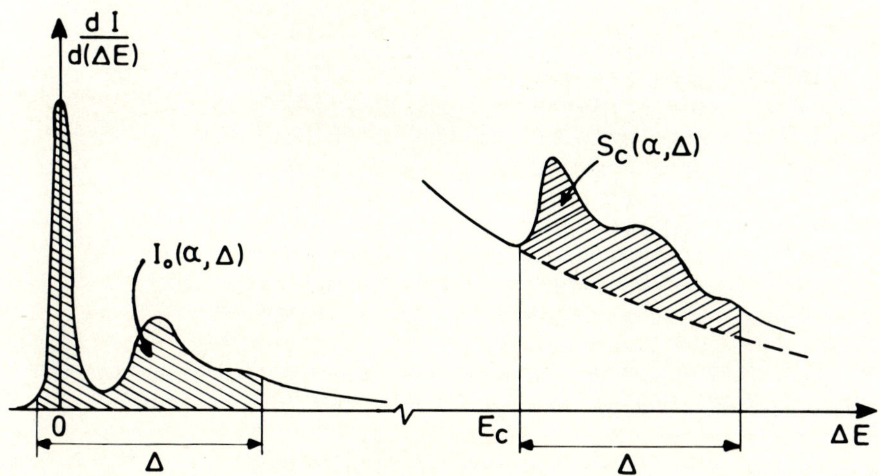

FIGURE 1. The characteristic signal $S_C(\alpha,\Delta)$ is defined, for a given angle α of collection, as the integration over an energy window Δ of the characteristic spectrum, i.e., the difference between the recorded spectrum and the extrapolated background.

$$S_C^A(\alpha,\Delta) = \sigma_C^A(\alpha,\Delta) \cdot I_o(\alpha,\Delta) \cdot N_A \cdot t$$

where $\sigma_C^A(\alpha,\Delta)$, is the partially integrated cross section for the excitation of this particular core level.

$I_o(\alpha,\Delta)$ is the integration over Δ of the low loss region (Figure 1).

t is the thickness of the analyzed volume, which can be experimentally derived from a Kramers Krönig transformation of the low-loss region of the spectrum (Trebbia, 1979; Colliex and Trebbia, 1979; Colliex and Trebbia, 1978).

The relative atomic concentration $\frac{N_A}{N_B}$ of two species A and B detected in the same spectrum can then be straightforwardly estimated:

$$C = \frac{N_A}{N_B} = \frac{S_C^A(\alpha,\Delta)}{S_C^B(\alpha,\Delta)} \times \frac{\sigma_C^B(\alpha,\Delta)}{\sigma_C^A(\alpha,\Delta)}$$

The magnitude of the characteristic signal $S_C(\alpha,\Delta)$ is very sensitive to experimental parameters such as:

- E_o : Primary energy of electrons
- J : Incident flux
- T : Total time of irradiation per unit area
- α_o : angular width of the illuminating beam
- α : angular width for the collection of scattered electrons
- β : acceptance of the spectrometer
- η : transmittance of the spectrometer
- δ_E : energy resolution of the spectrometer
- Δ : energy width of signal integration.

The most of these parameters is generally set up before a run of experiments, or is a built-in characteristic of the instrument.

But the measurement of S_c (α, Δ) is also correlated, for each experiment, to the estimation of the background. This is an operator - dependent parameter which must be considered with sufficient attention for minimizing the total uncertainty on the relative atomic concentration C, for example.

2. The Problem of Background Stripping

The general method for background stripping is to define before the threshold of excitation E_c a fitting region of interest, or fitting window, which is typically 50 or 100 eV wide and where a given mathematical model law (in this case a power law $A\ (\Delta E)^{-R}$ (Egerton, 1975) is adjusted to the raw data.

It has been already emphasized by several authors (Egerton, Joy, Maher, Trebbia) that this fitting window must be large enough to insure reasonable precision, but not too wide, however, because the power law model is only valid on a restricted energy range ($\sim$ 200 eV) (Trebbia, 1979; Trebbia and Colliex, 1979).

Once the fit is made, the power law is then extrapolated under the characteristic spectrum (Figure 1) on an energy window which must be (Trebbia, 1979) of the order of 100 eV: Figure 2, for example, shows that the relative variation of the signal determination, when two different fitting windows are used, passes through a minimum (5.48% in that case) for Δ = 90 eV.

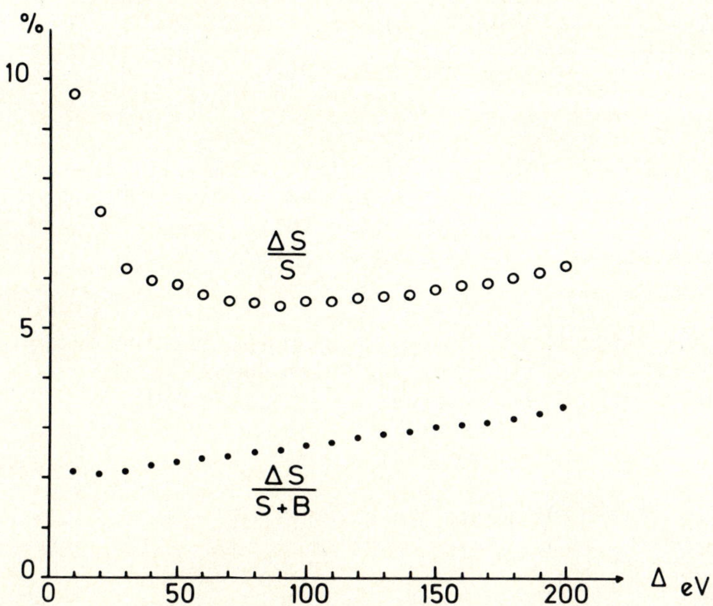

FIGURE 2. Two fitting windows (77 and 100 eV wide) have been used in the procedure of the program ANAL for the determination of the characteristic signal relative to the excitation of the 2p electrons in Titanium. (E_O =50 kV, α = 34 mrad). The relative variation with Δ of the difference between the two computed signals is minimum for Δ = 100 eV.

It is not the purpose of this paper to develop the well-known theories of statistical fluctuations inherent to each measurement nor to examine all the possible fitting procedures. The reader interested in this particular point will find in Bevington (1969) a rather complete discussion of these problems.

We shall only define some quantities and properties:

-Any counting experiment agrees with a Poisson distribution, which determines the probability, when measuring the quantity Y (X_i), to observe the value Y_i. The standard deviation is defined as

$$\sigma_i = \sqrt{Y(X_i)}$$

-Fitting procedures are generally based on the method of "maximum likelihood" which involves minimizing the quantity

$$\chi^2 = \sum_i \left[\frac{1}{\sigma_i^2} \; (Y_i - Y(X_i))^2 \right]$$

where Y_i is the experimental value at the point X_i, and Y (X_i) is the computed value by the fitting function. In our case, the whole set of the X_i points represent the fitting window

$$(X_i \equiv \Delta E, \; Y_i \quad I(\Delta E), \; Y(X_i) = A \cdot X_i^{-R}).$$

- The method of least squares fitting consistently underestimates the area under an experimental curve by an amount approximately equal to the value of

$$\text{Area (data)} \simeq \text{Area (fit)} + \chi^2_{min.}$$

and the validity of this equation can be considered, together with the magnitude of the reduced

$$\bar{\chi}^2 = \frac{\chi^2}{\nu} = \chi^2_\nu$$

(ν : number of degrees of freedom), as a test of the goodness of the fit $\chi^2_\nu > 1$).

-The uncertainty in measuring an area A may be estimated as:

$$\sigma_A \simeq \sqrt{A}$$

If $\chi^2_\nu > 1$, the uncertainty has to be multipled by χ^2_ν

$$s^2_A \simeq \sigma^2_A \; \chi^2_\nu$$

-When one substracts two areas,

$$S_c \; (\alpha,\Delta) = I \; (\alpha,\Delta) - B \; (\alpha,\Delta)$$

The statistical uncertainty in S_c (α,Δ) is according to the general rules of propagation of errors:

$$\sigma^2_{S_c} = \sigma^2_I + \sigma^2_B$$

$$= I \; (\alpha,\Delta) + B \; (\alpha,\Delta) \cdot \chi^2_\nu$$

$$\text{where } I(\alpha,\Delta) = \int_{E_c}^{E_c+\Delta} I(\Delta E)\, d(\Delta E)$$

$$\text{and } B(\alpha,\Delta) = \chi^2_{min.} + \int_{E_c}^{E_c+\Delta} Y_{Fit}(\Delta E)\, d(\Delta E)$$

3. The Different Fitting Methods Adapted to EELS

Several methods have been already proposed for the extrapolation of the background under the characteristic spectrum:

<u>a. Visual Extrapolation</u>. The simplest one is based on Egerton's remark (Egerton, 1975) that in Log-Log coordinates, the background before the ionization threshold can be approximated as a straight line (i.e., obeys to a power law $A.(\Delta E)^{-R}$) . Although it seems rather easy to determine visually the best line passing through experimental data, such a method is not advisable for two reasons: (1) In Log-Log coordinates, significant dispersions of data are considerably smoothed in such a way that it is almost always possible to fit any data with a straight line. (2) Moreover, this method provides no information about accuracy of the fit and induced uncertainty in extrapolation and integration of the characteristic signal.

<u>b. Program ANAL</u>. We have proposed a method of least square fitting (Trebbia, 1979; Trebbia and Colliex, 1979) (program ANAL) which optimizes independently the two parameters A and R by minimizing the quantity:

$$\sum_i (Y_i - Y(X_i))^2$$

Although this program has been intensively used, giving rather good results (Trebbia, 1979; Trebbia and Colliex, 1979), it suffers major inconveniences:

-Its execution is rather long since the two parameters are handled independently.

-Precision in the determination of coefficient R is arbitrarily limited to one digit after the decimal point.

-It does not take into account the fact that a variance due to statistical fluctuations is attached to each experimental data point. Since a counting experiment agrees with a Poisson distribution, this experimental variance is:

$$S_i^2 = Y_i$$

and varies with i.

c. Egerton's Formulae. Egerton has proposed very recently (1980) a new direct method of determination of A and R by using the following equations: $R \simeq 2\ \mathrm{Log}\,(I_1/I_2)/\mathrm{Log}(E_2/E_1)$

$$A = (I_1 + I_2)\ (1-R)/(E_2^{1-R} - E_1^{1-R})$$

where the fitting window E_1, E_2 is divided in two halves of respective integrals I_1 and I_2.

This approach is very interesting because these formulae give rather accurate values $\frac{\Delta R}{R} \simeq 1\%$ in minimum computing time. The precision on A is, however, dependent on the precision on R and no values of σ_A and σ_R are provided.

Starting from these Egerton's formulae, we have written a new program of least square fitting, the CURFIT program, that we describe now.

d. Program CURFIT. This program is directly derived from the FORTRAN CURFIT program described in Bevington (1969).It has been translated in the Flextran interpreter language in order to be executed by our TRACOR-NORTHERN multichannel analyzer. It is handled by a PDP 11-05 minicomputer as a subroutine which can be directly linked to the software that we have already written for EELS purposes (Trebbia, 1979; Trebbia and Colliex, 1979). This program can be divided into five main parts:

(1) For each experimental data point Y_i, the attached value S_i is imputed. Then, starting from Egerton's formulae, a first couple of parameters (A, R) is determined and the quantity

$$\chi^2 = \sum_i \left[\frac{1}{S_i^2} \ (Y_i - Y(X_i))^2 \right]$$

is estimated.

(2) According to the so-called 'gradient search method,' the program determines in the three-dimensional space (χ^2, A, R) the direction of the steepest descent in order to minimize as quickly as possible the χ^2 value by a new choice of (A, R). Actually, the trajectory followed in the (A,R) space by the program CURFIT can be compared to a torrent path starting from a mountain and looking for the best way to reach the lowest point in a flat valley. This ravine search is very efficient as far as the difference in level is not too small.

(3) When the relative variation of χ^2 between two values of the couple (A, R) tends to vanish out, the gradient search method is no longer able to determine the direction of the steepest descent, and then the program CURFIT uses the method of linearization of the fitting function by expanding it to the first order:

$$Y(X) = Y_o(X) + \frac{\partial Y_o(X)}{\partial A} \cdot \delta A + \frac{\partial Y_o(X)}{\partial R} \cdot \delta R$$

(4) The program is completely executed when the relative variation of χ^2 is smaller than an arbitrary predetermined value, 10^{-4}, 10^{-6}, 10^{-8} , for example. The result of the fitting is then displayed on a TV screen superimposed on the initial spectrum. The program provides the following final results: A, σ_A, R, σ_R χ^2, χ^2_ν. A comparison of the final value of χ^2 with a standard table enables to say whether the final fitting function describes with sufficient accuracy the unknown parental distribution of an ideal experiment performed during an infinite counting time.

(5) If the answer to this question is positive, then the program CURFIT computes for each point $Y(X_i)$ of the fitting curve a standard deviation

$$\sigma_i = \sqrt{Y(X_i)}$$

and provides an exhaustive list of all the experimental points Y_i which can be considered as spurious, that is:

$$|Y_i - Y(X_i)| > k\,\sigma_i \qquad k = 1, 2, 3.$$

Taking into account the number of degrees of freedom of the fit and the statistical distribution of the parental population, it is in fact possible to predict the probability of having such a number of spurious experimental points and this comparison between prediction and reality is a second test of coherence.

Since the program CURFIT uses Egerton's formulae as a starting point of search, and handles simultaneously the two parameters A and R, its execution is significantly faster than ANAL's one (a few minutes against half an hour).

It must be also noted that this program CURFIT can be modified in a very simple way for taking into account more than two parameters, at the cost, however, of a longer time of execution.

4. Preliminary Results.

On Figures 3 and 4, we have reported energy loss spectra recorded on a same Fe_xGd_{1-x} evaporated alloy, x ranging between 0.85 and 0.88 (Trebbia, 1979; Trebbia and Colliex, 1979). Figure 3 is related to the L_{2-3} signal of iron (i.e., excitation of the 2p electrons) and the Figure 4 to the N_{4-5} signal of gadolinium (excitation of the 4d electrons).

From these two spectra, we have performed a set of tests of the program CURFIT.

a. <u>Verification of the unicity of the fitting solution.</u> The first question to be answered is: Are we sure that looking for the minimum χ^2 value in the (A, R) space, we will find a unique minimum, and is this minimum a relative or an absolute one?

For this purpose, we have made eight different fits of the background shown on Figure 3 with the fitting window number two (i.e., 100.8 eV wide) starting at an arbitrary (A, R) initial value rather far away from the expected final value.

On Figure 5 we have sketched in the (A, R) plane the eight different trajectories that the program CURFIT followed when minimizing χ^2 step by step up to the final (A, R) value located by a star. As we have explained it in

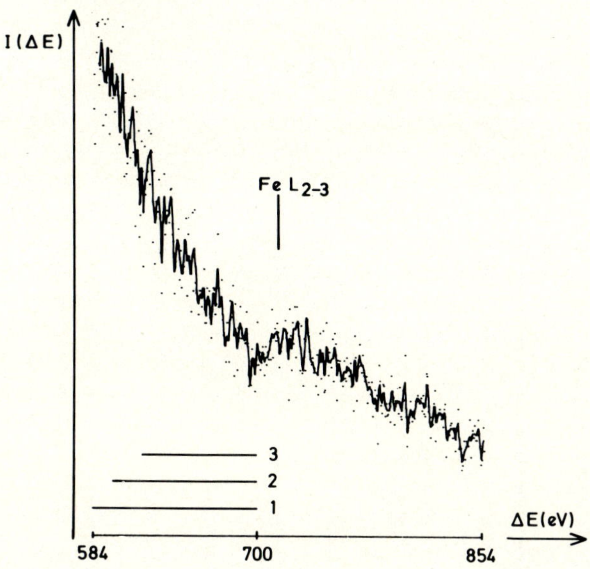

FIGURE 3. Energy loss spectrum of a thin (550 Å) evaporated $Fe_{0.88}Gd_{0.12}$ alloy in the vicinity of the L_{2-3} edge of Iron.

(E_o = 50 kV, α = 10 mrad).

Dot points: raw data.

Solid curve: after a three points smoothing procedure. Three different fitting windows (1, 2, and 3) were used by program CURFIT for the determination of the background.

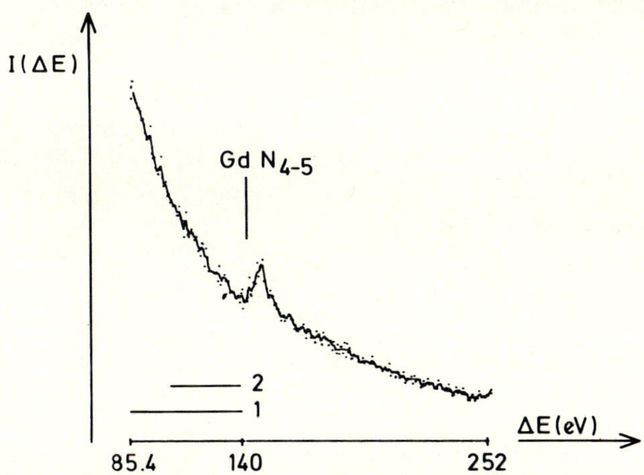

FIGURE 4. N_{4-5} edge of Gadolinium recorded on the same sample as Figure 3. (E_o = 50 kV, α = 3 mrad). Two different fitting windows.

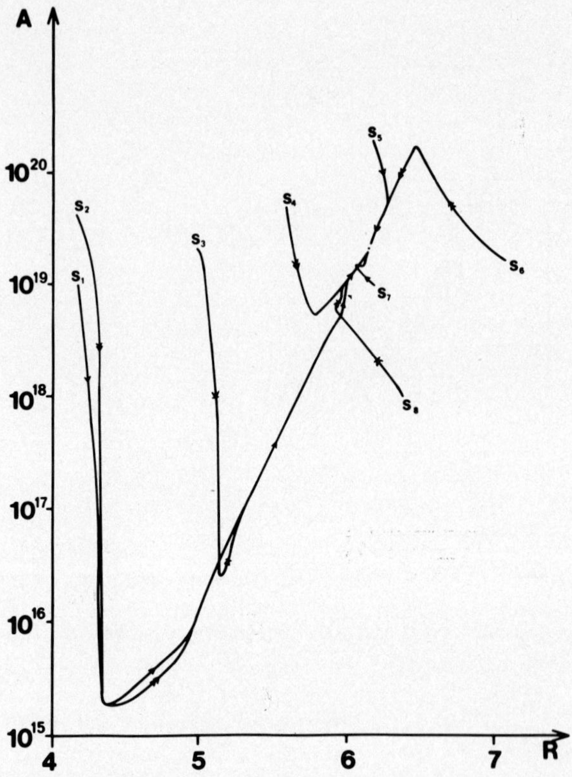

FIGURE 5. Ravine search path followed by the program CURFIT when searching the lowest value of χ^2 in the (A, R) plane. (Example of Figure 3, fitting window n° 2).

Section 3d, these trajectories are orthogonal to χ^2 iso-values curves. A detailed analysis of these trajectories reveals that:

- since χ^2 is exponentially dependent on R, this parameter is almost always optimized in the good direction.

- all the trajectories converge to the same point ($A = 2.127 \times 10^{19}$, $R=6.1636$), within, respectively, 4 and 5 significant digits precision, at the condition that the program CURFIT is not interrupted before the relative variation of χ^2 is smaller than 10^{-8}.

- several minima of χ^2 have been revealed to be only relative and the program succeeded in finding the absolute one.

- note also that the map of Figure 5 has been drawn at a very large scale. In practice on a concrete case, and with sufficiently accurate initial values for the starting point (Egerton's formulae), the trajectories are almost straight lines.

<u>b. Influence of the fitting window width.</u> We have tested how the final value (A, R) is dependent on the fitting window width. Concerning the spectrum of Figure 3, we performed a CURFIT fit using three distinct fitting windows labeled 1, 2 and 3 (of respective widths : 115.5, 100.8, and 79.1 eV), and starting from (A,R) value given, in each case, by the Egerton's formulae. In the same manner, we made also two independent fits on the background shown on Figure 4. The results are summarized in Table 1 and will be discussed in Section 4d.

<u>c. Influence of a smoothing procedure.</u> A three-point smoothing procedure of raw data does not, in principle, induce any strong distortion of the data distribution except that it increases the uncertainty in the estimates of the width and position of eventual peaks. This procedure can be well-suited for data distributions that we know agree with a given mathematical law, but that cannot be fitted with a sufficiently small χ^2 value. This is due to the fact that "data represent a random sample from the parent population and that an exact description of the data merely represents an estimate of the parent distribution" (Bevington, 1969).

Smoothing data strongly reduce the minimum value of χ^2 (χ_ν^2 is generally smaller than 1 with no special physical significance) and, by the fact, the difference between Area (data) and Area (fit) (see Section 2).

We have, therefore, tested the influence of smoothing data on our fitting method by repeating the five fits described in the precedent section. The results are gathered in Table 2.

<u>d. Discussion of the results.</u> In Tables 1 and 2, the subscripts E and C refer to results given respectively by Egerton's formulae and program CURFIT.

TABLE 1. Test of program CURFIT on Figures 3 and 4.

Fitting Window Width (eV)	ν	A_E	R_E	χ^2_E	A_C	R_C	χ^2_C	Area (fit_E)	Area (fit_C) + χ^2_C	Area (data)	
115.5	164	$4.859\ 10^{18}$	5.9316	228	$7.999\ 10^{18}$	6.0127	217	19182	18945	18946	Fig. 3
100.8	143	$2.133\ 10^{19}$	6.1600	193	$2.127\ 10^{19}$	6.1636	182	15492	15278	15277	Fig. 3
79.1	112	$4.731\ 10^{18}$	5.9268	137	$5.174\ 10^{19}$	6.3002	127	10863	10685	10685	Fig. 3
52.8	87	$1.458\ 10^{7}$	2.0254	144	$1.398\ 10^{7}$	2.0214	89	97807	95604	95599	Fig. 4
34.8	57	$2.071\ 10^{7}$	2.0964	133	$1.430\ 10^{7}$	2.0264	62	53990	52156	52158	Fig. 4

TABLE 2. Test of program CURFIT on Figures 3 and 4 after smoothing data.

Fitting Window Width (eV)	ν	A_E	R_E	χ^2_E	A_C	R_C	χ^2_C	Area (fit_E)	Area (fit_C) + χ^2_C	Area (data)	
115.5	164	$4.633\ 10^{18}$	5.9240	83	$7.436\ 10^{18}$	6.0000	77	19216	18972	18973	Fig. 3
100.8	143	$1.854\ 10^{19}$	6.1381	69	$1.973\ 10^{19}$	6.1506	64	15518	15296	15298	Fig. 3
79.1	112	$3.585\ 10^{18}$	5.8837	58	$3.766\ 10^{19}$	6.2499	51	10890	10696	10700	Fig. 3
52.8	87	$1.447\ 10^{7}$	2.0236	84	$1.403\ 10^{7}$	2.0220	30	97882	95644	95642	Fig. 4
34.8	57	$1.979\ 10^{7}$	2.0869	87	$1.389\ 10^{7}$	2.0201	20	53992	52162	52163	Fig. 4

Egerton's formulae are obviously a very good basis for a direct fitting of the background. At the scale of Figure 5, it is quite impossible to separate Egerton's values and CURFIT ones.

Even when one compares χ_E^2 and χ_C^2 obtained for the spectrum of Figure 3, it seems that Egerton's fit is rather good.

But the test of stability of integrals reveals that Egerton's formulae induce an Area (fit $_E$) of the background always greater than the actual Area (data). This is of importance because an overestimation of the fitted background means that, after extrapolation under the characteristic spectrum, the resultant signal will be significantly underestimated. According to this point, the CURFIT results

$$\text{area}(\text{fit}_C) + \chi_C^2$$

are excellent for the two spectra.

It can also be drawn out from the results obtained for Figure 3 an interesting consideration on the influence of the fitting window width. As it decreases, R_E seems to have quite erratic values, but R_C increases. This behavior can be interpreted if the true R value is strongly dependent on the energy loss ΔE.

It is already well-known that R increases with ΔE from 2 to 6 or 7 in the energy range 100 eV $\rightarrow$ 1000 eV. But this ΔE dependence of R seems to be stronger and stronger as ΔE increases: on a wider fitting window, the mean value obtained for R will be smaller than the one obtained on a more reduced window.

This result, which cannot be observed by the simple use of Egerton's formulae, is in perfect agreement with previous experiments (Trebbia, 1979) we performed on Aluminium (Figure 6): For four different values of α, we have plotted in Log-Log coordinates the spectrum of Aluminium between 140 and 700 eV. These spectra can be considered purely as backgrounds since no characteristic signal is present in this energy range.

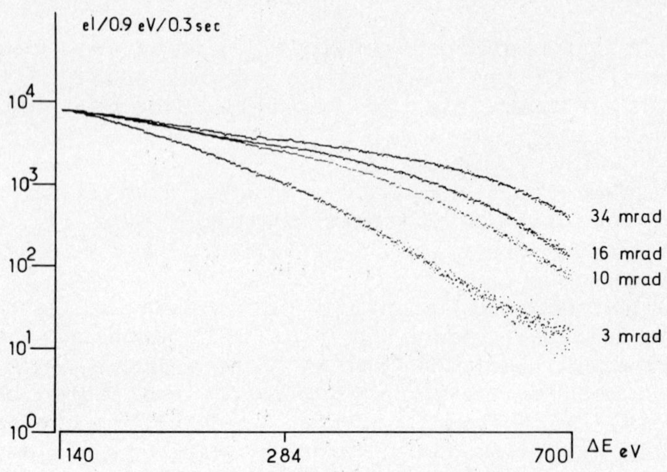

FIGURE 6. Aluminium background recorded between 140 and 700 eV for four different values of α. The curves are arbitrarily normalized at 140 eV. (E_o = 50 Kv, log-log coordinates).

From this figure, it is obvious that the absolute value of the mean slope increases with ΔE for any value of α.

Such a variation of the background intensity with ΔE might be described by a more refined model:

$$I\ (\Delta E) = A \cdot e^{-R_1\ (\mathrm{Log}(\Delta E))} \cdot e^{-R_2\ (\mathrm{Log}(\Delta E)^2}$$

Three parameters (A, R_1, R_2) instead of two would be handled by the program CURFIT for finding the χ^2 minimum value in the (χ^2, A, R_1, R_2) hypersurface.

It seems, however, necessary to confirm this behavior of R (ΔE), either by doing independent experiment on "round-robin" samples, or by computing over large energy ranges the decay of oscillator strengths and atomic cross sections.

A comparison of Tables 1 and 2 shows that smoothing data does not perturb in any way our preceding conclusions or previsions: $-\chi_E^2$ and χ_C^2

are considerably reduced to smaller values, that is, χ_ν^2 is <1. This simply reflects the fact that smoothed data cannot be considered as a random sample from the parent distribution because the observed variance of the smoothed data is much smaller than the predicted one.

- the test of stability of integrals reveals that Area (fit_E) is overestimated by an amount approximately independent on smoothing, while Area (fit_C) + χ_C^2 is always in very good agreement with the true Area (data).

- the variation of R_C with the fitting window width is of the same order of magnitude in Tables 1 and 2, though smoothing data slightly distorts the (χ^2, A, R) surface and leads to smaller final values (A, R).

To the question: Is smoothing data useful?, we may answer "no" so far as the raw data variance is not too high; i.e., the raw data can be reasonably considered as a random sample from the parent distribution. This point must be checked after each fitting procedure by listing the "spurious data" (see Section 3d).

5. Further Developments.

The formula giving the apparent ratio of the concentration of two atomic species

$$C = \frac{S_C^A(\alpha,\Delta) \cdot \sigma_C^B(\alpha,\Delta)}{S_C^B(\alpha,\Delta) \cdot \sigma_C^A(\alpha,\Delta)}$$

enables us to make direct microanalysis experiments if the partially integrated cross sections are known either by previous measurements on standards or by <u>a priori</u> calculations of the oscillator strengths.

But the value of C obtained by this method has no real physical significance if the confidence interval associated to it cannot be determined. This is a rather general problem inherent to all indirect measurements in which statistical fluctuations phenomena are involved.

Because the ratio

$$\frac{S_C^A}{S_C^B} = \frac{I_C^A - B^A}{I_C^B - B^B}$$

of the differences of independent Poissonian variables cannot be regarded as a Gaussian nor a Poissonian random variable, we may not have any direct information on the confidence interval attributed to C.

Ancey, Bastenaire, and Tixier (1977) found an elegant solution to this problem by deriving, for x-ray microanalysis experiments, a second-degree equation whose roots are the limits of the confidence interval with a confidence level (95% or 99%) a priori chosen.

Moreover, these authors give a statistical definition of the detection limit, taking into account the two main risks:

	Estimation of C	Though real value of C
risk 1	> 0	= 0
risk 2	= 0	> 0

It is of fundamental interest to elaborate similar equations specially adapted to EELS problems and to compare their derivations with previous estimations (Isaacson and Johnson, 1975; Colliex and Trebbia, 1979).

REFERENCES

Ancey, M., Bastenaire, F., and Tixier, R. (1977). J. Physics D. 10:817. "Microanalyse et Microscopie a Balayage," Les Editions de Physique, p.323 (1979). (English translation to be published).

Bevington, P. R. (1969). "Data Reduction and Error Analysis for the Physical Sciences." McGraw Hill, New York.

Colliex, C., and Trebbia, P. (1978). Proc. 9th Int. Cong. Electron Microscopy, p. 268. Toronto, Ontario.

Colliex, C., and Trebbia, P. (1979). In "Microbeam Analysis in Biology" (C. Lechene and R. Warner, eds.), p. 65. Academic Press, New York.

Egerton, R. F. (1975). Phil. Mag. 31:199.
Egerton, R. F. (1978). Ultramicroscopy 3:243.
Egerton, R. F. (1980). "An Automated System for Energy Loss Microanalysis." (Private communication before publication in the proceedings of the EMSA meeting to be held in California).
Ferrier, R. P. (1980). Microprobe Conference. Seattle, Washington.
Isaacson, M., and Johnson, D. (1975). Ultramicroscopy 1:33.
Isaacson, M. (1979). Ultramicroscopy 4:193.
Joy, D. C., and Maher, D. M. (1979). In "Introduction to Analytical Electron Microscopy" (J.J. Hren, J.I. Goldstein, and D.C. Joy, eds.), pp. 223-59. Plenum Press, New York.
Trebbia, P. (1979). Doctoral Dissertation. Universite de Paris--Sud.
Trebbia, P., and Colliex, C. (1979). J. Microsc. Spectrosc. Elect. 4:281.
Trebbia, P., and Colliex, C. (1979). In "Proceedings EMAG, 1979" (T. Mulvey, ed.), p. 333.

DISCUSSION

SPEAKER: P. Trebbia.

OTTENSMEYER: When you extrapolate and compare the areas under Egerton's estimation and your estimation, I notice that the difference is sometimes less than 1%.

TREBBIA: Yes, but as I discussed, this very small difference can induce noticeable effects when you extrapolate on a rather high energy window of integration, say 100 eV or more.

OTTENSMEYER: How do you know which extrapolation is true? Basically, I was just wondering if you had just taken a background curve without any characteristic peak, estimated the extrapolated curve from a fitting region, and then compared it to what it actually was further down.

TREBBIA: I think that Egerton's procedure basically neglects the statistical character inherent to all counting measurements. Statistical methods, like a χ^2 test for example, can help you for deciding which extrapolation is the less inaccurate.

Concerning the second point of your question, Figure 6 shows such backgrounds recorded on our experimental device. This log-log plot makes evident the fact that any fitting with a power law model, either with Egerton's procedure or CURFIT's one, is only valid on a reduced energy range, and that the two extrapolations will be both overestimated.

Actually, the point is: Are the results of Figure 6 due to artifacts in our experimental device or to physical effect? In a few months we shall be able to answer this question because we asked Peter Rez to calculate the generalized oscillator strength related to the experiment reported on Figure 6, and because we will make again such measurements with the new STEM machine which will be soon delivered in our laboratory.

FERRIER: Yes, but in your case, you could do it actually, but in the case of a biological sample, you are in a bit of a mess actually, because you do not know what the hell to use for the generalized oscillator strength. But in your case, you could do it actually and check how well his predictions agree with your semi-empirical methods which you are plotting here. That would be interesting.

TREBBIA: Yes.

SOMLYO: For your unknown biological sample, should one first deconvolute the multiple scattering effect before doing the background substraction?

FERRIER: If you actually integrated over, say 100 eV, the effect of the convolution of the low energy end of the spectrum is not going to be that marked at first order. If you are doing the stripping over, say 100 or 120 eV, would you agree with that?

TREBBIA: Yes, I would agree.

SOMLYO: This 100 eV as the optimized window is for a specific case. You may have to use a different window, depending on which sample you are studying.

FERRIER: I think all you are interested in is over what range in the low-energy region is there a significant change in intensity. Once it is convoluted, if you go out, for example, to 100, integrating underneath the curve would be that the influence of that plural scattering is not that large and to first order you could neglect it.

SOMLYO: How long does it take on your computer?

TREBBIA: It is a TRACOR Northern machine with a PDP 1105 and it takes about 5 minutes.

ELECTRON ENERGY-LOSS ANALYSIS IN BIOLOGY: APPLICATION TO MUSCLE AND A PARALLEL COLLECTION SYSTEM

H. Shuman
A.V. Somlyo
A.P. Somlyo

Pennsylvania Muscle Institute
University of Pennsylvania
Philadelphia, Pennsylvania

INTRODUCTION

Electron energy-loss analysis (EELS) is a method showing great potential for the characterization of biological materials (Isaacson and Johnson, 1975; Colliex and Trebbia, 1979; Isaacson and Crewe, 1975; Johnson, 1979; Joy and Maher, 1979; and Fejes, 1978) through the following three major approaches: (1) elemental analysis based on the measurement of absorption edges due to core-shell excitations (Egerton, 1978; Leapman, 1979; and Maher, 1979); (2) detection of chemical bonding states and nearest-neighbor relationships revealed in the low (0-10eV) energy loss region of the spectrum and the near absorption edge and extended edge (EXELFS) fine structure (Isaacson, 1972; Johnson, 1979; Joy, 1979; Leapman, 1979; Shuman, in press); (3) imaging of compositional contrast through elemental or molecular mapping at, respectively, the absorption edges (Isaacson, 1975; Joy and Maher, 1979; Simon and Ottensmeyer, 1979; and Hui, 1979) or fine structure features (Hainfeld and Isaacson, 1978). Loss of low atomic number elements due to radiation damage has been quantitated (Egerton, 1980), and the potential for single-atom discrimination through EELS has been anticipated (Isaacson and Utlaut, 1979).

ISBN 0-12-362880-6

The value of methods capable of obtaining in situ compositional information about subcellular regions has been demonstrated by energy-dispersive electron probe analysis (EDS), a method now well established in biology (for review, see Hall, 1979; Hutchinson, 1979; and Lechene and Warner, 1979). EDS, however, provides only atomic (elemental), but not molecular (nearest neighbor) information, and it does not detect elements $Z<11$ (except with windowless detectors). EDS is also more limited than EELS by low geometric detection efficiency and by the relatively low fluorescence yields of low Z materials. Furthermore, the computerized fitting routines required for separation of overlapping peaks that occur in biological x-ray spectra, such as the overlap of the K and Ca K lines, are not readily adapted for x-ray mapping and reduce the sensitivity of Ca-measurements in the presence of physiological concentrations of potassium (Shuman, Somlyo, and Somlyo, 1976; Somlyo, Shuman, and Somlyo, 1977). In view of the importance of calcium as a regulator of muscle contraction and our interest in this system (Somlyo, Shuman, and Somlyo, 1977; Gonzalez-Serratos et al., 1978), we have begun to explore the use of EELS for the detection of Ca through its L-edge spectrum. We shall show that this approach, as compared to K-edge or K x-ray detection, improves the detection levels for minimal detectable concentrations.

The use of a doubly corrected spectrometer with high energy resolution (approximately 0.5 volts) and operated in a serial collection mode also enabled us to make some observations on the electron energy-loss fine structure (EXELFS). Serial collection of the spectrum, however, is associated with higher radiation dose than parallel collection and, therefore, radiation damage. In view of the necessity of minimizing radiation damage in biological specimens, the desirability of a parallel collection system has been recognized (Isaacson and Johnson, 1975) and occasionally realized by recording on photographic plate (Curtis and Silcox, 1971; Ottensmeyer, 1979). However, we are not aware of any published method showing parallel data collection by means that readily lend themselves to digital processing. Therefore, in this communication we will also demonstrate a method for parallel collection with a Silicon Intensified Target (SIT) Vidicon (Shuman, in press).

EXPERIMENTAL ARRANGEMENT

The spectrometer used for these experiments (Shuman, 1980) is a second-order, aberration-corrected magnetic prism, and is mounted on a Philips field emission gun equipped EM 400. A schematic drawing of the system is shown in Figure 1. The object plane (OP) of the spectrometer coincides with the differential pump aperture of the EM400, which is also the back focal plane of the microscope imaging lenses. Either the diffraction pattern or image of the specimen at the OP is then dispersed at the spectrometer's image plane (IP_1) with an energy dispersion of 4.18 μm/eV (at 80keV). The optics of the spectrometer are such that the zero-loss electrons transmitted through the specimen are focused to a point at IP_1 independent of the scattering angle β , while energy-loss electrons are focused to a straight line (perpendicular to the dispersion and above the optic axis) also at IP_1. For serial collection, a pair of knife edges, forming an energy-selecting slit, are placed at IP_1. The electrons were detected with a CaF(Eu) scintillator and photomultiplier, and the signal digitized with a Teledyne 4707 voltage to frequency converter. The magnet ramp generator and a 50Mhz counter for the V-F

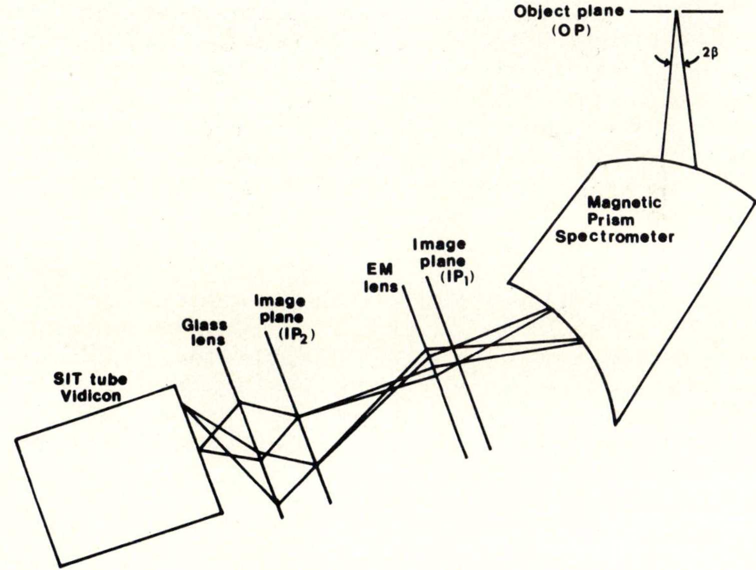

FIGURE 1. Schematic diagram of the EELS parallel Collection Optical system.

converter output were built onto a Unibus interface card for a PDP11/34 computer. The 11/34 is further used for data storage and manipulation. For parallel collection measurements with a Silicon Intensified Target (SIT) Vidicon detector the spectrometer dispersion was increased by a factor of 6 by magnifying the dispersion plane at IP_1 with a round EM lens (focal length = 21.4 mm) onto a phosphor-coated fiber optic bundle placed at IP_2. A pair of f/1.4 lens were used to image the spectra from the phosphor onto the SIT tube's photo cathode. (The SIT tube and associated electronics were on loan from PAR).

RESULTS AND DISCUSSION

Elemental Analysis

The detection and quantitation of calcium, as mentioned earlier, is an important problem in determining the mechanism of excitation-contraction coupling and contractile activation in muscle. Although EDS x-ray analysis has provided useful information on calcium concentrations in muscle, low collection efficiency and Ca K overlap with the K-K β peak can make this analysis difficult. Preliminary measurements with EELS indicate a substantial improvement for Ca detection. Figure 2 shows an EELS and an EDS spectrum of, respectively, the Ca K-edge and the $K\alpha\beta$ x-ray peak taken simultaneously from an evaporated film of CaO. The serially collected EELS Ca spectrum was taken with 20eV resolution to increase the signal. The estimated number of electrons and the number of x-rays in the two spectra are, approximately equal. Figure 3 shows the EELS spectrum of the Ca L_2 and L_3 edges and an EDS spectrum taken from the same area of a CaO film. The scale of the EELS spectrum was compressed by a factor of 512, otherwise the spectra were acquired under identical conditions (total time and probe parameters). The minimum detectable mass of Ca is clearly lowest for the EELS Ca $L_{2,3}$ edge measurement, while the Ca K-edge and Ca K x-ray peak have roughly equivalent minimum detection masses in fair agreement with previous calculations (Isaacson and Johnson, 1975).

The energy resolution for the parallel collection system was estimated from the FWHM of the zero-loss peak to be 5eV. An energy-loss spectrum of a thin crystallite of $CaSO_4$ obtained by parallel collection is shown in Figure 4. The

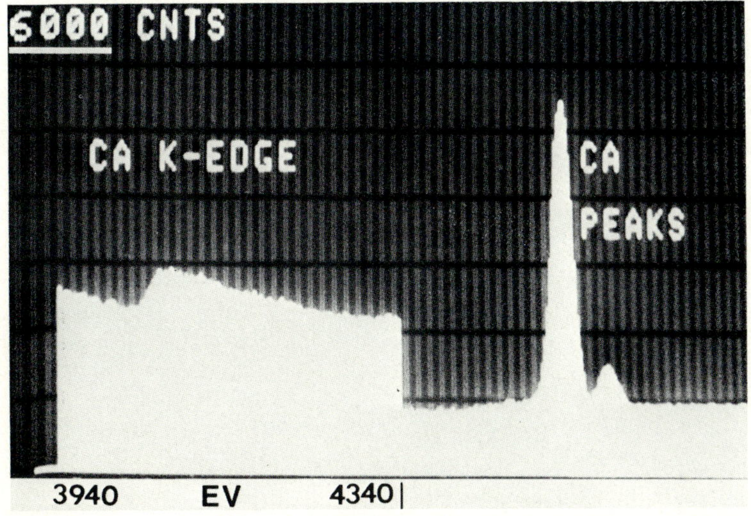

FIGURE 2. Ca K-edge and Ca Kαβ peak from evaporated CaO film, taken simultaneously in 100 sec. EELS spectrum shows the energy-loss region 3940 < E < 4340 eV, and was taken with a half-acceptance angle referred to specimen of $\beta = \sim 125$mrad, and 20eV resolution.

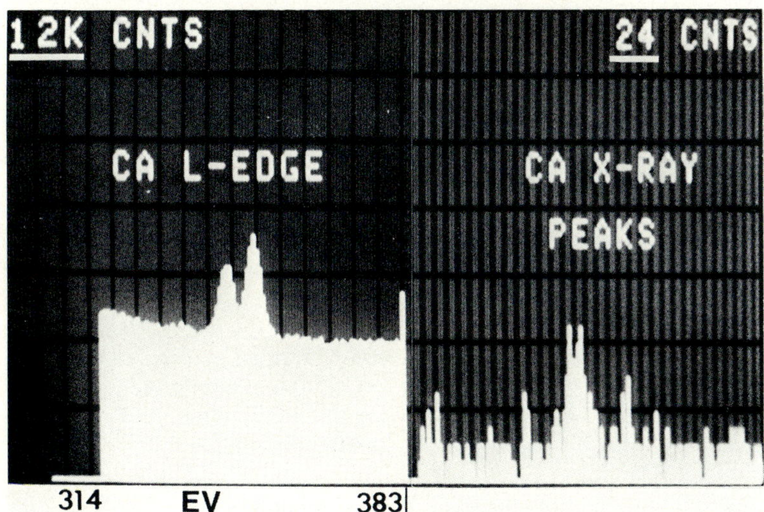

FIGURE 3. Ca L-edge and Ca Kαβ peaks from CaO film taken from the same area, each for 250 sec. The EELS spectrum covers the energy-loss region 323<E<383 eV at 2eV resolution and β = 17mrad.

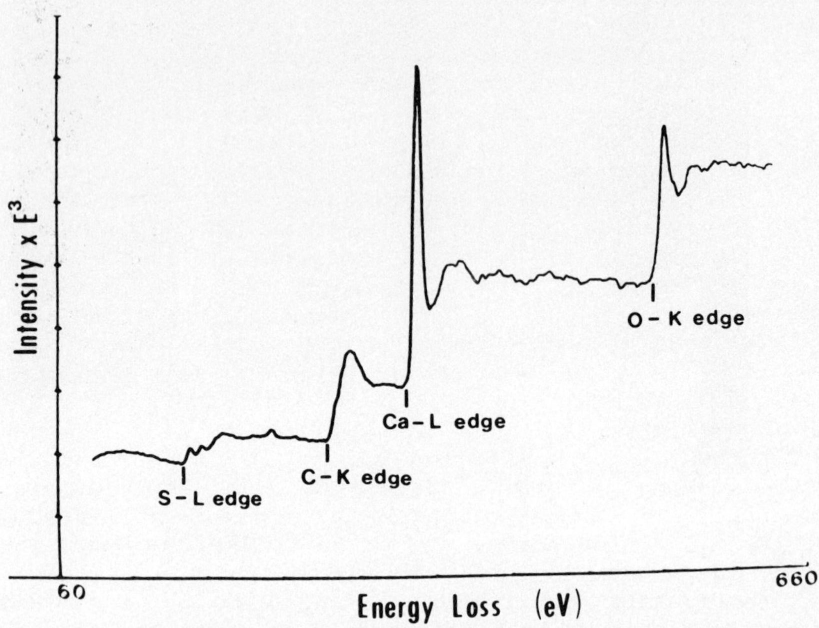

FIGURE 4. Parallel recorded energy-loss spectrum of $CaSO_4$. The intensity is scaled by E^3 to compensate for the energy-loss dependent decrease in signal.

intensity is multiplied by E^3 to compensate for the rapid fall off of signal with increasing energy loss. The two fine structure peaks at the sulfur L-edge characteristic of sulfate are 9eV apart and clearly separated. The 5eV separated edges at the carbon K-edge appear as a slight shoulder while the 3.5eV splitting of the calcium L_2 and L_3 edges is not visible, consistent with a 5eV Rayleigh resolution. The $CaSO_4$ spectrum was obtained in 40 sec. with 2nA beam current, and there are 10×10^6 counts/channel in the spectrum. Although the energy resolution presently obtainable with the SIT detector is worse than with serial collection, the higher collection efficiency leads to lower minimum detectable mass with parallel collection.

For trace elemental measurements, as in biological specimens, the minimum detectable concentration is a more important parameter. EELS and EDS measurements were performed on the terminal cisternae (TC) of frog striated muscle. Many previous EDS spectra have been taken for this system and its calcium concentration is well-known (Somlyo,

Shuman, and Somlyo, 1977; Gonzalez-Serratos et al., 1978). Specimens were ultrathin cryosections of rapidly frozen bundles of frog semitendinous muscles. Detailed description of preparatory techniques have been published (Somlyo and Silcox, 1979). Serially collected EELS spectra were taken from a 400Å spot at the center of a frog TC and again from a neighboring region of cytoplasm, known to contain very little Ca. The spectra were scaled to match in the energy-loss region preceding the Ca $L_{2,3}$ edges and subtracted. The resulting spectrum is shown in Figure 5. The number of electrons detected in a 46eV region under the edge is estimated to be $S = .41 \times 10^6$ electrons while the number of electrons in the same region of the unsubtracted TC spectrum was $S + B = 20.0 \times 10^6$ electrons. The signal-to-noise ratio for the EELS spectrum can be estimated to be $S/N = S/(S + B)^{1/2} \cong 90$. The x-ray determined concentration and estimated standard deviation of [Ca] in this TC was obtained from the EDS spectrum to be Ca = 120mmol/kg dry wt. ± 4.6 (x-ray). So that for the EELS measurement, assuming σ_c (EELS) = Ca/(S/N)(EELS), [Ca] = 120 mmol/kg ± 1.3 (EELS).

For calcium, at least, the EELS of the L-edges collected serially has a three times smaller error and consequently a three times smaller MDC. Figure 6 shows a spectrum obtained in the parallel mode in 40 sec. with 2nA beam current from a 100nm diameter area of TC. The calcium concentration in this

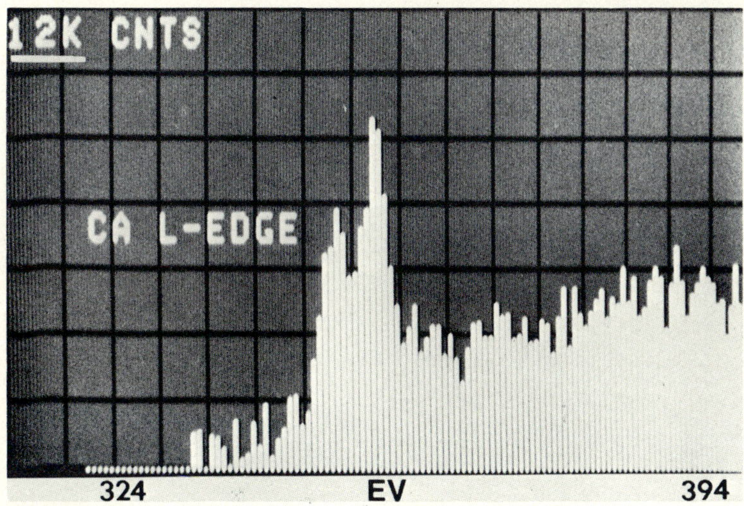

FIGURE 5. Background subtracted EELS spectrum from terminal cisterna in freeze-dried, ultrathin cryosection of frog semitendinous muscle taken in 100 sec., for the energy region 333eV to 393eV with β = 8mrad.

area is expected to be 100mM/kg dry weight (Somlyo, Shuman, and Somlyo, 1977). A neighboring cytoplasmic region was analyzed with identical parameters. The difference spectrum of TC-cytoplasm is also shown in Figure 6. The mass thickness of the TC is higher (Somlyo, Shuman, and Somlyo, 1977), so a fraction of the K-edge remains. The Ca L-edge at 346eV is visible above the remaining background noise. There are 1.2×10^8 counts in the 8 channels centered around the Ca L-edge peak above background in the differences spectrum. The signal-to-noise ratio is $S/N \cong 90$ for an estimated concentration error of $\pm$ 1mM/kg dry wt. Thus, using a smaller fraction of the available spectrum and 1/6 the analysis time, the parallel recording system provides roughly the same minimum detectable concentration as the serial system.

We also compared measurements of the P concentration of the cytoplasm of frog semitendinous muscle through EDS and serial EELS spectra of simultaneously obtained phosphorus K

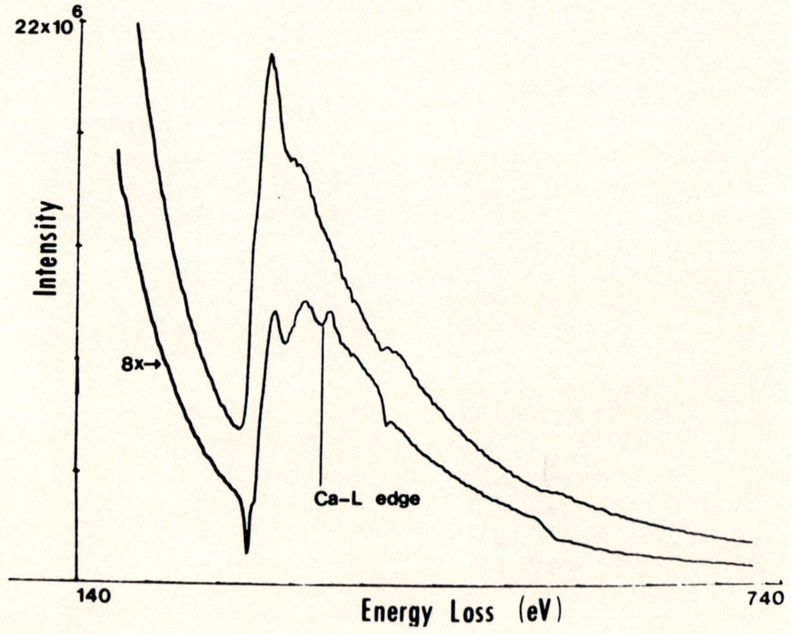

FIGURE 6. EELS spectrum of terminal cisterna and difference spectrum of TC minus cytoplasm. The difference spectrum is scaled by a factor of 8.

x-ray peaks and the K-edge (2145eV) energy losses. The EELS spectrum was taken with an energy resolution of 20eV and $\beta = \sim 125$mrad. A subsequent EELS spectrum of extracellular space of approximately equivalent mass thickness and known to contain a minimal amount of P was used to subtract the cytoplasm background. The number of electrons detected in a 240eV region under the P K-edge was estimated to be S = $.10 \times 10^6$ electrons, while the number of electrons in the same region of unsubstracted cytoplasm was S and B = 5.6×10^6 electrons or S/N = 42. The x-ray determined concentration of cytoplasmic phosphorus and error were P = 212mmol/kg dry wt. ± 5.9 (x-ray); so that for EELS, the concentration would be P = 212mmol/kg ± 5.0 (EELS).

Fine Structure: Near Edge, Extended Edge, and UV Region

The ability to distinguish chemically different atomic species, or to determine molecular composition on a microscopic scale is perhaps the most exciting potential of EELS. Figure 7 shows two superimposed EELS spectra of the sulfur $L_{2,3}$ edge at ∿ 165eV energy loss. The dotted spectrum was obtained from thin crystals of elemental sulfur deposited from ethyl alcohol solution, while the solid spectrum is for crystals of Na_2SO_4. The L-edge of pure sulfur is broad, as expected from theoretical prediction for atomic sulfur, while the edge for the sulfate has three strong peaks presumably due to the four coordinated oxygen atoms. The phosphorus L-edge shows a similar effect. Figure 8 shows the P $L_{2,3}$ edge at ∿ 132eV energy loss for a thin film of phosphoric acid. The multipeak structure is again presumably due to the coordinated oxygens.

The fine structure observed past the absorption edge (EXELFS) can also provide chemical information. Figure 9 shows a parallel recorded spectrum for the K-edge of magnesium in $MgSO_4$. The oscillations in intensity visible after the K-edge are due to the reflection of low energy electrons, produced during ionization, from atoms surrounding the Mg atom. Analysis of this EXELFS can be used to measure the immediate environment of the ionized atom (Johnson and Csillag, 1980).

The low energy-loss region of the spectrum corresponds to visible and U.V. light absorption (Isaacson, 1972). Ultimately, the information available from optical spectroscopy should be obtainable in EELS with the spatial

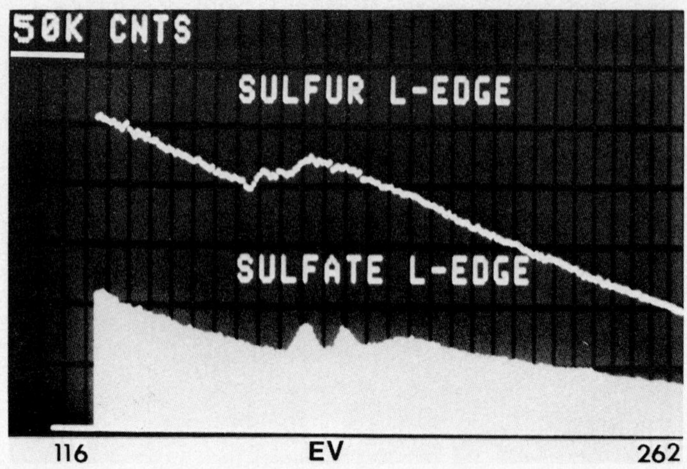

FIGURE 7. EELS spectra of sulfur (dotted line) and sulfate (Na_2SO_4) over energy region 125eV to 262eV with β = 8mrad. The carbon support film background was subtracted in both cases.

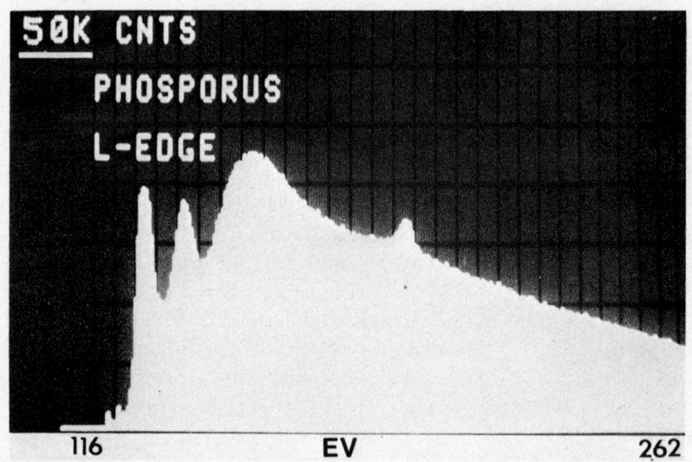

FIGURE 8. Phosphate (H_3PO_4) $L_{2,3}$ edge at 132eV and L_1 edge at 189 eV. Spectra were collected for 100 secs, at β = 8mrad.

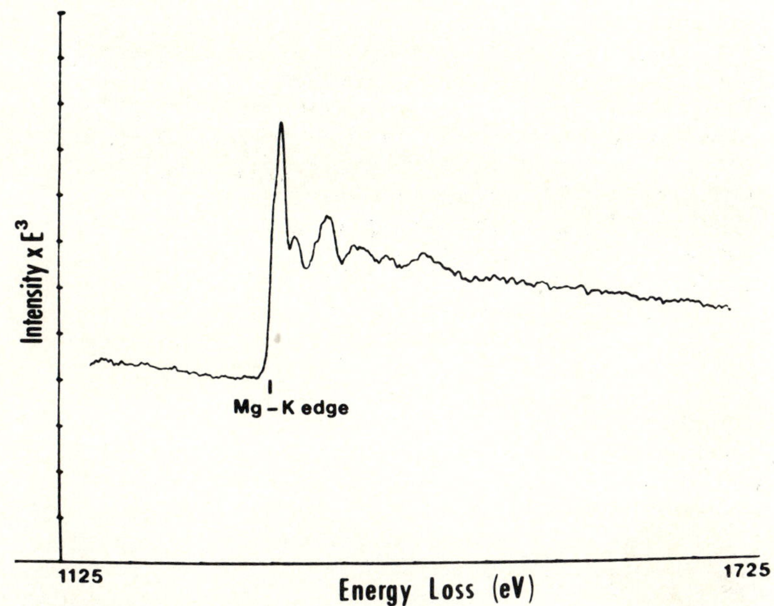

FIGURE 9. EELS spectrum of Mg K-edge from $MgSO_4$. Intensity scaled by E^3.

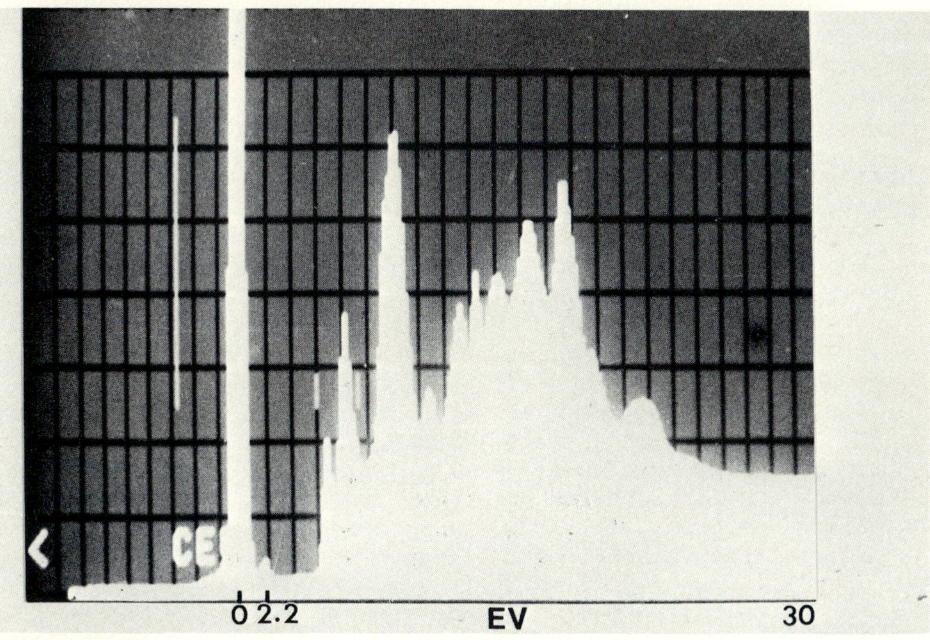

FIGURE 10. EELS spectrum of KCl. The F=center, energy-loss peak at 2.2eV is indicated.

resolution associated with electron microscopy. The optical properties of alkali halides provide a simple example of this correspondence as shown in Figure 10. The electron energy-loss spectrum taken from a small (1000Å diameter) crystallite of KCl over the energy region of 0-30eV shows many prominent features related to its optical spectrum. The small peak at $\sim$ 2.2eV energy loss is due to the ionization of an F center and accounts for the magenta color of radiation-damaged KCl.

CONCLUSION

For elemental analysis of low concentrations of calcium in thin sections of muscle, electron energy-loss spectrometry of L-edge losses is more sensitive than energy-dispersive x-ray analysis. This is true even in the case when the EELS data is collected with the information waste associated with serial collection. With parallel collection of EELS spectra, the predicted minimum detectable concentration of calcium is .23-mmol/kg for a 250 sec collection and for a cryosection of $\sim$1000Å thick ($\sim$2.5 x 10^{-6} gram/cm^2 dry). For thicker sections the shape of the background near the carbon K-edge becomes thickness sensitive, making accurate background subtraction far more difficult. Even this difficulty may be overcome with deconvolution of multiple scattering effects (Ray, 1979).

Quantitation from the K-edge losses of the elements Na to Ca is not as severely affected by thickness dependent variations in background. Our preliminary results with the P K-edge spectra agree with the prediction (Isaacson and Johnson, 1975) that EELS in the serial collection mode is as efficient for P analysis as EDS.

Measurements of the carbon K-edge for various nucleotide bases showed dramatic variations of near edge fine structure (Isaacson and Johnson, 1975). The P and S $L_{2,3}$-edges for the phosphate and sulfate ligands also show fine structure. It is almost certain, however, that in order to measure L-edge fine structure, multiple scattering effects must be eliminated with deconvolution procedures (Ray, 1979) or by using ultrathin sections or films.

REFERENCES

Colliex, C., and Trebbia, P. (1979). In "Microbeam Analysis in Biology" (C. P. Lechene and R. R. Warner, eds.), p. 65. Academic Press, New York.

Curtis, G. H., and Silcox, J. (1971). Rev. Sci. Inst. 42:630.

Egerton, R. F. (1978). Ultramicroscopy 3:243.

Egerton, R. F. (In press). Ultramicroscopy.

Fejes, P. L. (1978). Proceedings of a Specialist Workshop in Analytical Electron Microscopy. Cornell University. Ithaca, New York.

Gonzalez-Serratos, H., Somlyo, A. V., McClelland, G., Shuman, H., Borrerro, L. M., and Somlyo, A. P. (1978). Proc. Natl. Acad. Sci. 75:1329.

Hall, T. A. (1979). In "Microbeam Analysis in Biology" (C. P. Lechene and R. R. Warner, eds.), p. 185. Academic Press, New York.

Hui, S. W. et al. (1979). Proc. 37th EMSA Mts., p. 512.

Hutchinson, T. C. (1979). Biochem. Biophys. Acta. 58:115.

Isaacson, M. S. (1972). J. Chem. Phys. 56:1803.

Isaacson, M. S. (1972). J. Chem. Phys. 58:1813.

Isaacson, M. S., and Crewe, A. V. (1975). Ana. Rev. Biophys. Bioen. 4:165.

Isaacson, M. S., and Johnson, D. (1975). Ultramicroscopy 1:33.

Isaacson, M. S., and Utlaut, M. (1979). Proc. 37th EMSA Mtg., p. 524.

Johnson, D. E. (1979). In "Introduction to Analytical Electron Microscopy" (J. J. Hren, J. I. Goldstein, and D. C. Joy, eds.), p. 245. Plenum Press, New York.

Johnson, D. E., and Csillag, S. (1980). EMSA 38th, 118.

Joy, D. C., and Maher, D. M. (1979). Science 206:162.

Leapman, R. D. (1979). Ultramicroscopy 3:413.

Lechene, C. P., and Warner, R. R. (1979). In "Microbeam Analysis in Biology" (C. P. Lechene and R. R. Warner, eds.). Academic Press, New York.

Maher, D. M. (1979). In "Introduction to Analytical Electron Microscopy" (J. J. Hren, J. I. Goldstein, and D. C. Joy, eds.), p. 259. Plenum Press, New York.

Ottensmeyer, F. P. (1979). Proc. 37th EMSA Mtg., p. 380.

Ray, A. B. (1979). Proc. 37th EMSA Mtg., p. 522.

Shuman, H. (1980). Ultramicroscopy 5:45.

Shuman, H. (In press).

Shuman, H., Somlyo, A. V., and Somlyo, A. P. (1976). Ultramicroscopy 1:317.

Shuman, A. V., and Silcox, J. (1979). In "Microbeam Analysis in Biology" (C. P. Lechene and R. R. Warner, eds.), p. 535. Academic Press, New York.
Somlyo, A. V., Shuman, H., and Somlyo, A. P. (1977). J. Cell Biol. 75:828.

DISCUSSION

SPEAKER: Andrew Somlyo

FERRIER: On the question of the photodiode arrays, I think probably you are going to actually end up with most systems having optical coupling into charge coupled device, but what I will say actually is that in the CCD at least, or a photodiode array, you can anneal out electron beam damage at least three times so that the cost, if you like the picture, does go down a good deal actually, and it works out cheaper than a photographic plate.

SOMLYO: When there was still hope that we would buy a system with the Reticon, the engineer thought that we could anneal it. However, on this particular Reticon array, unfortunately, annealing did not work.

FERRIER: These were CCDs actually and you can bring those back.

SOMLYO: But you still think that for practical purposes optical coupling is best? We haven't found a detector that can take 80 Kv electrons.

FERRIER: The question is whether you can become essentially quantum noise limited. If you use direct electron bombardment, you have got so many electron-hole pairs created, that you have got overkill by a long way, but our calculations indicate that it is going to be hard to get down toward the quantum noise limit with optical coupling into either a CCD or a photodiode array.

SOMLYO: Have you considered slowing down the electrons and then going with low-velocity directly on the array?

FERRIER: That does not help. Well, yes, you can gain a bit that way, the dynamic range is limited, but of course the point is you just interrogate.

SOMLYO: Well you can interrogate at variable speeds the different channels. That may be one way to get around it.

FERRIER: I think you can certainly use the CCD for direct imaging to look at say, the fine structure in the EXELFS, because there the dynamic range is not that bad and you could use it for that purpose, but I suspect that the final versions for everyday use will involve optical coupling and Dale Johnson has been working on this as well.

SOMLYO: One nice thing that I could see from using the television camera is that since you can see where your slit is, for some experiments the main beam can be intercepted by a slit. It also makes alignment easier.

STERN: I was curious in some of the data you have shown that the L-edge is split and the intensity seems to be the L-23 edge with smaller intensity than the L_1 edge. Isn't that unusual? Do you have an explanation?

SOMLYO: No, I do not. This is why I asked Bob the question on the calculation of the L-2,3 intensity, because when Christian Colliex's came through, he indicated to us that there was disagreement among the theoreticians about the shape of the L-2,3 edge.

STERN: Is there no question about the quantitative comparison experimentally?

SOMLYO: That is so straightforward that there is no question.

STERN: What was the determining factor to your resolution in the parallel collection on this?

SOMLYO: In the parallel collection mode it is: (1) the finite size of the Vidicon element which is 25 microns; and (2) gain variations. We can improve it to some extent by increasing the dispersion of the spectrometer with a magnetic lens, but the one we have now is not sufficiently strong.

FERRIER: Are you talking about indirect bombardment? I think it is certainly true to say that the charged coupled devices even used in the optical mode, have much better gain uniformity over channels than devices based on P-N junctions.

SOMLYO: Are the charge couple devices two-dimensional?

FERRIER: Yes.

SOMLYO: But you see, the channel-to-channel variations could be reduced very much by increasing the dispersion with a lens and this could be relatively simple to do. Henry's choice of a Vidicon, rather than CCD, I think, was due to the fact that there is a commercially available system used for optical spectroscopy which uses the Vidicon and it seemed much more convenient for us to adapt this system.

SOME ASPECTS OF ELECTRON SPECTROMETER DESIGN

M. Isaacson

School of Applied and Engineering Physics
Cornell University
Ithaca, New York

INTRODUCTION

Over the last decade, there has been a considerable increase in interest in being able to analyze the energy lost by electrons being transmitted through thin specimens in an electron microscope, e.g., Isaacson and Johnson (1975), Colliex et al. (1976), Jouffrey et al. (1978), Joy and Maher (1979). Along with this interest have come discussions as to the method that is best suited for analyzing these energy-loss electrons. This analysis generally takes place by installing a system of electric and/or magnetic fields in which the transmitted electrons get dispersed in space according to their energy or momentum. The devices which produce this dispersion are called spectrometers, and in order to compare different spectrometers, we generally define some "figure of merit" which is supposed to be an indication of the performance of the device. The situation is further complicated in electron microscopy, since several other electron optical elements can be inserted between the sample and the "spectrometer" and the "figure of merit" can apply to the entire system. In that case, the optical performance of the spectrometer itself is sometimes obscured.

It is not the purpose of this paper to compare various spectrometer systems, but rather to present those aspects of spectrometer design which might be helpful to the user of the device. For discussions of instrumental artifacts observed in electron energy-loss spectra, the reader is referred to the article by Joy and Maher (1980). To keep this discussion

ISBN 0-12-362880-6

brief, I will concentrate entirely on spectrometers based upon uniform field magnets, since because of their simplicity the majority of spectrometers attached to electron microscopes are of this design.

ABERRATIONS

It is easy to see the principle of operation and the optical aberrations of uniform field magnetic spectrometers by considering the case of a 180° magnet (i.e., a magnet in which the primary electrons undergo a 180° deflection). From the Lorentz force equations one finds that the radius of curvature of an electron in a uniform magnetic field is proportional to its momentum and the deflection is perpendicular to the direction of the field. In other words, we get deflection in a plane perpendicular to the field direction, no deflection in the plane parallel to the field direction. For an electron of any energy, the radius of curvature in the plane perpendicular to the uniform field direction is given by

$$R = \frac{1}{300B}\sqrt{T^2 + 2Tmc^2} \tag{1}$$

where R is the radius in cm, B is the magnetic field in gauss, T is the kinetic energy of the electrons in eV and mc^2 = 511 KeV, the rest energy of the electrons. To simplify our discussion, we will consider the non-relativistic case where $T \ll mc^2$, then R is proportional to $\sqrt{T}$. If we have electrons of energy T and $T-\Delta T$ which leave point O in the uniform magnetic field, then they have a different radius of curvature in the field and after a 180° deflection they are separated by a distance $2\Delta R$ (Figure 1A),

$$\frac{\Delta R}{R} = \frac{1}{2}\,\frac{\Delta T}{T} \tag{2A}$$

Expressed another way,

$$\frac{\Delta R}{\Delta T} = K_1\,\frac{R}{T} \tag{2B}$$

where K_1 is a constant depending upon the geometry (one half in this case). This quantity is sometimes called the "dispersion" and expressed in microns per volt.

This would be the only quantity of interest, if it were not for the optical aberrations of the uniform magnetic field. All electrons of a given energy do not get focussed

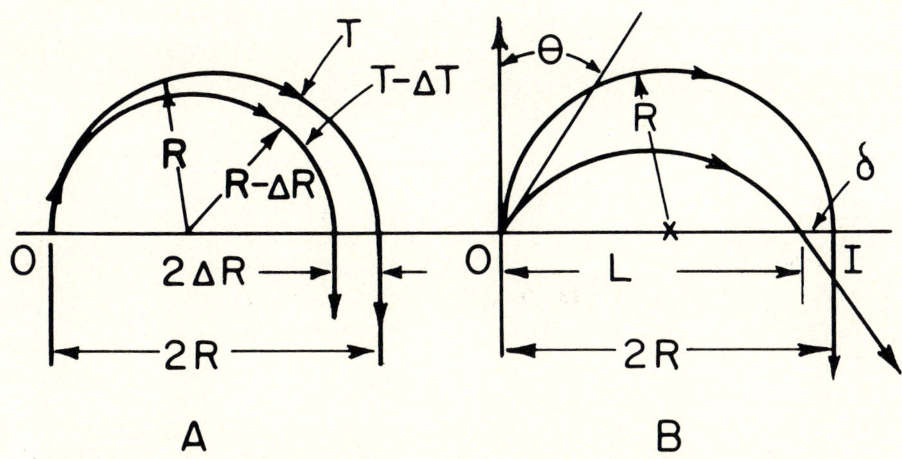

FIGURE 1. Aberrations in a 180° deflecting magnet. The direction of the uniform magnetic field is pointing into the page. Electrons of energy T and T-ΔT are separated by about 2ΔR in A. In B, electrons of energy T that enter the field at different angles with respect to the vertical axis (θ) are separated by about $R\theta^2$ at the magnetic field exit (to second order in θ).

to the same point (see Figure 1B). Electrons of energy T which leave point O at different angles with respect to the axis will be separated by a distance $\delta = 2R-L$ at the exit of the field. For a ray at an angle θ to the central ray[1]

---

[1]In the literature, the central ray is the ray that travels along the curved optical axis of the system midway between the magnetic poles. Aberrations are described as deviations from the path of the central ray. Other paths are generally evaluated as a Taylor series expansion about this central ray. (See, for example, Enge, 1967; Brown et al., 1967; Steffan, 1965.) The plane in which the central ray travels is called either the median plane or the first principal section (after Cotte, 1938).

$$\begin{aligned} \delta &= 2R[1-\cos\theta] \text{ exactly,} \\ &\cong R\theta^2 + \cdots \text{ for small} \\ \delta &= K_2 R\theta^2 \end{aligned} \tag{3}$$

K_2 is another constant that depends upon the geometry (it is unity in the case treated here).

Therefore, if we want to distinguish electrons separated by an energy ΔT, but which could have started out in the field at angles $\pm\theta$ with respect to one another, then we need, at least, that $\Delta R \approx \delta$. Putting in equations (2) and (3) we get that

$$(T/\Delta T) \cdot \theta^2 = K_1/K_2 \cong \text{unity} \tag{4}$$

That is, the product of the resolving power times the solid angle of acceptance of the spectrometer is of the order unity. The only way to reduce K_1/K_2 is to reduce the aberrations of the spectrometer. Equation (4) is a general property of all spectrometers with geometrical aberrations (although it has not yet been shown to be rigorously true). (Isaacson and Crewe, 1975.)

The above discussion was only meant to show the existence of geometrical aberrations within a uniform field. In fact, the situation is even more complicated in a magnet in which the source and the detector are located outside the field. Then we must get the electrons into and out of the field. Since in the real world we can't go abruptly from no field to a finite uniform field, there must be a region of finite spatial extent where the field goes from zero to the value well inside the magnet (the fringe field region)(Figure 2). Because this fringe field is not uniform, the electrons can get deflected in a direction parallel to the uniform field direction within the magnet. This has some advantages, since it is solely due to this fringe field that we can get such deflection (and, therefore, focussing in two directions). (See Figure 3.) As is common in the literature, we define the z direction to be parallel to the direction of the uniform field within the magnet and the y direction to be the direction perpendicular to the central ray orbit (i.e., we use a curvilinear coordinate system in which the direction of y changes along the path). (See, e.g., Steffen, 1965.)

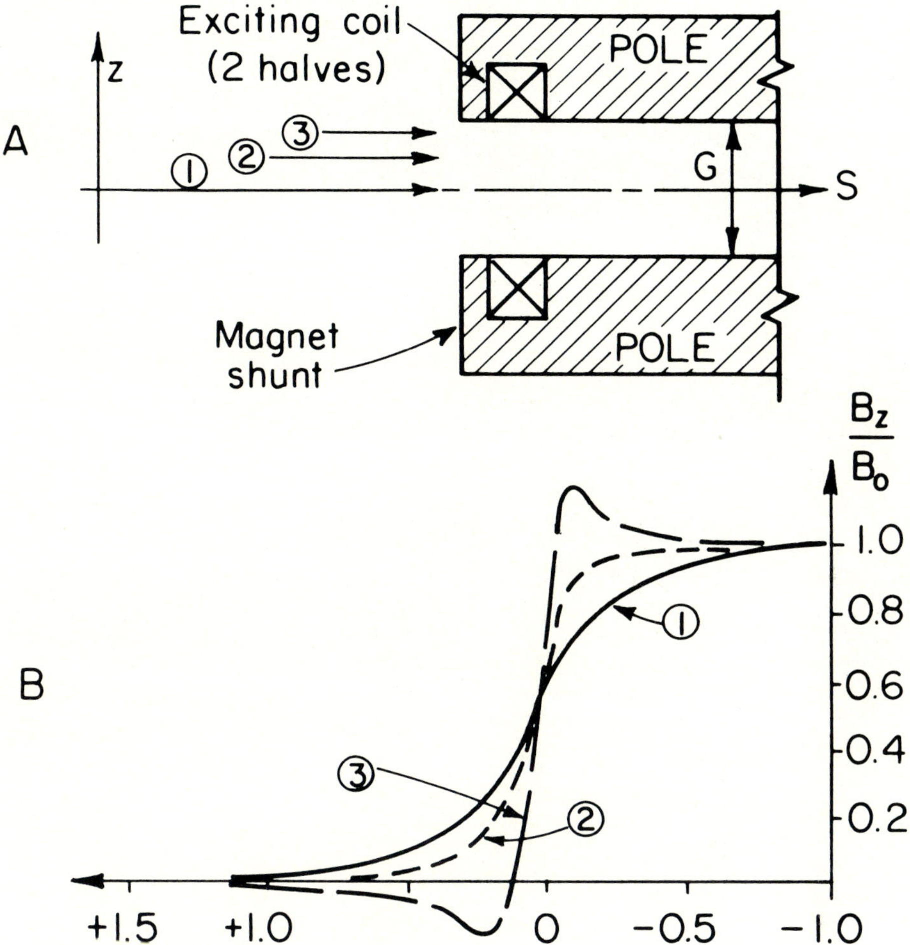

FIGURE 2. A. Schematic of a section perpendicular to the median plane of a dipole magnet showing the end configuration. This section is sometimes called the "second principal section." The gap between the poles is G and the electrons enter from left to right (in the S direction). z = 0 is midway between the two poles. B. the z component of the magnetic field along paths perpendicular to the field boundary and at different distances from the median plane. B_z/B_o = 1 well inside the magnet. Note that the fringe field can reverse sign if one gets close enough to the exciting coils.

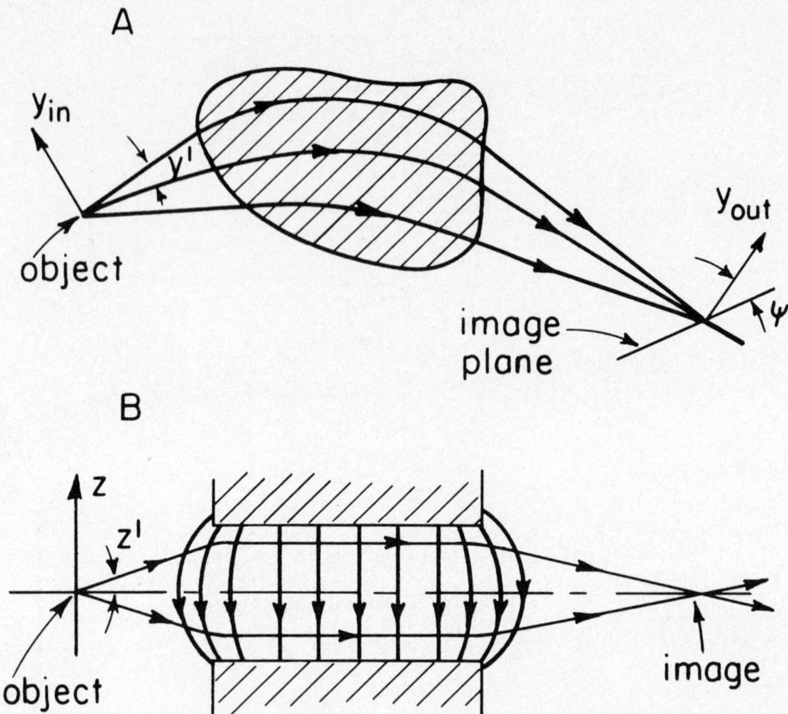

FIGURE 3. Ray paths in a uniform field magnet when the object and image are outside the field. A. Ray paths in the median plane (the first principal section). The field is directed into the paper. B. Ray paths in a curvilinear plane perpendicular to the median plane (in the z direction). The plane is perpendicular to the median plane and passes through the central ray path.

The question now is how to describe the optical aberrations in this non-uniform field--uniform field magnet system. Since the most common method of making this uniform field is with two flat magnetic poles separated by a constant distance G (the gap), the system has a mirror plane of symmetry (along z = 0, the median plane) (see Figure 3). It is known that any mirror-symmetric magnetic-field distribution can be written as a sum of mirror-symmetric multipole fields as shown in Figure 4 (e.g., Steffan, 1965; Rose and Plies, 1973).

$$B = B_2 + B_4 + B_6 + B_8 + \cdots = \sum_{n=1}^{\infty} B_{2n} \tag{5}$$

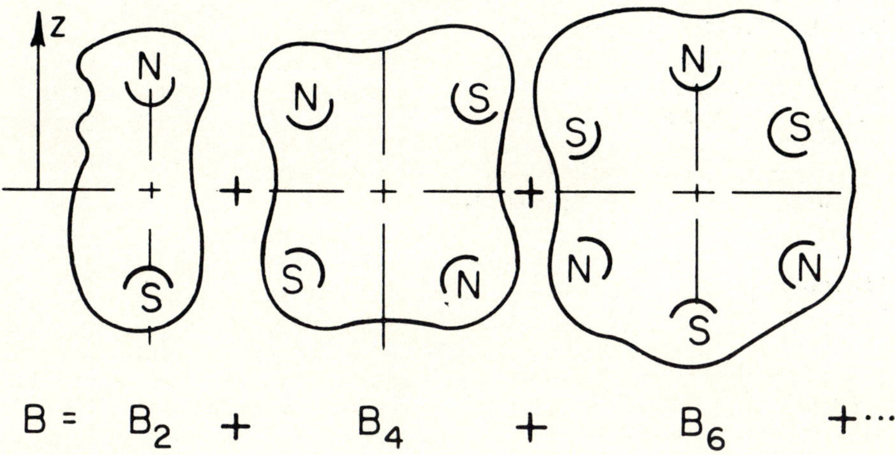

FIGURE 4. Multipole fields. The magnetic field shown schematically in Figures 2 and 3B can be constructed by expanding the field in multipoles about the plane of mirror symmetry (the median plane where z=0). For simplicity, we show the three lowest-order multipoles.

where $B_2(n=1)$ is a pure dipole field, $B_4(n=2)$ is a pure quadrupole field, $B_6(n=3)$ is a pure sextupole field, etc. The z component of the field for the dipole is constant as a function of y and z, for a quadrupole it is proportional to y (the distance from the quadrupole center), and for a sextupole it is proportional to y^2-z^2 (the total sextupole field, B_6, is proportional to the square of the distance off axis).

If a beam now enters the magnet with an angular spread of θ around the central ray, then one can show (in a fashion similar to our simple example) that the smear in the y direction at an exit plane outside of the magnet will be given by

$$\Delta y \cdot R = A_1 \theta + A_2 \theta^2 + A_3 \theta^3 + \cdots \quad (6)$$

where R is the radius of curvature of the central ray within the uniform field. The first term results from a field proportional to the distance the incident ray was off the axis of the central ray, the second term results from a field proportional quadratically to the distance off axis, etc. Thus, A_1 is the result of a quadrupole (B_{2n}, n=2) field and A_2 is due to a sextupole field (B_{2n}, n=3), etc. In other words, a 2n pole produces a θ^{n-1} aberration. That is, we can use quadrupole, sextupole, etc., fields to correct the aberrations of a real uniform-field magnetic spectrometer. This has been discussed extensively in the literature (e.g., Steffan, 1965; Enge, 1967; Wollnik, 1967; Parker, Utlaut, and Isaacson, 1978).

There are several ways of approaching this correction. But why do we want to do it in the first place? It is simply this: if we desire very good energy resolution (i.e., high resolving power), then we cannot allow a large solid angle of illumination of the spectrometer ($\pi\theta^2$), since $(T/\Delta T)\,\theta^2 \approx$ unity in an uncorrected spectrometer. A high resolution and high luminosity are mutually incompatible. We want the spectrometer to have a large angular acceptance, since we would like to collect as many of the energy-loss electrons as possible. Therefore, to keep a good energy resolution while increasing the angular acceptance requires us to remove some of the aberrations, i.e., reducing K_1/K_2. As an example, if $K_1/K_2 \approx 1$, then for 1/2 eV resolution with 100 KeV electrons, we must restrict the spectrometer to an angular acceptance of about 2.2 mrad.

If there were no lenses between the sample and the magnet, such an angle would only allow us to collect 10% of those electrons losing an energy of around 300 eV in passing through the sample and even less of those electrons losing more energy (e.g., see Isaacson, 1978). If we could reduce K_1/K_2 by 100 so that our angular acceptance increased to 22 mrad., that fraction would increase to more than 50% collection efficiency for 300 eV losses to almost 20% for 1500 eV losses. If there were lenses between the sample and the spectrometer, we might approach this increased collection by just demagnifying the angular spread of the scattered electrons before they enter the spectrometer. This has been discussed before in the literature (e.g., Johnson, 1980; Crewe, 1977; Egerton, 1980). Our concern here is to see how we can get increased collection by correcting the aberrations of the spectrometer. Then any angular demagnification we use makes the system that much better.

DESIGN

We can correct these aberrations simply by using the appropriate multipole fields. Thus, we can add quadrupoles, sextupoles, etc., at the entrance or exit of the magnet or even within the magnet. Since we are considering only uniform field magnets, we will only consider additional multipoles outside the magnet (or in the fringe field region). These multipoles can be made either by coils or by cutting shapes in the pole faces at the exit and entrance. Since the multipole fields have arisen because of the fringe field region, it seems reasonable to eliminate their effect by shaping that region. Figure 5 shows how this "shaping" accomplishes this effect. If we tilt the face such that the face normal is at an angle to the incident-beam path, then the excess field (or missing field) is proportional to the distance off axis, i.e., it acts just like a quadrupole (5A). Similarly, if we fashion the ends as circular arcs, then the excess field (or missing field) is proportional to the square of the distance off axis, just like a sextupole (5B). And if we fashion higher-order curves, we can expect to mimic higher-order multipoles.

Such shaped poles have been used for over a decade now in complex magnets for high-energy physics experiments (e.g., Enge, 1967; Brown, 1967; Keller and Hicks, 1979). Although

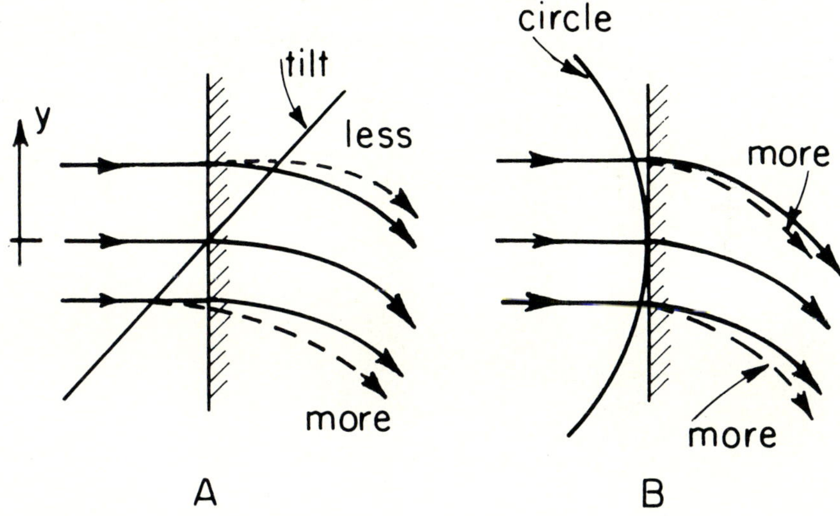

FIGURE 5. How we make a 2n pole magnetic field by cutting the shapes of the pole boundaries. It is only the end shape that matters. A. The excess (or missing) field is proportional to the distance off axis. (A quadrupole.) B. The excess (or missing) field is proportional quadratically to the distance off axis. (A sextupole.)

in those cases, the high-energy resolution was not necessarily the prime quantity of interest. It has only been recently that these shaping techniques have been used to try to eliminate some of the disturbing aberrations in spectrometers used with electron microscopes (Crewe, Isaacson, and Johnson, 1971; Ottensmeyer, 1980; Shuman, 1980). Let us look in more detail at how these aberrations affect the energy resolution. It is necessary to view the smearing in the two principal directions (i.e., perpendicular to the median plane, z, and in the median plane, y). For a point object[2],

$$\Delta y_{\text{image}/R} = C_{1y}\, y' + C_{2y}\frac{\Delta T}{T} + C_{3y}\, y'\frac{\Delta T}{T} + C_{4y}\, y'^2 + C_{5y}\, z'^2 + C_{6y}\frac{(\Delta T)^2}{T} + \cdots \tag{7A}$$

$$\Delta z_{\text{image}/R} = C_{1z}\, z' + C_{2z}\, y'z + \cdots \tag{7B}$$

where R is the radius of curvature of the central ray and y', z' denote the slope of the rays in the y and z directions, respectively.

Image smearing in the median plane means reduced energy resolution since that is the energy dispersion plane, while image smearing in the z direction just results in a line image. Note that the presence of C_{5y} means that electrons entering the spectrometer off the median plane can affect the energy resolution (this aberration results in a curved image).

Not only do these aberrations reduce the collection efficiency for a given energy resolution, but because they are not generally symmetrical in the two principal directions (z and y), the spectrometer transmission function can have an undesirable tail. An example of this is shown in Figure 6, where we compare a measured energy distribution (Egerton, 1978) with what we calculate for the same magnet (taking into

---

[2]In fact, the uniform field magnet is a momentum-dispersion device. We have been treating energy and momentum dispersion as interchangeable, although rigorously there is a difference at relativistic energies.

$$\frac{\Delta P}{P} = \frac{1}{2}\frac{\Delta T}{T}\left[\frac{1 + T/mc^2}{1 + T/2mc^2}\right]$$

is the correct equivalence between momentum resolution ($\Delta P/P$) and energy resolution ($\Delta T/T$). For 100 KeV electrons, the term in brackets is 1.089.

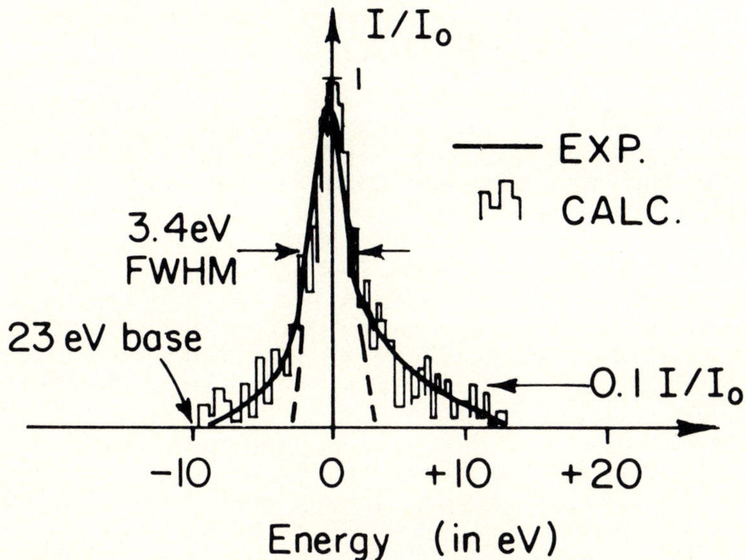

FIGURE 6. The effect of aberrations on the spectrometer transmission function. The histogram has been calculated using second-order matrix theory and assuming a 1.5 eV half-width Gaussian function as the source distribution. The experimental curve is from Egerton (1978) using a 4.7 mradian half angle of acceptance into the spectrometer. The area outside the dashed line is equal to the area within the peak.

account the source distribution). Note that although the full width at half maximum is 3.4 eV, half of the intensity is distributed in a tail that extends out to 10-15 eV. This would severely hamper measurements in which the shape of peaks were important.

For first-order focus, C_{1y} and C_{1z} are zero (we can tolerate a line image in z direction, so C_{1z} need not be zero). C_{3y} is the dispersion term (equivalent to K_1 in equation 2), and we would like this to be large; C_{3y} is the tilt of the image plane ($\tan \psi = C_{3y}/C_{2y}C_{1y}$) where C_{1y} is the angular magnification. C_{6y} is the second-order dispersion (i.e., the different energy electrons are not quite dispersed linearly along the image plane).

For $\Delta T/T = 0$, and first-order focus, the image shift becomes:

$$\Delta y_{image/R} = C_{4y}(y')^2 + C_{5y}(z')^2 \tag{8A}$$

$$\Delta z_{image/R} = C_{2z}y'z' \tag{8B}$$

Thus, if C_{4y}, C_{5y} have the same sign we get an elliptical aberration pattern, whereas if C_{4y} and C_{5y} have opposite signs, the pattern is a hyperbola. An example of the abberation pattern observed from a magnet with C_{4y} and C_{5y} of opposite signs is shown in Figure 7. The pattern was obtained by rocking the incident beam in angle about a fixed point at the spectrometer object and detecting those electrons that passed through a slit oriented perpendicular to the median plane. If we had eliminated the second-order aberrations, C_{4y} and C_{5y}, then that picture would have consisted of a uniformly illuminated spot (i.e., all electrons would have passed through the slit irrespective of the rocking angle).

The correction of C_{3y} implies that the image plane is not tilted with respect to the central ray path. This is important if one wishes to record simultaneously the energy-loss spectra, for then one would like to place the detector perpendicular to the exit beam path and not at some large angle.

Various attempts have been made to reduce the second-order aberrations (C_{3y}, C_{4y}, and C_{5y}) in magnets attached to electron microscopes. In 1971, Crewe, Isaacson, and Johnson constructed a magnet with shaped ends to eliminate C_{4y}. They experimentally demonstrated a reduction of C_{4y} by more than 20 over the equivalent

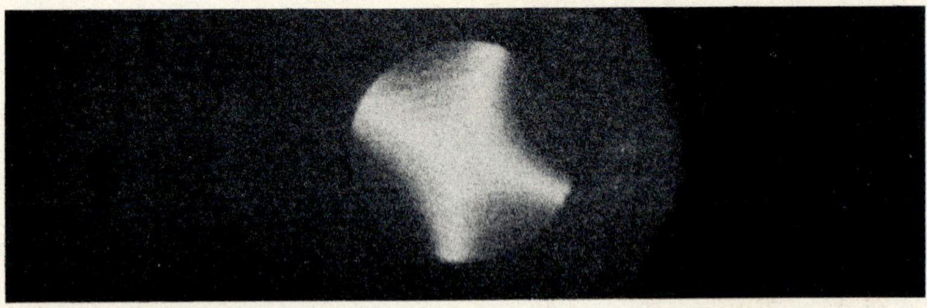

FIGURE 7. Aberration pattern observed in the VG Microscopes, Ltd., model HB5 Scanning Transmission Electron Microscope with the spectrometer that is supplied with the instrument. The hyperbola pattern indicates that C_{4y} and C_{5y} have opposite signs.

straight-edge magnet. In 1980, H. Shuman constructed a magnet with circular pole pieces in which he attempted to reduce C_{3y}, C_{4y}, and C_{5y} (see his paper in this volume). And more recently, other systems with aberration correction have been constructed elsewhere (University of California at Berkeley O. Krivanek, private communication , Cornell University, University of Toronto Ottensmeyer, 1980 , and University of Illinois K. Ramamurtii, private communication). A cut-away schematic of the system being instituted at Cornell is shown in Figure 8. For correction of the second-order angular aberrations C_{4y} and C_{5y}, one generally must have opposite curvatures on the entrance and exit faces. In the Cornell instrument, quadrupoles and sextupoles will be incorporated at the entrance and exit of the magnet as "trimming coils."

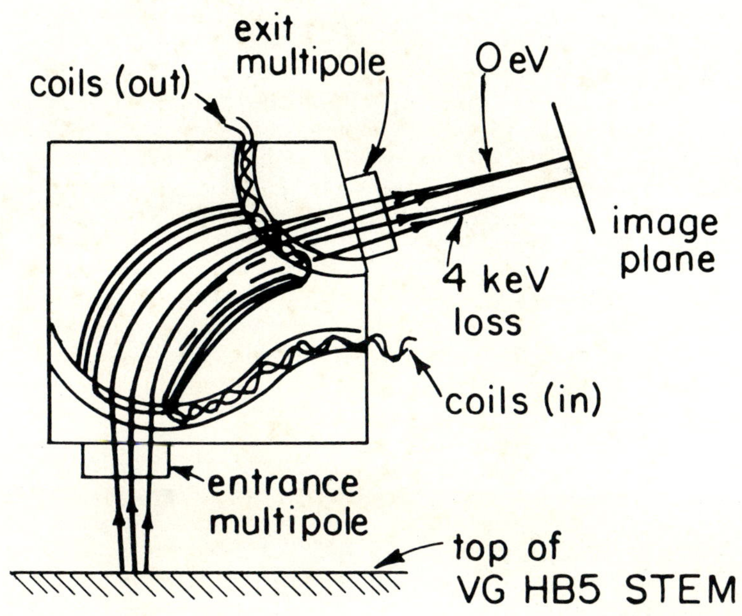

FIGURE 8. Schematic of the Cornell High Performance Electron Spectrometer to be attached to the HB5 STEM. The dispersion is about 2μ/eV at 100 KeV, the image plane will have zero tilt, and the expected energy resolution is about 1/2 eV at 20 mradian acceptance.

CONCLUSION

The initial systems that have been constructed indicate promise. However, the exact evaluation of these spectrometers from the experimental data given in the literature can be difficult. This is due to the use of pre-spectrometer lens optics to demagnify the angular spread into the spectrometer magnet and to other deleterious effects on the energy resolution such as AC magnetic fields, ground loops, etc.

In parallel with developing spectrometer aberration correction has been the trend towards simultaneous recording of the electron energy-loss data. (Discussed at the Second Cornell Workshop on Analytical Electron Microscopy, P. Fejes, ed., 1978 .) This push has come about in an effort to increase the efficiency of detection of the energy-loss electrons. For example, using the standard methods of sweeping the spectrum across a slit or an aperture, one is only accumulating data for a short fraction of the total time necessary to record the spectra. Some examples of simultaneous recording systems can be found in this volume (Johnson, 1981; Shuman et al., 1981). Insofar as optimizing the spectrometer system for use with these simultaneous recording methods, one would like to magnify the image plane of the spectrometer to match the detector element resolution and to have a flat focal plane perpendicular to the exit beam ($C_{3y}=0$) (in order to have equal resolution over a wide range of energy losses).

It appears that electron energy-loss analysis in the electron microscope is at a threshold. As the bugs are worked out of the new instrument design, and the quantitation methods are put on a sounder basis (see, e.g., Maher, 1979), we shall see an explosion of the technique just as we did with electron probe x-ray analysis during the time period between the first Battelle Conference on Microprobe Analysis (T. Hall et al., eds., 1974) and the present one.

ACKNOWLEDGMENTS

This work was supported by the National Science Foundation through the Cornell Materials Science Center and the National Research and Resource Facility for Submicron Structures. I would like to thank the organizers,

T. Hutchinson and A. Somlyo, for inviting me to participate in the Conference and in particular T. Hutchinson for his patience.

REFERENCES

Brown, K. L., Belbeoch, R., and Bounin, P. (1967). Rev. Sci. Inst. 34:481.
Brown, K. L. (1967). SLAC Report #75.
Colliex, C., Cosslett, V. E., Leapman, R. D., and Trebbia, P. (1976). Ultramicroscopy 1:301.
Cotte, M. (1938). Ann. Phys. (Paris) 10:333
Crewe, A. V. (1977). Optik 47:299.
Crewe, A. V., Isaacson, M., and Johnson, D. (1971). Rev. Sci. Inst. 42:411.
Egerton, R. F. (1978). Ultramicroscopy 3:39.
Egerton, R. F., (1980). Optik 56:363.
Enge, H. (1967). In "Focussing of Charged Particles" (A. Septier, ed.), Vol. 2, p. 203. Academic Press, New York.
Hall, T., Echlin, P., and Kaufmann, R., eds. (1974). "Microprobe Analysis as Applied to Cells and Tissues." Academic Press, New York.
Isaacson, M., and Crewe, A. V. (1975). Ann. Rev. Biophys. and Bioeng. 4:165.
Isaacson, M., and Johnson, D. (1975). Ultramicroscopy 1:33.
Isaacson, M. (1978). In "Scanning Electron Microscopy/1978." (O. Johari, ed.), Vol. 1, p. 763. SEM Inc., AMF O'Hare.
Johnson, D. (1980). Ultramicroscopy 5:163.
Johnson, D. (1981). In this volume.
Jouffrey, B., Kihn, Y., Perez, J. P., Sevely, J., and Zanchi, G. (1972). Proc. Ninth Int. Cong. on Elec. Microscopy (J. M. Sturgess, ed.), p. 292.
Joy, D., and Maher, D. (1979). Science 206:162.
Joy, D., and Maher, D. (1980). In "Scanning Electron Microscopy/1980" (O. Johari, ed.), Vol. 1, p. 25. SEM Inc., AMF O'Hare.
Keller, J. H., and Hicks, W. W. (1979). J. Vac. Sci. Tech. 16:1921.
Maher, D. M. (1979). In "Introduction to Analytical Electron Microscopy" (J. Hren, J. Goldstein, and D. Joy, eds.), p. 259. Plenum Publishing, New York.
Ottensmeyer, F. P. (1981). In this volume.
Ottensmeyer, F. P. (1980). Proc. 38th Ann. EMSA Mtg., p. 270.
Parker, N. W., Utlaut, M., and Isaacson, M. (1978). Optik 51:333

Rose, H., and Plies, E. (1973). In "Image Processing and Computer-Aided Design" (P. W. Hawkes, ed.), p. 374. Academic Press, New York.

Shuman, H. (1980). Ultramicroscopy 5:45.

Shuman, H., Somlyo, A., and Somlyo, A. (1981). In this volume.

Steffen, K. G. (1965). "High Energy Beam Optics." Interscience, New York.

Wollnik, H. (1967). Nucl. Inst. Meth. 52:250.

DISCUSSION

SPEAKER: M. Isaacson.

HALL: Would you or anybody be prepared at this point to say qualitatively what the prospects are for studying chemical bonding in biological material in view of the radiation damage that occurs?

ISAACSON: It depends on the questions you are asking, on how small a volume you want to do things from. Several questions have not been answered. One knows, in lowering the temperature of the sample, one can increase radiation resistance from the point of losing material. There are some experiments that show that when one is looking at certain types of energy loss, it does not change much at all when one goes down to nitrogen temperature. There are no experiments that I know of that indicate what happens when one goes, for example, close to helium temperature in terms of whether these peaks or whether these bonds are frozen in. So then one might have certain prospects, but one can make estimates of the damage or the number of coulombs per square centimeter which one needs to break certain of these bonds that can destroy some of the spectra and then go backwards and say, how many atoms or what type of volume do I need in order to be able to detect that? It certainly turns out that under certain conditions it is not crazy to think of being able to do it. It certainly might very well be possible at the 10-100 nm level if in fact there is an increased radiation resistance. In other words, if the spectra stayed for longer dose at very low temperatures, then that is a whole different story.

FERRIER: Ray Egerton has done some recent work, where. . .

ISAACSON: He is looking at mass loss only.

FERRIER: No. He was looking at the energy spectrum as far as I know.

ISAACSON: He monitors the elemental edges.

FERRIER: That is in an organic PMMA.

ISAACSON: But still what we have found is that, in looking at a variety of different amino acids, there is surely mass loss. In other words, you keep the material around, but whether or not the bonding is anything related to what it was is another story.

FERRIER: That is right.

SOMLYO: Isn't it true that if one wants to work on biological systems, for example, one might first do pure compounds using maybe the diffraction mode for an analysis in a large area with small angles to minimize radiation damage? If you want to do true biological microanalysis in a cell, then you have very heterogeneous systems, and I do not think that the theoreticians are prepared to help us solve those. But in terms of damage (not to use jargon), won't it depend on what happens at liquid helium temperature whether the damaging event is correlated or not? In other words, is there the possibility for reversibility when you trap the excited state?

ISAACSON: The easiest thing is to do it. You can mull about it for a long time. In terms of damage each material is very different, so one is very hard put to make statements of a general nature.

FERRIER: We have done it in chlorine copper phthalocyanin, and there you can study it. We did diffraction studies.

HUTCHINSON: In most of our studies we were looking at chlorine loss in PVC which follows an exponential decay curve, but it is primarily the escape parameter that we are concerned with. When we start to think about RNA and DNA, etc., there is no escape parameter. No diffusion occurs in a matrix. Once you make the ionization event occur, mass is lost.

ISAACSON: But you change the confirmation. . .

HUTCHINSON: Yes, exactly.

ISAACSON: In which case the bondings change. It depends on the parameter you are after.

SOMLYO: If you are an opportunist, you will ask for answers which your method hopefully can give you.

FERRIER: Mike, in terms of the final result, as you say you can shape the pole pieces and simulate quadrupole and sextapole, would your idea then be to do this to your main machine?

ISAACSON: This is just a precaution because when you add a sextupole outside, you essentially have a system with a variable curvature. So you do the best you can to get the right curve, then you use the sextapole coils as trimming coils to compensate for your machining or calculation errors.

FERRIER: But roughly speaking, what tolerance are you talking about as far as your machinist is concerned?

ISAACSON: Oh, not very good tolerances, because we are talking about curvature tolerances of a few percent--5 or 10% maybe.

SOMLYO: I noticed no one has mentioned here the fields in the electron microscope that contribute practically but not theoretically to the aberrations in the magnet. That is a practical problem.

ISAACSON: It is very difficult, because all of this focussing and things we are talking about are all due to what basically is happening outside the magnet. So the sextupole effects depend on how wide you make the magnet, what is nearby (if you put material in front of it, the magnetic shunts), where the coils are, and things of that sort.

FERRIER: And homogeneity.

ISAACSON: Well, homogeneity on the inside one assumes.

FERRIER: You don't use metal chairs.

ISAACSON: You never use metal chairs.

SOMLYO: In electron microscopes, most of them unfortunately were not designed with the idea in mind that these stray fields should be minimized in the microscope.

SOMLYO: For dispersion can you still use the sextapole or would you add the quadrupole?

ISAACSON: The hexapole does not change it, so if you want added dispersion, you use a lens. The advantage of the lens is it has a tilted focal plane. So if you are clever, you match the post spectrometer lens focal tilt with whatever you happen to get there, so that you wind up with a focal plane which is flat. You can increase your dispersion in one plane with a quadrupole as well. It depends whether you are more prone to wind things Cw or CCw.

ELECTRON ENERGY-LOSS MICROANALYSIS WITH HIGH SPATIAL RESOLUTION, ENERGY RESOLUTION, AND SENSITIVITY

F. Ottensmeyer
D. Bazett-Jones
K. Adamson-Sharp

Ontario Cancer Institute
Toronto, Ontario

INTRODUCTION

The ideal in microanalysis is to have a quantitative, two-dimensional map of any desired element in the specimen of interest with a sensitivity of detection and a spatial resolution that is commensurate with single, or at least small, clusters of atoms. Compromises to this ideal have, of course, had to be accepted, since on the one hand the physical technique available may be inadequate or, on the other, the specimen is uncooperative by deteriorating before the data can be collected in biology often even before the specimen is put in the measuring instrument.

For high atomic number elements, x-ray microanalysis has served admirably with a potential resolution of about 5 nm (Goldstein, 1979) in thin specimens and an estimated potential sensitivity of 90 x 10^{-21} g for the detection of iron (Shuman and Somlyo, 1976). In practice, weak signals force the use of larger beam probes and thicker specimens, such that resolutions of 50 nm and bigger are more common. For low atomic number elements, the x-ray fluorescence yield and the efficiency of detection of the soft x-rays that are emitted decrease, resulting in still weaker signals.

A complementary technique to x-ray microanalysis is electron energy-loss analysis, the analysis and use of primary electrons that have suffered an energy loss in passing through the specimen and exciting electrons from

ISBN 0-12-362880-6

valence shells or energy bands of molecules and crystals or from lower-lying electronic states, including the inner shells of atoms. The cross section for such an interaction, a primary event, is higher the lower the atomic number of the element. Moreover, the detection efficiency of the electron spectrometer is higher the lower the energy loss to be analyzed.

Although electron energy-loss analysis was considered twenty years ago (Hennequin, 1960; Castaing and Henry, 1962), it was the advent of the transmission scanning microscopes and scanning attachments that created a renewed interest in the technique (Hren, Goldstein, and Joy, 1979). The ease with which magnetic spectrometers are built for scanning systems suggested a spot-by-spot mode of analysis and imaging, parallel to x-ray microanalysis. However, Castaing and Henry (1962) showed that entire images can be analyzed simultaneously with a more complex imaging energy filter, a prism-mirror-prism (PMP), installed in a fixed-beam electron microscope.

THE ELECTRON MICROSCOPE

The Filter

We have now installed our second such PMP system in a high-resolution Siemens Elmiskop 102 (Figure 1), after a prototype in a Siemens Elmiskop 1 (Henkleman and Ottensmeyer, 1973) indicated that aberrations in such a system were sufficiently small in practice not to affect the normal performance of the unmodified microscope. This finding proved to be true still in the higher resolution instrument.

The filter, inserted above the projector lens in the column of the microscope, consists of an electromagnet creating a triangular magnetic field, an entrance lens to adapt the optics of the microscope to the optics of the filter, an electrostatic mirror made from an electron gun, a slit system for energy band width selection, and an exit lens which focuses on the energy-loss spectrum at slit height or on the achromatic image within the triangular magnetic field.

Since the instrument was intended primarily for biological applications, simplicity of operation was one of the design imperatives. So if energy-selected images only

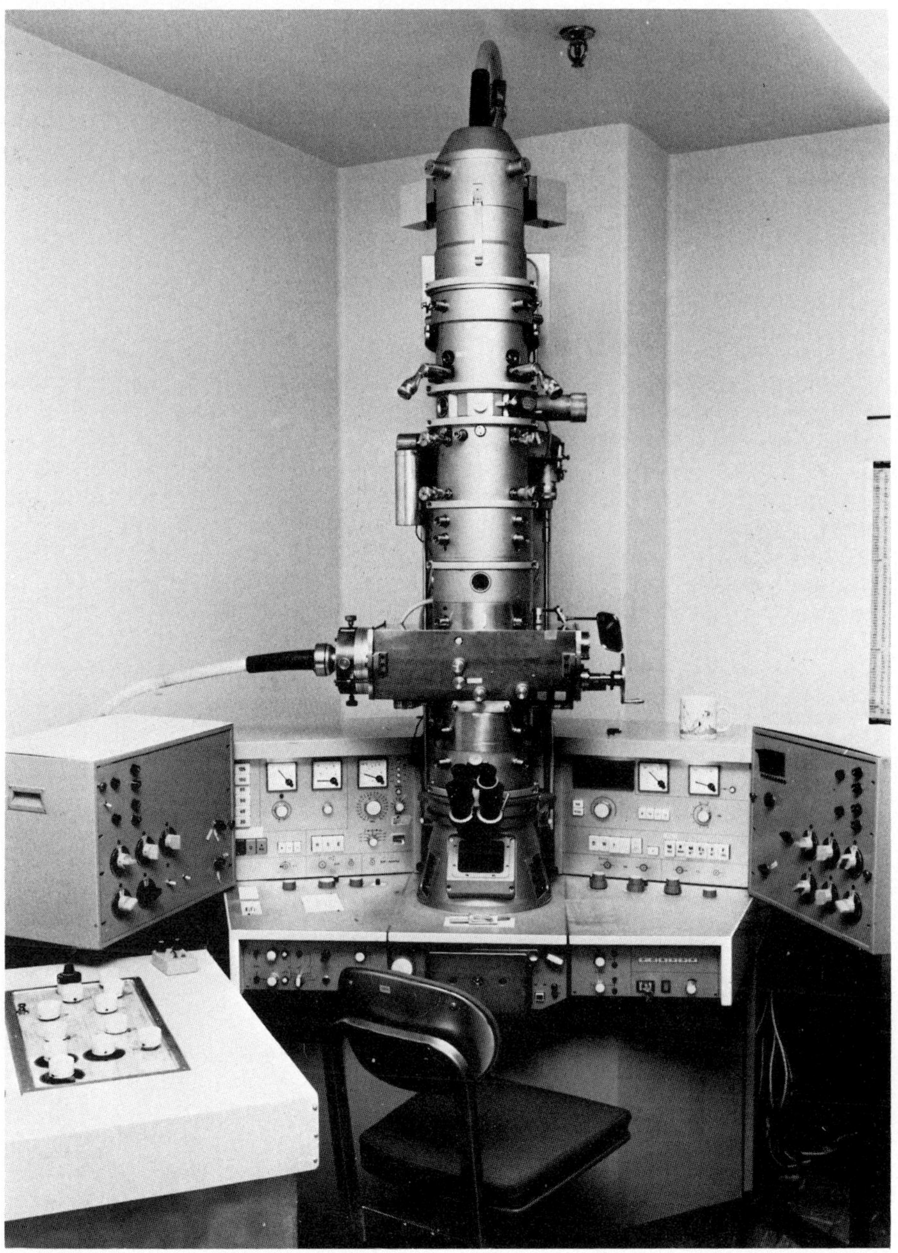

FIGURE 1. Siemens Elmiskop 102 fitted with PMP imaging energy filter. Controls for additional lenses and the magnetic prism are on the left and right panels. Energy-loss selection is controlled from the tank in the left foreground.

are desired, only the calibrated energy-selecting knobs need be used in addition to the normal operation of the microscope. If spectra also are desired, a two-way switch selects between two exit lens excitations preset for the spectral or image modes. For spectral observations, the energy-selecting slit is retracted. All other operations of the microscope--bright field, dark field, diffraction, etc.--are normal.

Specifications

The physical characteristics of the modified microscope are indicated in Table 1. Resolutions of around 0.3 nm were obtained, both with 0 eV electrons (elastic scatter) and with 150 eV loss electrons (phosphorus $L_{2,3}$-edge enhancement) (see below). The latter result indicates that while lower energy losses from molecular orbitals or plasmon excitations may be relatively delocalized (Castaing, 1975), the interaction of the imaging electron with an inner shell of the atom is localized to within 0.3-0.5 nm at least.

TABLE 1. Performance characteristics of the energy electron microscope.

Energy Loss	Magnification	Spatial Resol.	Comment
eV		nm	
	200,000 x	∿0.3	Measured by optical diffraction of duplicate exposures of a carbon film
0	80,000 x	∿0.5	
150	50,000x	0.3-0.5	Imaged thickness of phosphorus atom layers in viral membrane
Acceptance Angle		**Energy Resolution**	
m rad		eV	
15		15	Corresponds to full field of view at camera level
1.5		2	for 5 μA beam current
1.5		0.8	for 1 μA beam current

Minimum mass detected with a signal to noise of 4 ± 1 was 2×10^{-21}g phosphorus in the membrane of a leukemia virus using the $L_{2,3}$-edge of phosphorus (Adamson-Sharpe and Ottensmeyer, 1981).

Energy resolution and, therefore, energy selection in images is limited to 15 eV over the full field of view due to second-order aberrations of the prism-shaped magnetic field. For small fields of view (small acceptance angle of the spectrometer) a 2 eV resolution is typical for normal beam currents used. This limit, due to the Boersch effect, can be reduced to 0.8 eV if a small beam current is accepted. The latter limit is still not the spectrometer limit but is the energy spread in emission of electrons by the tungsten hairpin filament.

Applications

A typical spectrum recorded on film in 0.4 sec. and densitometered subsequently is shown in Figure 2. In the low-energy region, the 3 eV loss corresponding to the optical absorption of the prophyrin ring of the heme molecule is seen. The broad, high peak, due primarily to interband excitations and plasmon losses in the carbon support, then decays away at higher energies, interrupted only by stepped increases in intensity at the absorption edges of carbon, nitrogen, oxygen, and iron.

Thus, all the elements in the heme molecule (except for hydrogen) are identified. Quantification of the elements in the energy-loss spectrum is actively pursued in the laboratories of Egerton (1978), Maher (1979), Colliex (1980), and others.

The energy-loss spectrum contains a continuum of intensities at any energy due to events lying at lower energy but still excited at higher energy losses. Thus, an image taken above the absorption edge of an element contains information about the spatial distribution of that atom enhanced above a general image of the total structure. This is shown in Figure 3, a micrograph of murine leukemia virus taken with a loss of 150 eV to enhance the phosphorus contribution from the $L_{2,3}$-edge at 138 eV. A complete image is seen, rather than only a phosphorus distribution. On the one hand, this is a boon, since the spatial relationships of one structure to another are immediately apparent. On the other, it is a drawback, since a second image, a reference image, must be taken with an energy immediately below the absorption edge of the chosen element in order to obtain a measure of the contribution of the background intensity. The subtraction of the two images, suitably normalized, is then the net distribution map of that

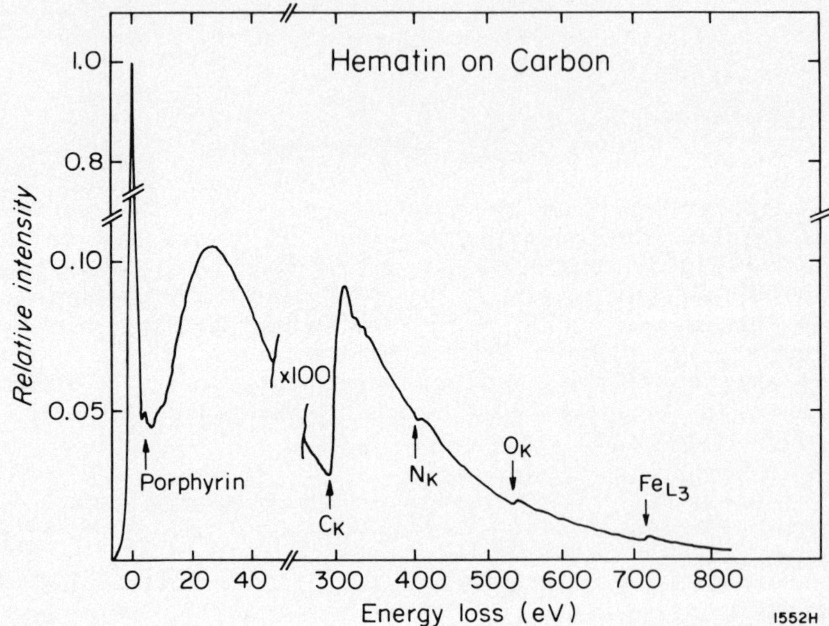

FIGURE 2. Energy-loss spectrum of hematin on carbon showing the molecular absorption of the porphyrin ring at 3 eV, the K-edges of carbon, nitrogen, and oxygen at 283 eV, 402 eV, and 532 eV, respectively, followed by the $L_{2,3}$-edge of iron at 720 eV.

element. This subtraction can be carried out photographically over the entire plate (Ottensmeyer and Andrew, 1980) or a small portion on each image may be digitized, spatially correlated, normalized, subtracted, and displayed on a suitable video system. This process permits immediate relative quantification over the digitized image. One of the virus images in Figure 3 was handled in this way (Figure 4).

An analysis of the membrane portion of such virus images has yielded both a measure of spatial resolution and of

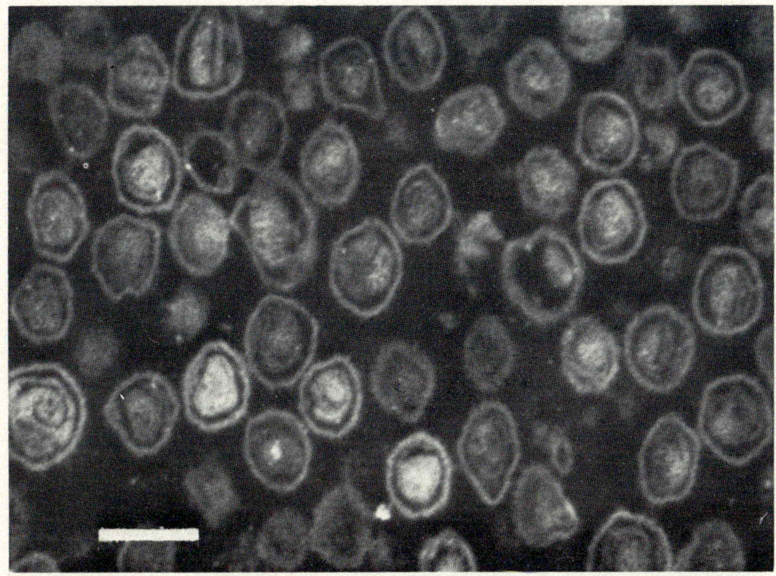

FIGURE 3. Section of purified, embedded, unstained murine leukemia viruses taken at an energy loss of 155 ± 7 eV above the phosphorus $L_{2,3}$-edge at 138 eV. Although the phosphorus signal is enhanced, the large background intensity at this energy in the electron energy-loss spectrum provides a complete overview of the morphology of the virus. The scale equals 0.2μm.

detection sensitivity in images taken with an energy loss. On the assumption that the phosphorus atoms in the phospholipid bilayer of the membrane form a thin sheet of atoms, a view of sections perpendicular to this view should yield a measure of the spatial resolution of the filter. This assumption has two provisos: that the atoms indeed form a uniform thin sheet and that the interaction between the inner shell of the phosphorus atom and the imaging electron is sufficiently localized. The observation that one-third of the measurements falls as low as 0.3-0.5 nm suggests that both provisos are met.

Since the phosphorus composition of the membrane is known, the minimum mass of phosphorus could be calculated as 2×10^{-21}g with a signal-to-noise ratio of 4 ± 1 (Adamson-Sharpe and Ottensmeyer, 1981).

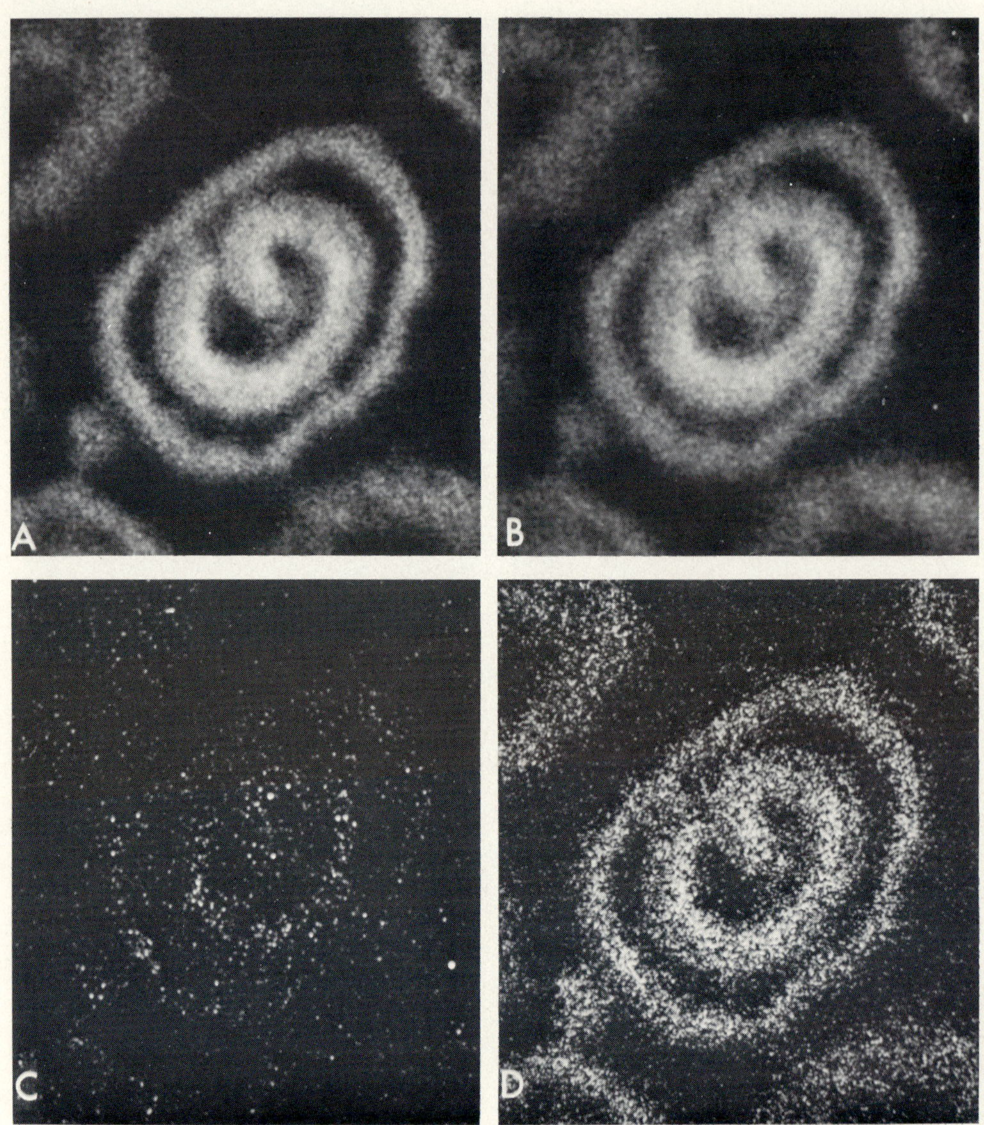

FIGURE 4. (a) Phosphorus-enhanced image at 155 eV loss of one of the viruses in Figure 3 digitized on a 512 x 512 raster. (b) Reference image at 105 eV of the same virus as in (a). (c) Numerical subtraction, after cross-correlation and normalization of the images in (a) and (b). The image represents the net phosphorus distribution in the virus section. (d) Same as (c) but contrast enhanced by digital processing.

Comparisons

On the surface, energy-loss microanalysis appears to be preferable to x-ray microanalysis from many aspects. The energy-loss event being prior to conversion into a fluorescent x-ray has a higher cross section than x-ray emission. Moreover, the spatial resolution and detection sensitivity shown here are one or two orders of magnitude better than in x-ray microanalysis to date. A closer examination, however, indicates a much more complementary relationship with a crossover between the techniques at an energy of 2000 eV, or perhaps even 1000 eV.

The intensities of the signals detected in the two cases (Joy, 1979) are given by:

$$I_{EEL} = I \cdot N \cdot \sigma_E \cdot \eta_\beta \cdot \eta_{\Delta E} \quad (1)$$

and

$$I_{X\text{-}ray} = I \cdot N \cdot \sigma_E \cdot \omega \cdot F \cdot D \quad (2)$$

where I is the incident number of electrons;

N is the number of atoms in the beam;

σ_E is the cross-section of excitation of an electron from energy shell E;

η_β is the fraction of characteristic loss electrons captured in the spectrometer acceptance angle;

$\eta_{\Delta E}$ is the fraction of characteristic loss electrons passed in the energy band width selected by the slit;

ω is the fluorescent yield;

F is the geometric efficiency of the x-ray detector;

D is the sensitivity of the x-ray detector to rays of different energy.

Choosing typical values for these parameters from Joy (1979) and Zaluzec (1979) and an energy window of 15 eV in our EEL spectrometer, the ratio of intensities $I_{EEL}/I_{X\text{-}ray}$ for the phosphorus K-absorption at 2.1 KeV is 3 in favor of EEL. If one considers the relative background intensities in the two spectra, the advantage is non-existent or in favor of x-ray microanalysis.

For the fluorine K-absorption at 685 eV, the ratio increases to 2.1 x 10^3, a clear advantage for EEL. However, EEL can also very easily use L-absorption at even lower energies, while x-ray microanalysis cannot usefully reach below 2 KeV. Thus, a comparison of the L-absorption of phosphorus at 140 eV in EEL with the K-fluorescence at 2.1 KeV in x-ray analysis yields a ratio of 3.3 x 10^4 (Table 2). Clearly, if intense low-energy absorptions are available, EEL analysis has an advantage, whereas x-ray microanalysis remains the technique of choice above 2 KeV.

TABLE 2. Comparison of Signal Intensities for Several Inner Shell Absorptions and Emissions for Electron Energy Loss and X-ray Microanalysis.

Energy shell considered in		I_{EEL} $I_{X\text{-}ray}$
EEL	X-ray	
P_K(2.1 KeV)	P_K(2.1 KeV)	3
F_K(0.7 KeV)	F_K(0.7 KeV)	2.1 x 10^3
P_L(0.14 KeV)	P_K(2.1 KeV)	3.3 x 10^4

The comparison of a scanning system in EEL analysis versus a fixed-beam system is more difficult since the usage determines the preference. Both systems are, in principle, capable of the same spatial resolution, the same energy resolution, and the same detection sensitivity. If spot analysis is desired, clearly the scanning system or a scanning attachment is desirable to achieve a small spot size. For imaging, the PMP system is probably preferable. It offers a large field of view over which the specimen grid can be scanned. More decisive, however, are the number of resolution elements (pixels) that can be recorded in one image. At a magnification of 50,000 X, an energy-loss image revealed structure at 0.3-0.5 nm resolution on a 9 cm x 6.5 cm plate. To sample this plate digitally without loss in resolution by the Nyquist criterion, one would have to sample at about 7.5 μm spacings, resulting in about 10^8 pixels per plate. Good scanning display systems today can reproduce about 100 x 100 pixels or 10^6 pixels per image. One photographic plate, therefore, contains the equivalent of 100 scan images.

Nevertheless, after image treatment--normalization, correlation, subtraction--would definitely be simplified if the information is already gathered digitally in a correlated fashion by simultaneous multiple-energy recording. This, however, is not yet taking place on any of the present scanning systems.

ACKNOWLEDGMENTS

This work was supported by the Ontario Cancer Treatment and Research Foundation, the National Cancer Institute of Canada and grant MT-3763 from the Medical Research Council of Canada.

REFERENCES

Adamson-Sharpe, K. M., and Ottensmeyer, F. P. (1981). J. of Microscopy.

Castaing, R. (1975). In "Physical Aspects of Electron Microscopy and Microbeam Analysis" (B. M., Siegel and D. R. Beaman, eds.), p. 287. Wiley, New York.

Castaing, R., and Henry, L. (1962). C. R. Acad. Sci. Paris 255:86.

Egerton, R. F. (1978). Proceedings. 11th Annual SEM Symposium, Vol. 1, pp. 13-23. SEM, Inc., Chicago.

Goldstein, J. E. (1979). In "Introduction to Analytical Electron Microscopy" (J. J. Hren, J. I. Goldstein, and D. C. Joy, eds.), p. 83. Plenum Press, New York.

Henkelman, R. M., and Ottensmeyer, F. P. (1973). J. Microsc. 102:79.

Hennequin, J. F. (1960). Diplome d'Etudes Superieures, Paris.

Joy, D. C. (1979). In "Introduction to Analytical Electron Microscopy" (J. J. Hren, J. I. Goldstein, and D. C. Joy, eds.), p. 223. Plenum Press, New York.

Maher, D. M. (1979). In "Introduction to Analytical Electron Microscopy" (J. J. Hren, J. I. Goldstein, and D. C. Joy, eds.), p. 259. Plenum Press, New York.

Ottensmeyer, F. P., and Andrew, J. W. (1980). J. Ultrastruc. Res., V. 72:336.
Shuman, H., and Somlyo, A. P. (1976). Proc. Natl. Acad. Sci. USA. 73:1196
Trebbia, P., and Colliex, C. (1980). In this volume.
Zaluzec, N. J. (1979). In "Introduction to Analytical Electron Microscopy" (J. J. Hren, J. I. Goldstein, and D. C. Joy, eds.), p. 121. Plenum Press, New York.

DISCUSSION

SPEAKER: F. P. Ottensmeyer

ORNBERG: I must say I am quite impressed by the pictures of phosphorus in the membrane, particularly after osmium fixation and then to dehydration. Corn, in a paper in Science a few years ago, showed that 25% of phosphorus lipid is extracted during most procedures. Have you looked at any artificial systems where you might be able to see domains within a phospholipid domain such as within a membrane within one of the leaflets?

OTTENSMEYER: No, I haven't. Basically we thought initially of making artificial membranes and not using osmium fixation. But we never were able to get the bilayer through the process of dehydration and embedding, and still have something reproducible there. Osmium fixation, of course, allows you to go through that process. We were at the time worried about an osmium loss from the O_1 shell at 84 eV giving us a signal which could be interpreted as phosphorus. However, this osmium signal should be there in the reference as well as in the phosphorus peak image, and so it should subtract out to some degree.

SOMLYO: It is not intuitively clear to me that with a very strong Z-dependence of the background, given that osmium is 76, you have no apparent difficulty with the very high Z material in those regions.

OTTENSMEYER: We have looked at myelin sheaths around peripheral nerves and what surprised us is the lack of contribution to contrast from osmium in the reference image at 105 eV, the image that did not contain the phosphorus contribution. At 105 eV, you could just barely make out that the specimens were myelin sheaths, whereas as soon as you went above the phosphorus loss to 150 eV, the myelin pattern was very strong.

SOMLYO: How was the myelin prepared? You might try glutaraldehyde--urea embedding which really gets around this whole business of extracting the phospholipids. It is a little messy but can be done. Then what is left is the phosphorus. In terms of comparison with what is available in x-ray maps, I think the best that has so far been obtained is a result that Henry got from a stained catalase map showing 80 Angstrom. I was trying to analyze what Nigel sent to me; Nigel sent me some ribosomes which I tried to analyze. One could say that a globe-like dimer at 250 Angstrom could be resolved from a monomer next to it at about 120 Angstrom in the phosphorus x-ray maps. So from that point of view, the results seem still comparable.

LECHENE: With what efficiency in the x-ray map?

SOMLYO: The x-ray map is obtained by serial collection; therefore it would be less efficient. But the question is now, when one is dealing with DNA coils and such, what is the limiting radiation damage, with the dose that one has to deliver?

OTTENSMEYER: We have information on that on the nucleosome. We looked at the signal-to-noise ratio in net phosphorus image of the nucleosome over 30 nucleosomes and obtained an average signal-to-noise ratio of about 30. If you take two images one after the other with equivalent doses and do the subtraction on those to get an idea of the signal-to-noise, it is around 0 plus or minus 1. So the phosphorus does contribute very heavily to the signal. This signal-to-noise ratio of 30 in the nucleosome is derived from 140 base pairs or 280 atoms of phosphorus. If one assumes that one can still see the signal at a signal-to-noise ratio of 5, then the limit of detection is around 50 atoms of phosphorus. This is the same limit we obtained in the membrane analysis. So from two different points of view, this appears to be the limit, even though we are giving two doses so heavy that radiation-induced alterations are obviously there.

SOMLYO: I didn't mean that I was worried about the detection itself, or the loss of phosphorus, but rather how much spatial information is lost. I admit when I look at those images, I figure I am no longer looking at DNA, but rather phosphoric acid that was baked. The question is about spatial information really.

LECHENE: I would like to know, practically now, when one wants to do imaging by the scanning mode, what would be the highest resolutions and the total area on the average that can be practically analyzed, compared to parallel collection?

OTTENSMEYER: Well, it depends on the resolution that you offer. Supposing you are now talking x-rays versus . . .

LECHENE: No. I am talking EELS with parallel collection.

OTTENSMEYER: I can give you an answer for that. If you scan, say with 1000 by 1000 pixel points, that gives you 10^6 picture elements for whatever resolutions you want to choose. We can choose any magnification, but for example at 50,000 times in the TEM, we still have 3-5 Angstrom resolution, corresponding to 10^8 pixel points on the plate. So the 512 by 512 virus image that you saw is equivalent to a single field that you could scan and analyze at one time. On a single plate, we have 100 times as much in terms of area analyzed all at once. That is one difference. I don't know whether that is the answer that you are after.

FERRIER: I think it is very difficult to answer the question actually. You can get down to your 10

Angstrom areas by scanning if your specimen stays still long enough during the time you want to record it.

SOMLYO: Perhaps in some modes the limit is not at high magnification, but rather at lower magnification in terms of the spectrometer acceptance because of the large angles involved. Maybe either Dale (Johnson) or Mike (Isaacson) wants to comment on whether at a low magnification the energy resolution disintegrated, or depending on what energy resolution you are mapping, you can no longer pick up the edges. If I understand Claude Lechene's question, he was asking about the low magnifications where he was worried about energy resolution.

JOHNSON: You can unscan to any degree that you want to and go right back to good energy resolution for a big scanned area, even in the image mode. You put the same coils below the specimen that you have above and take the scan right back out again. The scan doesn't have to be a fundamental limitation to the size and the area.

SOMLYO: In the TEM mode, can you do the same thing?

OTTENSMEYER: The lowest magnification at which we comfortably work is 2000x. If we go below that, the lens configurations in the Siemens 102 change and we have to realign. So we haven't checked out how low one could go, though I see no fundamental problem yet.

FERRIER: But if one can arrange to do some background subtraction in some way, the fact that the edge isn't all that well defined doesn't necessarily limit one in terms of plotting maps and I think that's the real way one might be going.

LECHENE: From what Peter Ottensmeyer said now, the resolution you get is determined by his electron gun whose lower energy spread is something like 0.8 eV, an energy which is required when you are reading dyes. How would this compare practically with scanning, with all the high difficulties there are, not to lose the resolution in the spectrometer?

FERRIER: If you have a field emission gun, you can get down to 1/4 of an eV approximately.

LECHENE: But can you collect it?

FERRIER: Yes, you can certainly. If you want to optimize the signal, however, you may want to set it so that as far as your spectrum is concerned your are working at 5 eV resolution. That says nothing about your gun; actually, that says what you are getting out of your spectrometer.

SOMLYO: I think Peter Ottensmeyer also said that to get 0.8 eV, he dropped down the emission current.

FERRIER: I am surprised he got as low as that.

SOMLYO: What it comes down to then is the signal that one can get from the specimen you obtain.

OTTENSMEYER: The 0.8 eV was a demonstration of the present energy resolution. Moreover, it is not even limited to that value if one gets the higher brightness and lower energy spread of a field emission gun.

QUANTITATIVE ELEMENTAL ANALYSIS USING ELECTRON ENERGY-LOSS SPECTROSCOPY

D. Maher
D. Joy

Bell Laboratories
Murray Hill, New Jersey

INTRODUCTION

The localization and quantification of light to medium atomic number elements by the electron energy-loss technique are important to the field of analytical electron microscopy. With this technique elemental analysis can be achieved by identifying and quantifying the inner-shell ionization edges present in an energy-loss spectrum. When the technique is implemented in a scanning transmission electron microscope fitted with a suitable electron spectrometer, the expectation is that the focussed electron probe (diameter 1 to 50 nm) can be positioned or scanned locally relative to image features and a high spatial-resolution analysis of the specimen obtained. At present there are several aspects of quantitative elemental analysis which require careful consideration before the technique can be applied in a general materials sense. In this paper the basic features of electron energy-loss spectra are discussed and then an approach to quantitative analysis based on calculated cross sections is outlined. Certain limitations of quantification are considered and last some estimates of the detection sensitivity of the technique for the case of a carbon matrix are reviewed.

ISBN 0-12-362880-6

SPECTRAL ACQUISITION

Electron energy-loss experiments are done by allowing some fraction of the electron beam to enter a magnetic or electrostatic spectrometer which then deflects these electrons by an amount proportional to the energy transferred to the specimen. For a magnetic prism spectrometer the result is a linear dispersion of electrons (for details see Isaacson, 1981). In general a slit assembly is placed perpendicular to the focal plane of the linear dispersion and this is followed by a scintillator, light pipe, photomultiplier tube and counting electronics (Figure 1). Energy-loss spectra are then obtained by scanning the linear dispersion across the slit and the resulting intensities, in the form of pulses, are stored sequentially in a multi-channel analyzer or computer. The fraction of scattered electrons which the spectrometer accepts is determined by an aperture placed between the specimen and spectrometer and possibly by post-specimen lens optics (Crewe, 1977; Joy and Maher, 1978; Egerton, 1978a; and Johnson, 1980). For elemental analysis a spectrometer resolution δ in the range 5 to 10 eV is required and typically a resolution in this range will be controlled by the slit width W and linear dispersion D (in μm per eV) of the spectrometer. For example, if D is 4 μm per eV and W is set to 20 μm then a slit limiting resolution of 5 eV would be achieved (for details see Joy, 1979, and Isaacson, 1981).

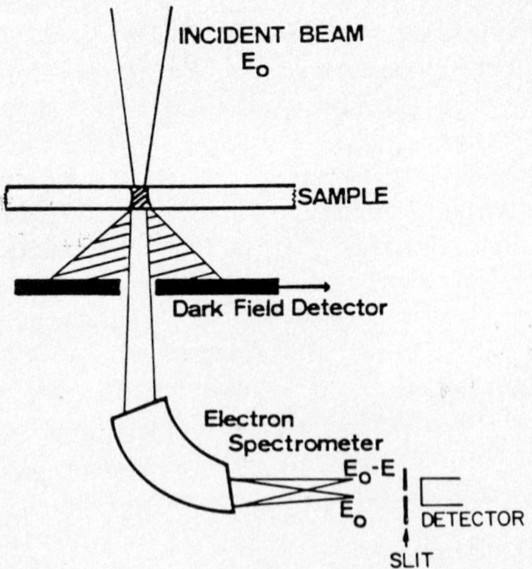

FIGURE 1. Schematic diagram of the basic components of an analytical scanning transmission microscope fitted with a magnetic prism spectrometer.

SPECTRAL FEATURES

In these experiments the energy-loss spectrum is recorded in the forward direction (i.e. $\theta \to 0$, where is measured with respect to the incident-electron beam direction) and within a semi-angle β subtended at the specimen (β is typically 1 to 15 mrad for 100 kV incident electrons). As shown in Figure 2, the spectrum is a plot of electron intensity versus energy lost, i.e., I_β (E) vs E, where I is counts per channel and E is the energy loss relative to the incident-beam kinetic energy E_0. Again with reference to Figure 2, spectra from thin specimens (typically < 100 nm thickness) exhibit a sharp peak at E = 0 (i.e., the zero-loss peak) followed by one or more broader peaks in the loss range up to E $\sim$50 eV. These broader peaks or so-called low-loss peaks are predominately due to the interaction of transmitted electrons with valence or conduction electrons, resulting in plasmon and/or single electron excitations. At higher energy losses, the excitations of inner-shell electrons are reflected in the spectrum by a rise in intensity at some

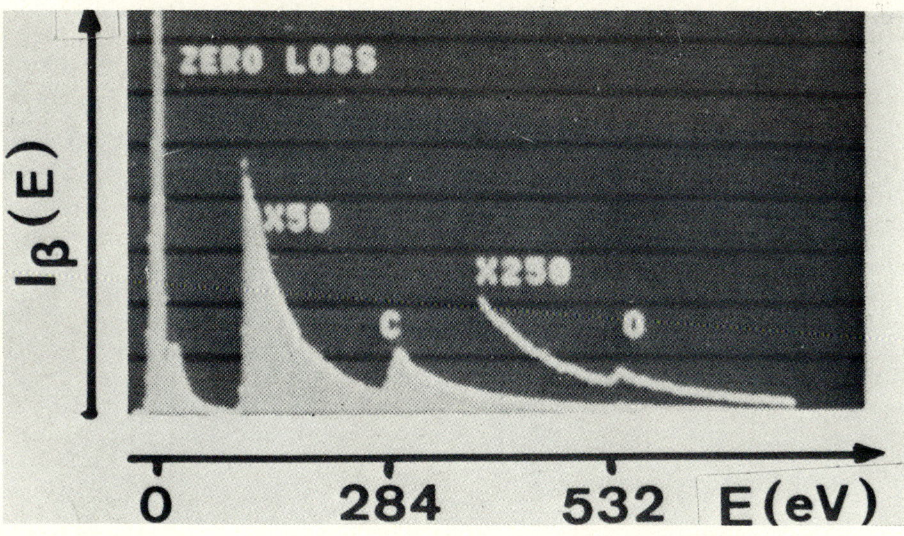

FIGURE 2. Experimental energy-loss spectrum from an amorphous carbon specimen ($\sim$ 20 nm thick): the spectrometer resolution was 5 eV; the acquire conditions were 0.2 sec per ch at 1 eV per ch; and a total gain change of X12.5 K relative to the low-loss condition was required to maintain approximately the same counting statistics for the K edges from carbon (284 eV) and oxygen (532 eV).

energy loss $E = E_k$ where E_k is the threshold energy for an edge. The threshold energy of an edge is, within a few eV, equal to the binding energy of an atomic shell k (k representing the type of shell: K, L, M, etc.). Since each element has unique inner-shell binding energies, it is possible to identify the presence of particular elements in a matrix of several elements from measurements of these threshold energies.

BINDING ENERGIES

In this paper our concern is with standardless quantification of light and medium atomic number elements and this implies using the K-shell edges for elements of atomic number Z = 3 through 15 and/or L-shell edges for Z = 11 through 30. The binding energies E_B for these edges are given in Tables 1 and 2.[1] As can be seen from the

TABLE 1. K-Shell Binding Energies for Element $3 \leq Z \leq 15$

Z	ELEMENT	E_K (eV)[a]
3	Li	54
4	Be	112
5	B	188
6	C	284
7	N	402
8	O	532
9	F	685
10	Ne	870
11	Na	1072
12	Mg	1305
13	Al	1560
14	Si	1839
15	P	2146

[a]From Larkins (1977).

[1]For a complete tabulation of inner-shell binding energies, see Larkins (1977).

TABLE 2. L_3- and L_2- Shell Binding Energies for Elements $11 \leq Z \leq 30$.

Z	ELEMENT	E_{L_3} (eV) [a]	E_{L_2} (eV) [a]
11	Na	31	31
12	Mg	51	51
13	Al	73	73
14	Si	99	100
15	P	135	136
16	S	164	165
17	Cl	200	202
18	Ar	249	250
19	K	294	296
20	Ca	346	350
21	Sc	402	407
22	Ti	456	462
23	V	513	521
24	Cr	575	584
25	Mn	640	651
26	Fe	708	721
27	Co	779	794
28	Ni	855	872
29	Cu	931	951
30	Zn	1020	1043

[a]From Larkins (1977).

tabulations elements from lithium to zinc can be investigated by an energy-loss experiment using a spectrometer that records losses out to $\sim$ 2500 eV. In principle, higher atomic number elements can be identified using M- and N-shell edges but, at present, routine quantification of these edges is only possible through the use of standards.

SPECTRAL DYNAMIC RANGE

The probabilities for inner-shell excitations are small compared to those associated with features in the low-loss region of spectrum, and therefore, by comparison, the net intensity of edges is also lower, possibly by several orders of magnitude. This is illustrated in Figure 2 which is an energy-loss spectrum from a thin amorphous-carbon film. The spectrum is divided into three regions. In the region below 50 eV, the zero-loss peak dominates and a single low-loss peak (at 24 eV) is present. A gain change[2], ΔG of ~ 50 X was required to observe the carbon K edge (at 284 eV) and a second gain change of ~ 250 X was imposed in order to observe the K edge from oxygen (at 532 eV). In the case of a high energy K edge, e.g., the phosphorous K edge at 2146 eV, a dynamical range of 10^5 to 1 relative to the low-loss region may be required to detect an edge above background. At present limitations set by the recording chain and memory capacity required for data collection and processing generally make it impossible to record and computer process an entire spectrum (say 0 to 2000 eV) at an optimum detector gain. Therefore a spectrum is recorded with a variable gain and, as is required for quantification, regions of interest are normalized to the low-loss gain condition using calibrated gain factors.

BACKGROUND MODELING

From the spectrum shown in Figure 2, it can be seen that edges ride on top of a decreasing background and, unlike x-ray peaks, edges are of indefinite extent (i.e., the edge intensity and background intensity converge at $E \rightarrow E_o$). Therefore, to obtain the true edge shape and the net edge intensity, the background prior to an edge first must be modeled and then extrapolated under the edge of interest. The range of energy losses over which the background can be extrapolated clearly must be finite (typically 100 to 200 eV) and this range is referred to as the energy window Δ. As proposed by Egerton and Whelan (1974), the background prior to an edge can be modeled by a power law, $A \cdot E^{-r}$, where A and r are constants whose values depend on the mass thickness of the specimen, β and E_k (e.g., Maher et al. 1979). Typical fitted values of r vary from 2 to 5. In practice background modeling and stripping is achieved by using an interactive computer program to: (1) fit a power law to the background over a moderate energy-loss range (~ 100 eV)

[2] In this experiment, the spectrum was recorded by digitizing the analog output from the photomultiplier tube. The gain of the PMT was varied by changing the high voltage applied to it.

prior to an edge at E_k using a least squares routine; (2) extrapolate the fitted model to $E_k + \Delta$; and (3) strip the extrapolated background intensity from the total intensity. These steps are illustrated in Figures 3a-c for a K edge from an amorphous carbon specimen. Extensive experiments by the authors have shown that an $A \cdot E^{-r}$ fit is almost always within the statistical channel-to-channel noise of the background prior to edges whose threshold energies extend from 100 eV to 1840 eV when the specimen is sufficiently thin. The accuracy of the background extrapolation is still a point of conjecture (e.g., Colliex et al., 1981). In this regard it is worthwhile pointing out that after the background has been removed, the limiting edge profile (i.e., for $E > E_K + 50$ eV) should follow an inverse power law, $B \cdot E^{-s}$, where B and s are again constants (see Figure 3d). Experimental values of s have been obtained for the K edge from amorphous carbon as a function of β (Maher et al., 1979) and these exponents compared to those derived from theory (Egerton, 1979). The two results agreed to within $\pm$ 5% which certainly lends support to the background extrapolation procedure outlined in this section.

EDGE SHAPES

The shape of inner-shell edges is important from the standpoint of both edge identification and quantitative analysis (Maher, 1979). Theoretical shapes based on an atomic model of the inner-shell excitation process for K, L, and M edges recently have been published by Leapman, Rez, and Mayers (1980). The qualitative features predicted by these calculations have been confirmed experimentally by a number of investigators. In terms of the present discussion, two shape classifications can be made: (1) edges which exhibit a sharp rise at threshold and have a characteristic "saw tooth" profile (see Figure 4a); and (2) edges which exhibit a delayed maximum with respect to threshold and have a "sleeping whale" profile (see Figure 4b). Shape classifications for K and L edges from light and medium atomic number elements are summarized in Table 3 where it can be seen that K edges for Z = 3 to 15 are of the saw tooth type whilst L edges for Z = 11 to 30 are a mixture of the two types. It must be kept in mind that in practice the detailed shapes of edges depend on the spectrometer resolution and solid state effects (e.g., Ray, 1979; Leapman and Grunes, 1980). The absence of a sharp rise at threshold and delayed maximum often makes it difficult to identify the second class of

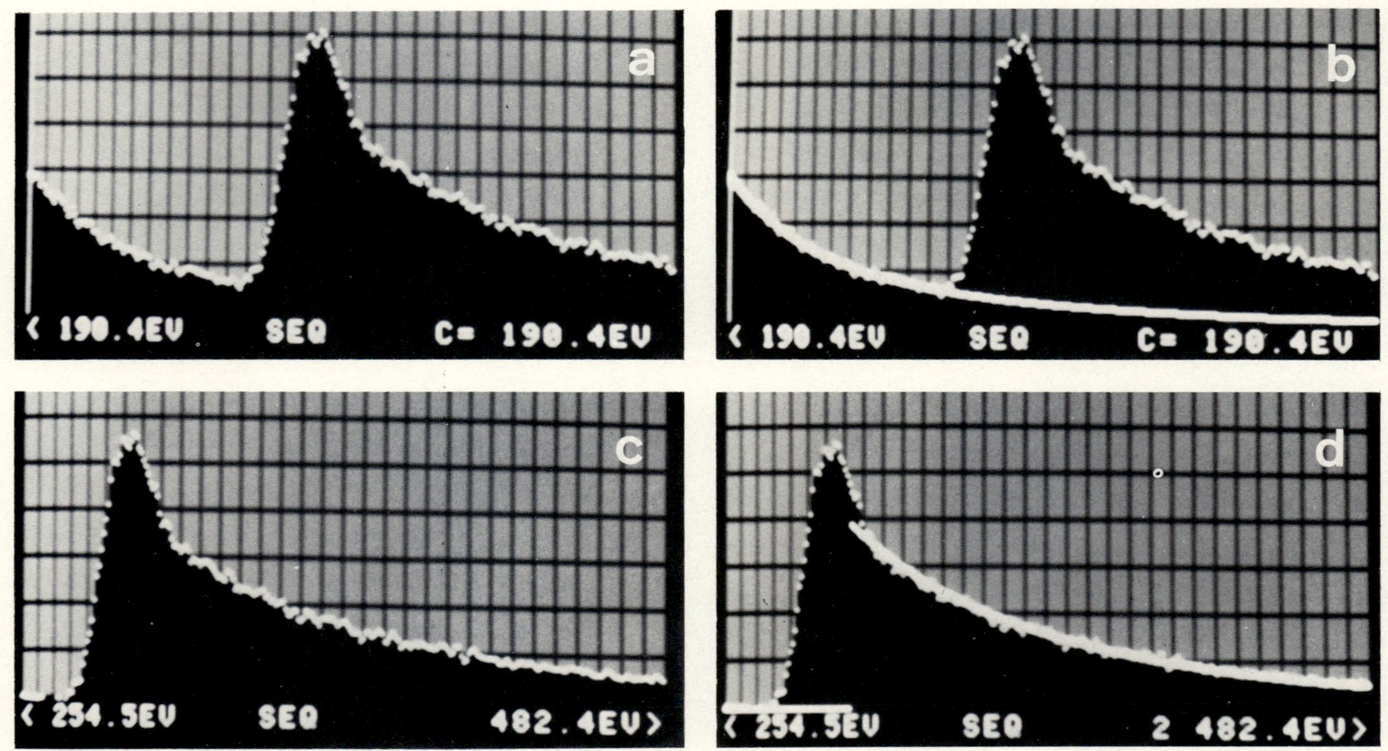

FIGURE 3. Illustration of the basic steps required to characterize an inner-shell edge: (a) raw spectral data of a K edge (E_K = 284 eV) from an amorphous carbon specimen ∿20 nm thick; (b) exponential background fit and extrapolation (solid curve); (c) edge profile after stripping off the extrapolated background; and (d) exponential fit to the limiting portion (i.e., $E > E_K + 50$ eV) of the stripped edge (solid curve).

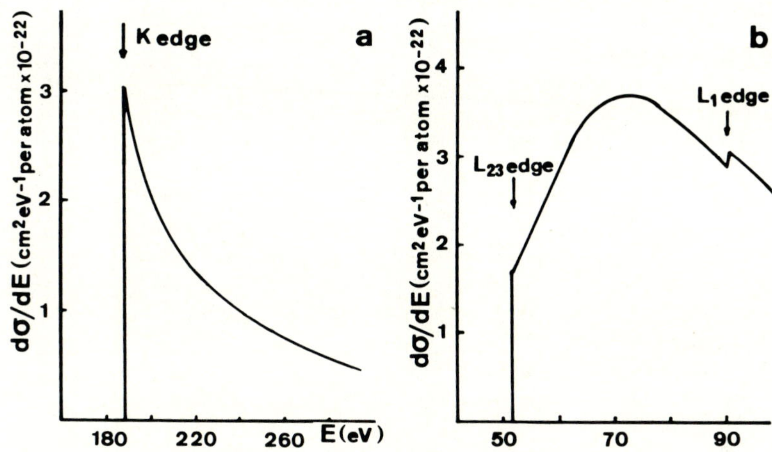

FIGURE 4. Theoretical edge profiles: (a) for the boron K edge (E_K = 188 eV) and (b) for the magnesium L edges ($E_{L_{23}}$ = 51 eV and E_{L_1} = 89 eV). In each case E_0 = 80 keV and β = 10 mrad (from Leapman, 1979).

TABLE 3. Shape Classifications of K and L EDGES for Elements $3 \leq Z \leq 30$[a].

CLASSIFICATION	EDGE TYPE	ELEMENTS
SAW TOOTH	K	Li(Z = 3) to P(Z = 15)
SAW TOOTH	L	K(Z = 19) to Fe(Z = 26)
SLEEPING WHALE	L	Mg(Z = 12) to Cl(Z = 17)
SLEEPING WHALE	L	Co(Z = 27) to Zn(Z = 30)

Note: Na(Z = 11) L edge is most probably SLEEPING WHALE and Ar(Z = 18) L edge is most probably SAW TOOTH

[a]From Leapman, Rez, and Mayers (1980).

edges, especially when they are riding on an appreciable background. The importance of these two shape classifications in quantification will become clear in the following section.

QUANTITATIVE ANALYSES

Quantitative analysis of inner-shell edges yields either the absolute amount N (in atoms - cm^{-2}) of an element present in a sufficiently thin specimen, or the relative amount of a particular element N^1 in a matrix of several elements N^j (j = 1, 2 etc.). The basic equation for deriving this information from an experimental spectrum is (Egerton, Rossouw, and Whelan, 1976; and Egerton, 1978b)

$$N = \frac{I_{k(\beta,\Delta)}}{I_{\ell(\beta,\Delta)}} \cdot \frac{GF}{\sigma_{k(\beta,\Delta)}} \quad (1)$$

where with reference to Figure 5 $I_k(\beta,\Delta)$ is the net integrated intensity under the kth edge measured for a semi-scattering angle β and over an energy loss Δ; $I_\ell(\beta,\Delta)$ is the integrated intensity under the low-loss region of the spectrum for the same scattering angle β and from the zero-loss peak up to Δ ; GF is a gain factor ($GF = \Delta G^{-1}$) relating I_k to I_ℓ; and $\sigma_k(\beta,\Delta)$ is a partial cross section (in cm^2 - $atom^{-1}$) for excitations from the kth inner shell to continuum states covering a range of scattering angles up to β and a range of energies from E_k to $E_k + \Delta$.

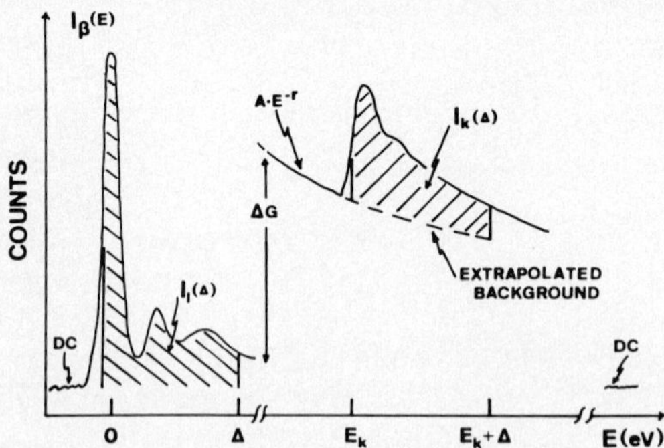

FIGURE 5. Schematic electron energy-loss spectrum of the signal intensity $I_\beta(E)$ versus energy loss E: DC represents the measured dark current and all other quantities are defined in the text.

Most often in quantitative analysis, the relative amount of a particular element is required. From equation (1) it follows that for a binary system the atomic ratio of element 1 to element 2 is given by

$$\frac{N^1}{N^2} = \frac{I_k^1(\beta,\Delta)}{I_k^2(\beta,\Delta)} \cdot \frac{\sigma_k^2(\beta,\Delta)}{\sigma_k^1(\beta,\Delta)} \cdot \frac{GF^1}{GF^2} \tag{2}$$

where the gain factors are determined with respect to $I_\ell(\beta,\Delta)$ and the type of edges can be mixed.

Therefore for quantitative analysis, it is necessary to evaluate the net-edge integral(s) and possibly the low-loss integral from the spectral data; to know the gain factor(s) and acceptance angle; and to derive values of $\sigma_k(\beta,\Delta)$.

In practice it is first necessary to take account of the dark current of the detector chain and either this signal is measured (as shown in Figure 5) and stripped from the spectrum or it is removed by electronic discrimination prior to spectral acquisition. Then the stripped edge count(s), and if required, zero-loss count, must be summed over an energy window Δ using suitable algorithms (e.g., Maher, Joy, and Mochel, 1978; and Joy, 1979). Gain factors and spectrometer acceptance angles must be obtained from appropriate calibration experiments (e.g., Maher, 1979). The partial cross sections for K edges can be estimated from a semi-empirical "efficiency-factor" method (Isaacson and Johnson, 1975; Isaacson, 1980) and, in principle, $\sigma_k(\beta,\Delta)$ for any type of edge can be estimated from standards (e.g., Maher, 1979) or calculated from theory (Leapman, Rez, and Mayers, 1978, 1980; Egerton, 1979, 1981).

The simplest approach to obtaining calculated values of $\sigma_k(\beta,\Delta)$ for K and L edges is to make use of a hydrogenic wave-function approximation (Egerton, 1979).[3] These calculations can be done analytically (the so-called SIGMAK and SIGMAL programs) and therefore they are easily incorporated into a software package for quantitative elemental analysis using electron energy-loss spectroscopy (e.g., Joy and Maher, 1981). More detailed calculations which are based on Hartree-Slater wave functions (Leapman et al., 1980) support the use of a "hydrogenic" model to obtain

---

[3]Graphed values of $\sigma_k(\beta,\Delta)$ for Be, B, C, N, O, and F can be found in Egerton (1981).

$\sigma_K(\beta, \Delta)$ for the light elements (see Table 3) provided a sufficiently large energy window is used, say 100 to 200 eV. The situation regarding calculations of $\sigma_L(\beta, \Delta)$ for the medium atomic number elements (see Table 3) using a modified "hydrogenic" model (Egerton, unpublished) is still not clear. For those L edges which exhibit a sharp intensity rise at threshold, this model should be a reasonable approximation and recent experiments on homogeneous compounds (e.g. V_3Si) support this hypothesis, at least in a relative sense (Maher and Joy, unpublished). In these experiments the atomic ratios N^1/N^2 are derived from L-edge intensities using $\sigma_L(\beta, \Delta)$ calculated by the SIGMAL program and K-edge intensities using $\sigma_K(\beta, \Delta)$ calculated by the SIGMAK program. The experimental and expected atomic ratios are within 10% agreement for energy windows > 75 eV (see Figure 6 for an example). This is the same relative accuracy which was achieved previously when quantifying homogeneous compounds using SIGMAK for K-edge to K-edge analyses (Joy et al., 1979). In contrast to these results, experimental work in progress by the authors suggests that a modified "hydrogenic" model may not be a very good approximation for those medium atomic number elements which exhibit a delayed

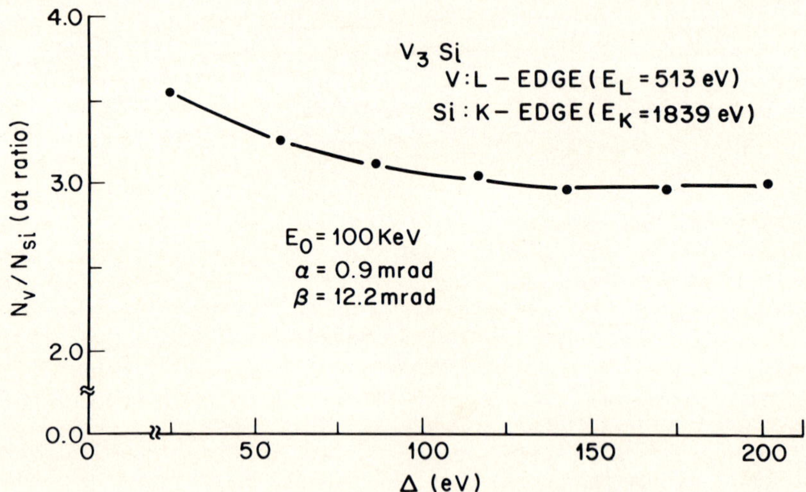

FIGURE 6. L-edge to K-edge test analysis of an amorphous V_3Si specimen ($\sim$40 nm thick) showing the variation of N_V/N_{Si} with Δ for spectral data collected in the CTEM mode at a spectrometer resolution of 8 eV.

maximum. This situation must be explored more systematically before any definitive conclusions can be reached. Clearly until the results of the more lengthy cross-section calculations based on Hartree-Slater wave functions are parameterized in a form which is suitable for typical electron energy-loss experimental conditions, the analytical calculations of $\sigma_L(\beta,\Delta)$ based on modified hydrogenic wave functions are certainly adequate.

When considering the application of quantitative analysis to general materials problems, it is important to examine the fundamental limitations of equations (1) and (2). One way to approach this is through the assumptions made in obtaining the partial cross sections. In this case there are three constraints which must be considered, namely: (a) the incident electron beam is assumed to be a plane wave (i.e., collimated illumination); (b) the inner-shell excitation process is atomic (i.e., solid-state, fine-structure makes a small contribution to the integrated edge intensity); and (c) incident electrons experience only a single elastic or inelastic scattering event in propagating through the specimen (i.e., the specimen is very thin). Obviously these are stringent constraints and in practice they are approximated by (1) setting the convergence semi-angle α of the incident beam so that $\alpha \lesssim 0.5\beta$ or selecting β so that $\beta \gtrsim 2\alpha$ (e.g., Isaacson, 1978); (2) choosing Δ sufficiently large, say 100 to 200 eV; and (3) using specimens whose thickness t is less than one total mean-free path for inelastic scattering (i.e., < 80 nm for carbon when E_0 = 100 keV) and minimizing Bragg diffraction for single crystal specimens (i.e., quasi-kinematical imaging conditions).

In practice, dynamical elastic and plural inelastic scattering events always occur to some degree and therefore in this respect equations (1) and (2) are approximations. Considerable effort has been put into defining limiting experimental conditions where the appropriate integrals for quantification are effected equally by elastic scattering (Egerton, 1978b; Rossouw and Whelan, 1979a, b, and 1980; Zaluzec, Hren, and Carpenter, 1980) and by plural inelastic scattering (Egerton, 1978b; Joy et al., 1979; Maher et al., 1979; Zaluzec, 1980; and Stephens, 1980).

Problems arising from elastic scattering are most important for single crystal specimens and as stated previously, the intensity in Bragg diffracted beams should be minimized in this case. For plural inelastic scattering the situation is complicated. Primary electrons can lose energy by multiple valence/valence and inner-shell/inner-shell electron excitations, as well as, by mixed valence/inner-shell or inner-shell/valence electron excitations. These plural excitations change both the energy and angular distribution of inelastic electrons relative to the single-scattering approximation. Therefore, for a given β and Δ, the integrals used for quantification can either decrease, remain unchanged or increase depending on t and E_k. The accuracy of equations (1) and (2) decreases with increasing t and experimental limits on the acceptable thickness which can be used in order to achieve a given accuracy have been reported for only a few materials (e.g., Joy et al., 1979; Maher et al., 1979; Zaluzec, 1980). The criterion proposed is to use the ratio of the integrated (or peak) intensities from the first low-loss peak (typically at $\sim$20 eV) to the zero-loss peak as a measure of the acceptable thickness. For ratios $\lesssim 0.2$ experiments indicate that equation (1) has a probable uncertainty of $\pm$ 20% whereas the value for equation (2) is $\pm$ 10% (Joy et al., 1979).

Edges in the same spectrum can be affected quite differently by plural inelastic scattering, and a quantitative understanding of this is essential for the future progress of the technique. In our experiments the intensity of the lowest lying edge with increasing thickness is always attenuated more rapidly for a given β and Δ than those edges which occur at higher energies (e.g., Maher, 1979). The computer simulations based on transport theory (Rez, 1979) show this effect; however, other experiments (Zuluzec, 1980) and model calculations (Stephens, 1980) are not completely in agreement with our results. The obvious solution to this problem is to carry out systematic experiments wherein spectra are taken as a function of thickness and then deconvoluted in energy and angle to obtain single scattering, inner-shell spectra (e.g., Colliex et al., 1976; Colliex, Gasinger, and Trebbia, 1976; Spence, 1977; Rez, 1979; Stephens, 1980); quantitative results from the raw spectral data then would be compared to those obtained from the deconvoluted spectra in order to set limits on the contributions of plural inelastic scattering to quantification.

This point may seem academic, but it is crucial even for the simplest quantitative analysis, i.e., assessing relative changes in composition within a given specimen or series of specimens, because the low-loss intensity and net edge intensities are equally affected by plural inelastic scattering only over a restricted range of experimental conditions fixed by t, E, β, and Δ.

BEAM CONVERGENCE

In the preceding section it was indicated that the constraint of collimated illumination (i.e., $\alpha \to 0$) was satisfied experimentally by collecting spectra under the condition where $\beta \gtrsim 2\alpha$. If a spectral analysis is made in a probe-forming system at high spatial resolution (i.e., point-to-point) then a focused convergent beam is required (typically α = 5 to 10 mrad). For a finite convergence angle, the α = 0 inelastic angular distribution is convoluted with the incident-beam angular distribution. As a consequence of this convolution, the angular distribution of inelastic electrons is broadened and the effective acceptance angle of a spectrometer generally will be less than the geometrical acceptance angle β.

For collimated illumination inelastically scattered electrons of energy E have a Lorentzian angular distribution I (θ) about the incident-beam direction (θ= 0) given by

$$\frac{I(\theta)}{I(\theta=0)} = \frac{1}{\theta^2 + \theta_E^2} \qquad (3)$$

where $\theta_E = E/2E_0$ (e.g., Joy, 1979). Moreover, the fraction η_β of the total inelastically scattered electrons having an energy E collected by a spectrometer accepting scattering events within a semi-angle β is (e.g., Isaacson and Johnson, 1975)

$$\eta_\beta = \frac{\ln(1 + \beta^2/\theta_E^2)}{\ln(2/\theta_E)} \qquad (4)$$

If the incident beam has a finite semi-angle of convergence α_{max}, then equation (3) is modified to

$$\frac{I(\theta)}{I(\theta=0)} = \int_0^{\alpha_{max}} \frac{I(\alpha)*\alpha d\alpha}{(\theta-\alpha)^2 + \theta_E^2} \qquad (5)$$

where I(α) is the normalized angular distribution of the incident-beam intensity (e.g., Isaacson, 1978) and * represents a two-dimensional convolution. Equation (5) can be solved analytically (Egerton, 1978b) and results for the carbon K edge and α_{max} = 1, 2, 3, and 10 mrad are shown in Figure 7 where it can be seen that a finite convergence angle broadens the angular distribution relative to the $\alpha = 0$ case calculated from equation (3).

Furthermore, for the case of a finite convergence angle, the fraction $\eta_{\alpha,\beta}$ of the total inelastically scattered electrons having an energy E collected by a spectrometer which accepts scattering events within a semi-angle β is given by

$$\eta_{\alpha,\beta} = \frac{\int_0^{\beta} I(\theta)\theta d\theta}{\int_0^{\theta_{max}} I(\theta)\theta d\theta} \qquad (6)$$

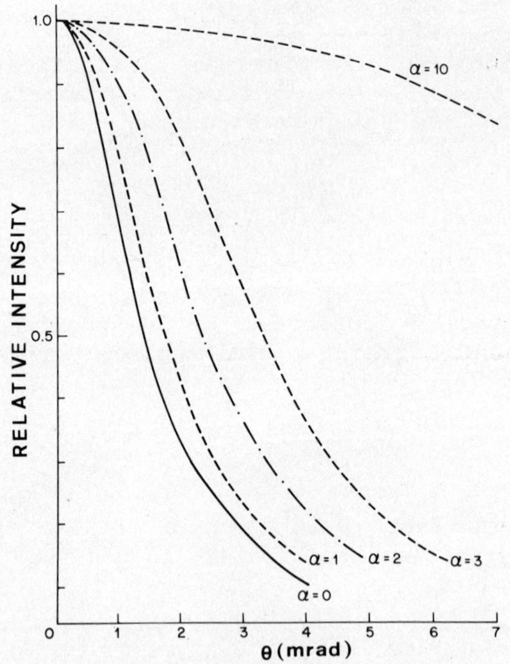

FIGURE 7. Relative angular distributions of intensities $I(\theta)/I(\theta = 0)$ for the carbon K edge (E_K = 284 eV and θ_E = 1.4 mrad) and assuming α = 0, 1, 2, 3, and 10 mrad.

where $I(\theta)$ is the normalized angular distribution obtained from equation (5) and θ_{max} is set equal to $\sqrt{2\theta_E}$. Equation (6) can be evaluated numerically assuming, as was done in evaluating equation (5), that the incident-beam intensity is constant for all $\alpha < \alpha_{max}$ and is zero for $\alpha > \alpha_{max}$ (i.e., there is no beam tailing). $\eta_{\alpha,\beta}$ is plotted as a function of α for the carbon K edge and three values of β in Figure 8. For $\beta > 2\alpha$ the angular collection efficiency for a finite beam convergence is approximately equal to the $\alpha = 0$ case whilst for $\beta \lesssim \alpha$ the two cases deviate markedly, thus demonstrating the point made previously that for this regime the effective acceptance angle of the spectrometer is smaller than the geometrical acceptance angle β.

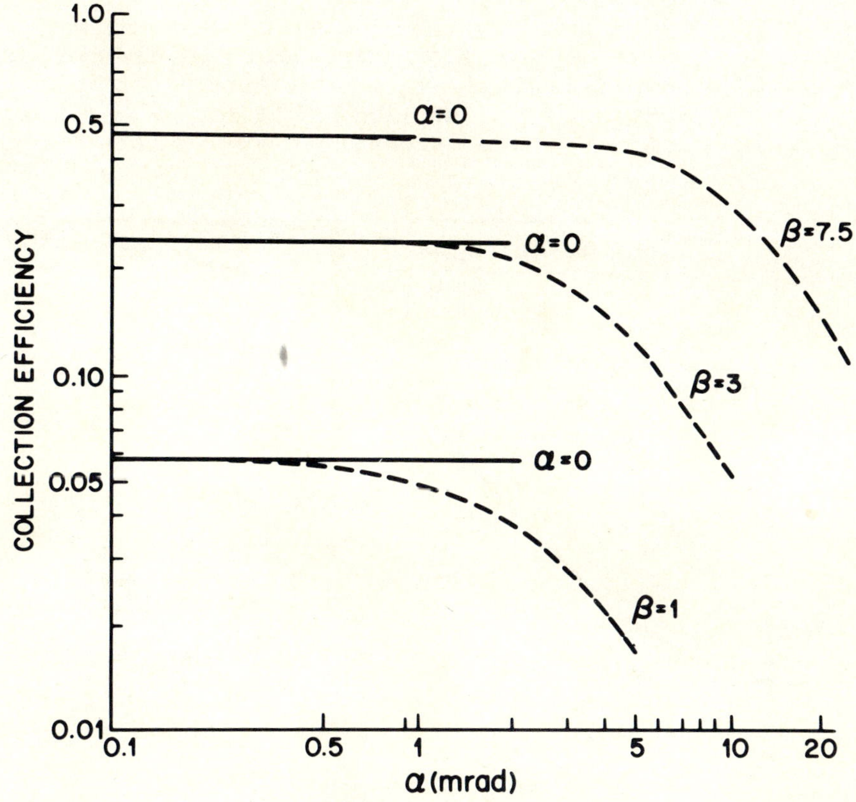

FIGURE 8. Angular collection efficiency $\eta_{\alpha,\beta}(\theta_E)$ versus α for the carbon K edge (E_K = 284 eV and θ_E = 1.4 mrad) assuming β = 1, 3, and 7.5 mrad.

For quantitative analysis, e.g., using equation (2), the appropriate cross sections are now $\sigma_k(\beta,\Delta)\cdot\eta(\alpha\ \beta\ \theta_E)$ where

$$\eta(\alpha,\beta,\theta_E) = \frac{\eta_{\alpha,\beta}(\theta_E)}{\eta_\beta\ (\theta_E)} \qquad (7)$$

Values of $\eta_{\alpha,\beta}/\eta_\beta$ are plotted as a function of β for the carbon K edge and three values of α in Figure 9. It can be seen from these results that if α is in the range 5 to 10 mrad and $\beta \leq \alpha$ then the cross-section $\sigma_k(\beta,\Delta)$ must essentially be corrected by anywhere from 10 to $\sim$ 80%. Such a correction is appreciable when compared to the probable uncertainty ($\pm$ 10%) expected when estimating relative atomic ratios under conditions where $\beta \gtrsim 2\alpha$.

Experiments designed to test this correction procedure are in progress (Joy, Maher, Farrow, Colliex, and Trebbia) and preliminary results for the carbon K edge are shown in Figure 10. It can be seen from these data that the corrected net edge intensities (i.e., $I_K(\alpha)/\sigma_K^*$ where $\sigma_K^* = \sigma_K(\beta,\Delta)\cdot\eta(\alpha,\beta,\theta_E))$ plotted as a function of α for β = 3.2 mrad and Δ = 100 eV are constant to within better

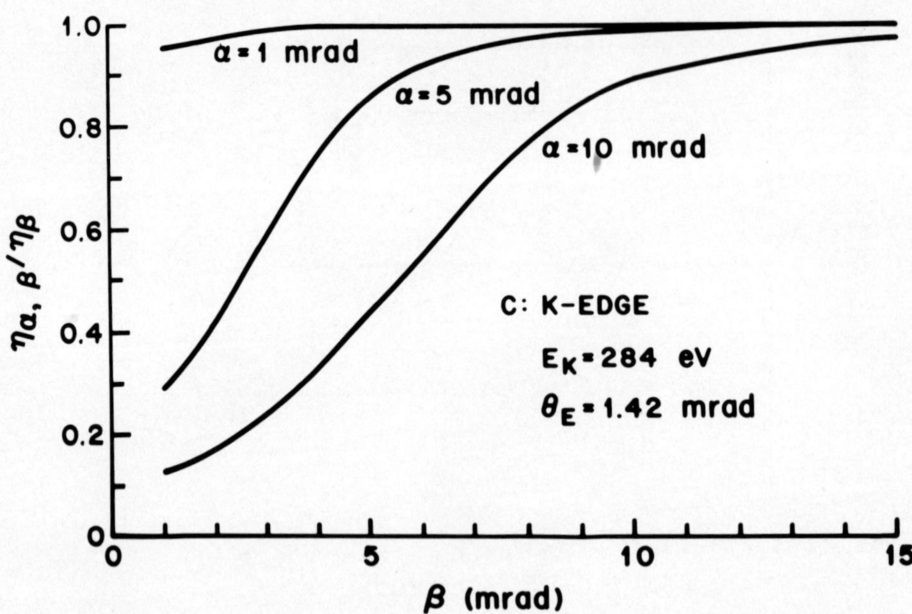

FIGURE 9. Relative angular collection efficiency $\eta_{\alpha,\beta}/\eta_\beta$ for the carbon K edge versus β assuming α = 1, 5, and 10 mrad.

than $\pm$ 10% for values of α in the range 0.9 to 14 mrad. The ability to achieve corrected values of I_k/σ_k which are independent of α and β is essential to the general application of the ratio method, especially when high spatial resolution is required. This result is encouraging and obviously further work is required in order to characterize the α and β dependence of other regions of the spectrum, including the low-loss region. It should be emphasized that because of the need to invoke a correction term which is a function of α, β, and θ_E, it will be necessary to maintain constant electron optical conditions when the energy-loss

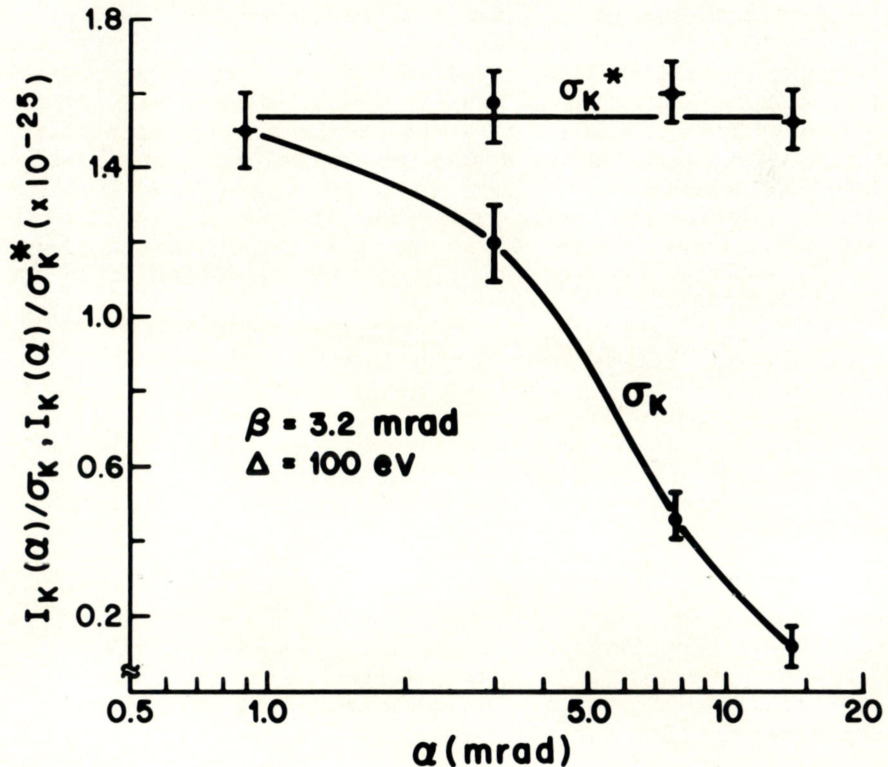

FIGURE 10. Experimental net K-edge intensities divided by: the hydrogenic cross section assuming $\alpha = 0$ (i.e., σ_K); and the corrected hydrogenic cross section taking α = 0.9, 3.2, 7.8, and 14 mrad (i.e. σ_K^*). The spectral data are from an amorphous carbon film ($\sim$20 nm thick) and they were recorded in the STEM imaging mode at 100K magnification using a 10 nm dia probe with E_o = 100 keV and a spectrometer resolution of 5 eV.

technique is being used even in its simplest form, i.e., estimating relative changes in N or N^i/N^j within a given specimen or series of specimens.

DETECTABLE LIMITS

A knowledge of the detection sensitivity of any analytical technique is important because it indicates what kind of experiments are feasible and because it sets an ultimate limit to the precision of the quantification schemes. In the case of electron energy-loss spectroscopy the detection sensitivity for a given element in the presence of one or more other elements is a function of many experimental parameters (e.g., the accelerating voltage, spectrometer acceptance angle, dwell time per channel, incident beam flux, etc.), as well as specimen-related quantities (e.g., atomic number of the element of interest, mass-thickness of the specimen, etc.). With so many variables a general expression for the sensitivity is not very useful, so instead it is usual to model the conditions of interest and then perform a very specific calculation (Joy and Maher, 1980).

As in any other counting experiment, the detection limit will be set by the ratio of the wanted signal (i.e., the characteristic inner-shell edge) to the noise (i.e., the background on which the edge is riding). Since the origin and magnitude of the background vary with energy loss, it is clear that the detection limit will also vary depending on the characterisitic loss being examined.

Consider the case of a thin biological section, which we will approximate as pure amorphous carbon of thickness 80 nm. Below the carbon K-edge the background arises from "plasmon" losses, non-characteristic, single electron excitations, etc., while above the K-edge the background is predominantly the "tail" of the edge.

In passing across the edge the effective background will increase by a factor of 5 to 10 times. The intensity of the background falls, on average, as E^{-4}, on the other hand the inner-shell cross-section of edges of elements in the K-series (i.e., Li at 55 eV to F at 680 eV) or the L_{23}-series (i.e., Al at 55 eV to Fe at 700 eV) will also fall, the rate depending on β and Δ . Taking a typical set of operating conditions appropriate for a TEM/STEM instrument

(E_0 = 100 keV, incident current density = 10 amp/cm^2 which represents 10^{-10} amps into a 30 nm probe, a counting time of 0.2 sec/channel, a spectrometer resolution of 10 eV and β = 3 mrad) the calculations show that about 10^3 atoms of lithium would be detectable (i.e., about 0.03 at %). For chlorine which has its L_{23}-edge just below the carbon K-edge, the detection limit rises to about 2000 atoms (0.06 at %), while for calcium which has its L_{23}-edge just above the carbon K-edge, the detection limit is 6000 atom (0.18 at %). For fluorine and iron which have edges at about 700 eV the corresponding figure is around 0.5 at %.

These detection limits are encouraging, since a volume of only $\sim 10^{-17}$ cm^3 is being sampled, and they show that the technique is sensitive for most analyses above the trace-elemental level. It should be pointed out, however, that when radiation damage is an important consideration these calculations are rather optimistic. The dose used in the above estimates, i.e., 10^5 electrons/nm^2, is substantially higher than the 10^2 electrons/nm^2 usually quoted as being necessary to maintain sample integrity.

REFERENCES

Colliex, C., Cosslett, V. E., Leapman, R. D., and Trebbia, P. (1976). Ultramicroscopy. 1:301.

Colliex, C., Gasgnier, M., and Trebbia, P. (1976). J. Physique. 37:397.

Colliex, C., Jeanguillaume, C., and Trebbia, P. In this volume.

Crewe, A. V. (1977). Optic. 47:299.

Egerton, R. F. (1978a). Ultramicroscopy. 3:39.

Egerton, R. F. (1978b). Ultramicroscopy. 3:243.

Egerton, R. F. (1979). Ultramicroscopy. 4:169.

Egerton, R. F. J. Microscopy. In press.

Egerton, R. F., and Whelan, M. J. (1974). Proc. 8th Int. Cong. on Electron Microscopy. Australian Academy of Science, Canberra. 1:384.

Egerton, R. F., Rossouw, C. J., and Whelan, M. J. (1976). In "Developments in Electron Microscopy and Analysis" (J. Venables, ed.), p. 129. Academic Press, New York.

Isaacson, M. (1978). Proc. 11th Ann. SEM Symp. 1:763.

Isaacson, M. (1980). Proc. 38th Ann. EMSA Meeting. p. 110. Claitor's Publishing, Baton Rouge.

Isaacson, M. (1981). In this Volume.
Isaacson, M., and Johnson, D. (1975). Ultramicroscopy 1:33.
Johnson, D. (1980). Ultramicroscopy. 5:163.
Joy, D. C. (1979). In "Introduction to Analytical Electron Microscopy" (J. J. Hren, J. I. Goldstein, and D. C. Joy, eds.), p. 223. Plenum Press, New York.
Joy, D. C., and Maher, D. M. (1978). J. Microscopy. 114:117.
Joy, D. C., and Maher, D. M. (1980). Ultramicroscopy. 5:333.
Joy, D. C., and Maher, D. M. J. Microscopy. To be published.
Joy, D. C., Egerton, R. F., and Maher, D. M. (1979). Proc. 12th Ann. SEM Symp. 2:817. SEM, Inc., Chicago.
Larkins, F. P. (1977). Atomic Data and Nuclear Data Tables 20:312.
Leapman, R. D. (1979). Ultramicroscopy. 3:413.
Leapman, R. D., and Grunes, L. A. (1980). Phys. Rev. Lett. 45:397.
Leapman, R. D., Rez, P., and Mayers, D. (1978). Proc. 9th Int. Cong. on Electron Microscopy. 1:526. Imperial Press, Toronto.
Leapman, R. D., Rez, P., and Mayers, D. (1980). J. Chem Phys. 72:1232.
Maher, D. M. (1979). In "Introduction to Analytical Electron Microscopy" (J. J. Hren, J. I. Goldstein, and D. C. Joy, eds.), p. 259. Plenum Press, New York.
Maher, D. M., Mochel, P., and Joy, D. C. (1978). Proc. 13th Ann. Conf. Microbeam Analysis Society. 53A. Kyser, Ann Arbor.
Maher, D. M., Joy, D. C., Egerton, R. F., and Mochel, P. (1979). J. Appl. Phys. 50:5105.
Ray, A. B. (1979). Proc. 37th Ann. EMSA Meeting. p. 522. Claitor's Publishing, Baton Rouge.
Rez, P. (1979). Proc. 14th Ann. Conf. Microbeam Analysis Society. p. 117. Kyser, San Antonio.
Rossouw, C. J., and Whelan, M. J. (1979a). In "Developments in Electron Microscopy and Analysis" (T. Mulvey, ed.), p. 329. Inst. of Physics, London.
Rossouw, C. J., and Whelan, M. J. (1979b). J. of Phys. D. 12:797.
Rossouw, C. J., and Whelan, M. J. (1980). Proc. 7th European Congr. on Electron Microscopy (P. Brederoo and V. E. Cosslett, eds.). 3:58.
Spence, J. C. H. (1977). Proc. 35th Ann. EMSA Meeting. p. 234. Claitor's Publishing, Baton Rouge.
Stephens, A. P. (1980). Ultramicroscopy. 5:343.

Zaluzec, N. J. (1980). Proc. 38th Ann. EMSA Meeting. p. 112. Claitor's Publishing, Baton Rouge.

Zaluzec. N. J., Hren, J. J., and Carpenter, R. W. (1980). Proc. 38th Ann. EMSA Meeting. p. 114. Claitor's Publishing, Baton Rouge.

DISCUSSION

SPEAKER: Dennis M. Maher.

SOMLYO: Where does the L cross-section come from?

MAHER: L cross-sections can be calculated from either a hydrogenic wave-function approximation or a more detailed calculation based on Hartree-Slater wave functions.

FERRIER: I think in terms of the Born approximation that the interaction is small, and I think the fact that you have got a finite convergence angle essentially means that you just have to use a series of plane waves, and you are looking at the effect of that convoluted with the scattering. I think then in terms of solid state effects that causes the rearrangement, alters the intensity in the peak; but I think the first order is still taken care of in terms of the following cross-section.

MAHER: Experiments have shown that fine-structure intensity for the L-edge of silicon contribute, at most, 5% to the net edge intensity when $\beta = 3$ mrad and $\Delta = 100$ to 200 eV. Furthermore, the effect of beam convergence is to broaden the angular distribution of inelastically scattered electrons and for small collections apertures this must be taken into account.

FERRIER: It is right that it changes the distribution, but in your L shell and K shell where you got .9, the shapes are quite different in the spectrum; the K is very nicely hydrogenic, the L is not. Isn't part of your discrepancy the fact that you are summing a relatively sharp range of E?

MAHER: In the experiments we are talking about, the ratio

$$\frac{I_{L(\beta,\Delta)}}{I_K(\beta,\Delta)} \cdot \frac{\sigma_{K(\beta,\Delta)}}{\sigma_L(\beta,\Delta)}$$

has been measured as a function of Δ and the limiting value of is determined by the convergence of this ratio, typically $\pm$ 2% is obtained.

FERRIER: The L shell you have truncated quite a bit.

MAHER: For $\Delta > 75$ eV the functional form of $\frac{I_{L(\beta,\Delta)}}{I_{K(\beta,\Delta)}}$ is the same as $\frac{\sigma_{L(\beta,\Delta)}}{\sigma_{K(\beta,\Delta)}}$, however the two are not same for $\Delta < 75$ eV.

FERRIER: The cross-section obviously could have. I mean the hydrogenic one for the L may not be that good an approximation; it does not explain.

MAHER: Yes, I agree with you. The main point here is that for L edges (especially those which exhibit a delayed maximum) the cross-sections calculated from a modified hydrogenic model may not be adequate for quantification and this situation needs to be examined in more detail.

ISAACSON: What I don't understand is your starting out with an equation. I read it correctly, sort of as an approximate equal-to sign and what you really want to do is do the actual convolution and you're approximating that convolution by separating out these two factors.

MAHER: I didn't present any of the details regarding the calculations of η, but the approach is quite similar to that which you presented in your SEM paper in 1978.

ISAACSON: No, but this ratio that you are measuring that I intensity divided by the cross-section, this is a partial cross-section which is really a convolution of the convergence illumination, with the scattering distribution, with the spectrometer apertures and you're approximating that by. . .

MAHER: Since we are applying the ratio method to a pure element (i.e., silicon), the ratio:

$$\frac{I_{L(\beta,\Delta)}}{I_{K(\beta,\Delta)}} \cdot \frac{\sigma_{K(\beta,\Delta)}}{\sigma_{L(\beta,\Delta)}}$$

should be close to one, assuming a sufficiently thin specimen is used, etc.

ISAACSON: The question is, do you expect it to be valid to better than 10%?

MAHER: No.

ISAACSON: Then why are you surprised that you get something that is maybe 10% wrong?

MAHER: The fact that the ratio is 0.8 makes me feel that the problem should be examined in more detail.

ISAACSON: Have you tried it for aluminum?

MAHER: Yes, and we get the same answer.

THOMSEN: You mentioned you prefer to work in a quasi-kinematic mode of operation. What starts to happen to the K-L intensity moving more to the dynamic?

MAHER: Well, in the zone axis case (i.e., .001) the ratio is $\sim$ 0.4. Generally speaking, the K-edge intensities increase more than the L-edge intensities as you go towards a Bragg condition for the brightfield case. However, the situation is complicated and depends on specimen thickness θ_E, θ_B, and β.

FERRIER: I think that is the point actually, this material is single crystal.

MAHER: There is a very good reason why these experiments were done on single crystal samples, and that is because in practice we must work with single crystals.

LIMITATIONS TO THE SENSITIVITY OF ENERGY-LOSS SPECTROMETRY

Dale E. Johnson

Center for Bioengineering
University of Washington
Seattle, Washington

INTRODUCTION

With energy-loss spectrometry (ELS) increasing both in availability and in use as a microanalytical tool, it seemed useful, for the purposes of this conference, to summarize and discuss various limitations to the ultimate sensitivity of the technique. The potential advantages of ELS over other microanalytical techniques have been widely discussed (Isaacson and Johnson, 1975; Colliex, Cosslett, and Trebbia, 1976; Joy and Maher, 1980), with the limitations to the sensitivity of ELS somewhat less appreciated. These limitations are, however, clearly important to successful applications. For the purposes of this discussion, we classify the limitations as follows:

A. Fundamental Limitations

(1) background energy-loss events

(2) multiple scattering events

(3) radiation damage

B. Instrumental Limitations

(1) aberrations of the energy analysis system (including post-specimen optics)

ISBN 0-12-362880-6

(2) energy resolution through spatial dispersion

We ignore here several limitations which in principle can be reduced to negligible levels. These include: spurious background (for example, from slit scattering) specimen contamination, current and voltage instabilities, and stray A.C. magnetic fields.

The overall effect of factors A and B above is not only to limit the ultimate sensitivity of energy-loss spectrometry but also to determine the optimum operating conditions under which this ultimate sensitivity is achieved. Although the sensitivity of ELS under these optimum conditions may be significantly better than, for example, EDS, the optimum operating conditions (e.g., optimum specimen thickness) may also be significantly more stringent.

BACKGROUND ENERGY-LOSS EVENTS

Considering the excitation of particular energy level electrons in a specimen which results in an energy loss of interest = ΔE, it is an unfortunate fact that all other electrons of the specimen for which $\Delta E >$ Ionization Potential I.P., can also absorb energy = ΔE from the incident beam. Thus for energy-loss events resulting from the excitation of low concentration elements, the potential background can be very large simply because the ratio of the number of excitable background to the number of excitable signal electrons is large.

On the fortunate side, for these background producing excitations, as the ratio ΔE/I.P. increases, the probability of the background energy electrons becomes more peaked at the large scattering angles (Bethe Ridge) (Inokuti, 1971). The latter allows discrimination against background events by the use of small angular acceptances. In such cases where the background scattering events are peaked at larger scattering angles an optimum acceptance angle exists which will optimize the signal to noise ratio ($\equiv P/\sqrt{B}$). This condition is indicated qualitatively as the first of four optimum operating condition curves shown in Figure 1.

In some cases, for example, the detection of the Ca L line (346 eV) in a carbon matrix, the main source of background from the C K line (285 eV) does not differ greatly

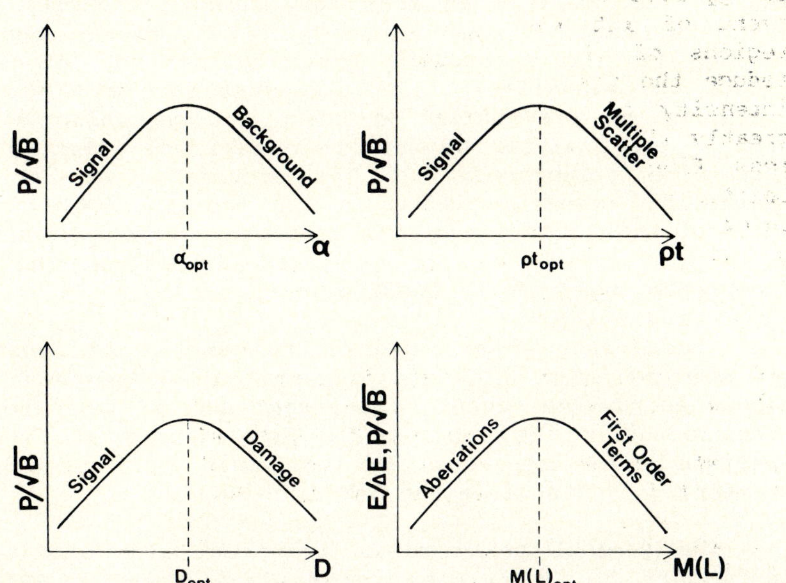

FIGURE 1. The qualitative behavior of four optimum conditions for ELS, with the limiting factors indicated for each. α = acceptance angle, ρt = specimen mass thickness, D = electron dose, and M(L) = post-specimen optics magnification (camera length).

in angular distribution from the signal peak and thus minimizing the effects of background through an optimum aperture angle may not be possible.

The variability of the energy-loss background signal (both in magnitude and shape) with changes in mass thickness also has important consequences for energy-loss mapping (i.e., contrast production through variation in a selected energy-loss signal). To be valid, an energy-loss map should reflect only compositional variation, which means that all effects of mass thickness variation must be eliminated. This is discussed in more detail in the next section.

MULTIPLE SCATTERING

Multiple scattering events limit the sensitivity of ELS in two main ways: (1) multiple <u>inelastic</u> events degrade the energy-loss spectrum in that they remove intensity from the event of interest and transfer it to higher energy-loss regions of the spectrum. This transfer of intensity may reduce the visability of an elemental edge by reducing the intensity of pre-ionization peaks and may also complicate greatly the analysis of both near edge and extended energy-loss fine structure. (2) The occurrence of any elastic scattering event in addition to the inelastic event of interest will reduce the effective beam current by scattering a large fraction of these electrons outside the angular acceptance angle of the spectrometer.

The effects of multiple scattering will be minimized if the specimen mass thickness is such that the probability of a single energy-loss event of interest and no other scattering event occurring is maximized. This maximum is found at a specimen thickness equal to one mean free path for all scattering events (Isaacson and Johnson, 1975).

The general behavior is indicated also qualitatively, in Figure 1. For $\rho t<\rho t_{opt}$ the signal-to-noise $(P/\sqrt{B})$ is limited by too little signal and for $\rho t>\rho t_{opt}$, $P/\sqrt{B}$ is limited by multiple scattering.

The fact that the energy-loss signal for any given excitation reaches a maximum and then decreases again as specimen mass thickness increases produces particular problems in energy-loss mapping. A simple example is shown in Figure 2 in which a specimen feature is assumed to have the projected mass density profile shown. Any energy-loss signal ($I_{\Delta E}$) of constant probability over this feature will, however, because of multiple scattering, map this feature into a distorted profile as shown. These "halo"-like features are common in energy-loss maps.

As discussed in the above section, varying mass thickness can affect both the magnitude and shape of the background while multiple scattering (especially elastic) can affect the magnitude of the background and the energy-loss peak by removing intensity from the beam. In the energy-loss analysis of a single region, the first effect is accounted for by fitting the background and extrapolating under the peak, and the second effect, by using a ratio of the peak area to, for example, a region of the background: This ratio is proportional to the concentration and independent of the amount of elastic scattering. Although rarely done in practice, these same two procedures are necessary for each point of an energy-loss map. As pointed out by

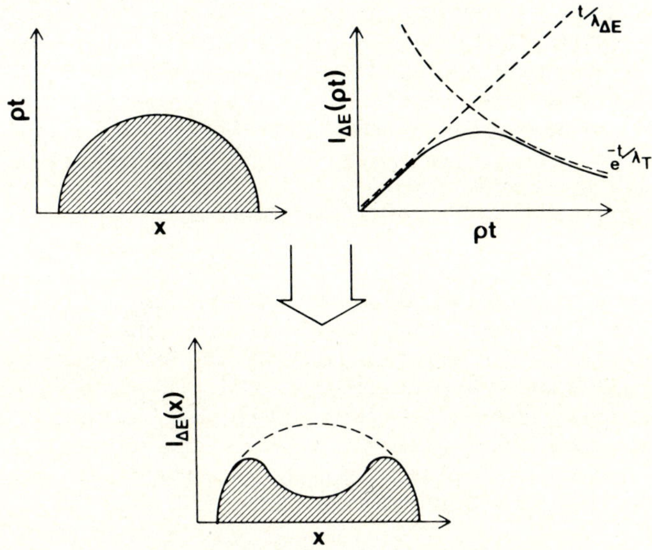

FIGURE 2. An illustration of the effect of multiple scattering on an energy-loss signal. An object is assumed to have a projected mass density distribution as shown in the upper left, with uniform probability of energy loss, ΔE. The net effect of increasing mass thickness (ρt) on an energy-loss signal is shown in the upper right, with the exponential decreasing term due to multiple scattering. ($\lambda_{\Delta E}$ = mean free path for the energy-loss region of interest, λ_T =mean free path for all scattering events.

Jeanguillaume et al. (Jeanguillaume, Trebbia, and Colliex, 1978), this may be approximated by obtaining energy-loss maps using at least two energy-loss intervals prior to the peak in addition to an interval containing the peak. This allows determination of both A and r in an assumed background dependence of the form $I_B(E) = AE^{-r}$.

Also it has been shown (Johnson, 1979) that if the change in shape of the background, with varying mass thickness, can be neglected, then the ratio of the contrast of a feature at a characteristic edge of an element to the contrast before the edge is a quantity that depends only on the concentration of that element. In this case only two energy-loss maps are needed since we need to account only for the effects on the magnitude of the background. This is in effect similar to the procedure described by Ottensmeyer (this volume). In using only two energy-loss maps, however, it must be constantly kept in mind that changes in background shape with varying mass thickness are not corrected for and in general, the use of at least three energy-loss intervals in forming an energy-loss map is preferred.

RADIATION DAMAGE

Of the limitations under discussion, the one least unique to ELS as a component of analytical electron microscopy is radiation damage. To the extent, however, that the high energy resolution of ELS is used to ask more detailed questions than just elemental composition (e.g., using near edge fine structure to determine chemical bonding states), then radiation damage becomes a more severe limitation than in, for example, EDS.

The situation is one in which the same events which carry information, the inelastic scattering events, are also with some probability capable of destroying the source of the information. The $P/\sqrt{B}$ in an energy-loss peak will initially increase with dose, and then decrease again, as for example, the number of undamaged molecules decreases.

The optimum dose will depend on the relative cross sections for the energy-loss event of interest and for damage events which eliminate the characteristic energy-loss events.

The maximum value of $P/\sqrt{B}$ will depend upon the magnitude of $\sqrt{I\tau}$ (I = beam current τ = irradiation time) at the optimum dose, and thus will increase as $\sqrt{A}$ (A = beam diameter) since $D = I\tau/A$. In other words, for a given beam current, the maximum $P/\sqrt{B}$ will be proportional to the beam diameter. In this sense, radiation damage can be considered as a limit to the spatial resolution of analysis: the more radiation sensitive the energy-loss event, the worse the spatial resolution possible. This limitation will be especially important in the use of near edge or extended energy-loss fine structure.

ABERRATIONS OF THE POST-SPECIMEN OPTICS-SPECTROMETER SYSTEM

As with any electron optical imaging device, the ultimate performance of an energy-loss spectrometer system will be limited by aberrations. Since energy-loss spectrometers typically lack cylindrical symmetry, second and higher order aberrations are present. In principle, spectrometers can be designed to eliminate these second order aberrations (Enge, 1967) and also, correction elements (e.g., sextupoles) can be added for aberration correction (Isaacson, this volume). In either case a higher order limitation then exists, and it now seems clear (Johnson, 1980) that the performance of an energy-loss spectrometer system can in general be increased by the use of post-specimen lenses in combination with the energy-loss spectrometer. These cylindrically symmetric lenses, lacking in second order aberrations, are used to form an image or a diffraction pattern of the specimen area analyzed at the spectrometer object plane. The size and angular divergence of this transformed object can then be varied by the lenses (at least two are required), so that for a given size and angular acceptance of the area analyzed, combined with the dispersion and aberration properties of the spectrometer, the optimum energy resolution is achieved.

This optimum behavior of the energy resolution as a function of the post-specimen lens operation (i.e., magnification, M or camera length, L) is also shown qualitatively in Figure 1. For small M(L) the angles of divergence are large and the energy resolution is limited by spectrometer and/or lens aberrations. For large M(L) first order terms dominate and the energy resolution is limited by the diameter of the image or diffraction pattern used as the spectrometer object. In Figure 1 we have plotted M(L) vs. $E/\Delta E$ but also indicated $P/\sqrt{B}$ on the ordinate, since $P/\sqrt{B}$ is clearly a function of the energy resolution.

ENERGY RESOLUTION THROUGH SPATIAL DISPERSION

Since the typical electron energy-loss spectrometer achieves high energy resolving power through spatial dispersion, an <u>efficient</u> (i.e., simultaneous detection of all energy events) detection system should consist of an array of detectors with the number of detectors equal to the number of discrete energy-loss intervals desired in the spectrum.

This is in contrast to a pulse height analysis system (e.g., EDS) in which the energy resolving power and simultaneous detection of all energy events is achieved with a single detector.

Since we are interested in measuring the electron intensity distribution only in the direction of dispersion, an array of "slit" detectors is needed in which the "slit" detector is narrow enough to provide the energy resolution required and long enough to collect all the intensity in the direction perpendicular to the direction of dispersion. This array of slit detectors can be formed by integration from a two-dimensional array of detection elements (e.g., a TV camera) (Shuman et al., this volume) with the advantage that the dispersion plane can be directly imaged. This may be useful in the observation of aberration and misalignment patterns. Such increased capability must be balanced against the added cost and complexity of a two-dimensional detector array. An alternative approach, using a one-dimensional array of photodiodes, is being evaluated in our laboratory and is shown schematically in Figure 3.

The parallel detection system consists of: a magnetic post spectrometer lens to increase dispersion, a phosphor conversion plate, coupling glass lenses, a dual channeltron image intensifier, and a cooled photodiode array (Reticon RL256C/17). The scanning of the array is controlled by a Kevex 7000 with the analog output of the array fed back into the Kevex, digitized and stored simultaneously with the acquisition of x-ray data. An example of a preliminary spectrum is shown in Figure 4.

An additional aspect of this energy resolution through spatial dispersion is that any other source of spatially varying intensity can mimic energy loss. These sources of spurious energy-loss structure include:

(1) Variations in specimen transmission within the area analyzed if the spectrometer images this area at the dispersion plane.

(2) Variations in the angular scattering distribution of the area analyzed if the spectrometer images the diffraction pattern of this area at the dispersion plane.

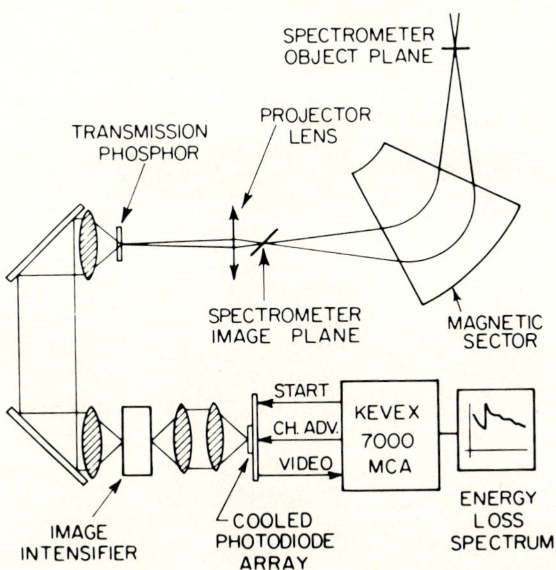

FIGURE 3. A schematic diagram of a parallel detection system under evaluation, which consists of: a magnetic post-spectrometer lens to increase dispersion, a phosphor conversion plate, coupling glass lenses, a dual channeltron image intensifier, and a cooled photodiode array. The scanning of the array is controlled by a Kevex 7000 with the analog output of the array fed back into the Kevex, digitized and stored simultaneously with the acquisition of x-ray data.

(3) Spatial variations in the dispersion plane intensity produced by post-specimen lens defocus and/or aberrations. This effect is analogous to phase contrast in a conventional EM image in which variations in the angular scattering distribution (produced by "interference" effects) appear as fringes only through defocus and/or aberrations.

The effects of (1) can be made negligible by using an area of analysis and total magnification such that the diameter of the image in the dispersion plane is $<< \delta E/D$. (δE = energy spread of the electron source, D = dispersion of the spectrometer).

The effects of (2) can be made negligible by using an angle of acceptance and a total camera length such that the diameter of the diffraction pattern in the dispersion plane is $<< \delta E/D$.

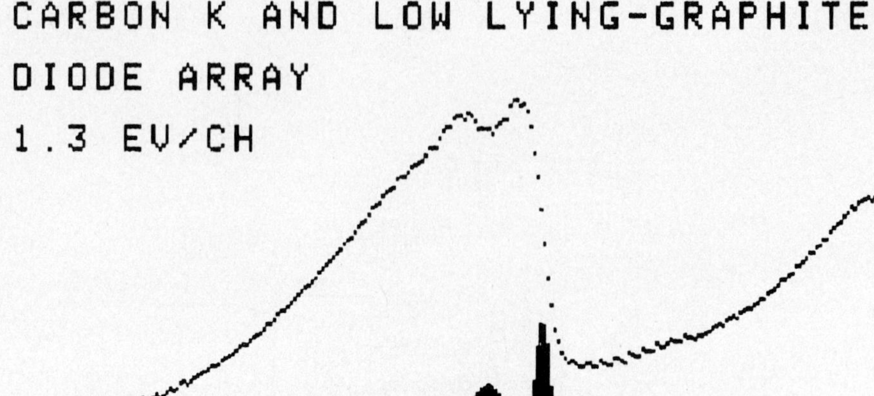

FIGURE 4. A preliminary spectrum obtained using only the diode array (without image intensifier) of the parallel detection system shown in Figure 3. The energy resolution was limited by the phosphor conversion plate. Each memory channel corresponds to a photodiode detector.

The effects of (3) can be made negligible by ensuring that at the best focus of the post-specimen optics, the angle of acceptance is chosen sufficiently small such that any intensity shift in the dispersion plane, due to aberrations, is $<< \delta E/D$.

The effects of (2) and (3) are most pronounced for specimens which diffract strongly, as the diffracted beams can appear as peaks in the energy-loss spectrum. The presence of such spurious peaks is most easily detected on the energy-gain side of the zero loss beam, but the absence of such "energy-gain" peaks is no guarantee of their absence also on the energy-loss side.

An example of such spurious structure is shown in Figure 5. The specimen is a thin polycrystalline Al foil, and low lying spectra are shown for two angles of acceptance. The area of analysis was imaged at the dispersion plane, and the spurious peak indicated was produced by the effect of lens aberrations on the diffracted beam. A similar spurious peak, however, would also have been produced if the diffraction

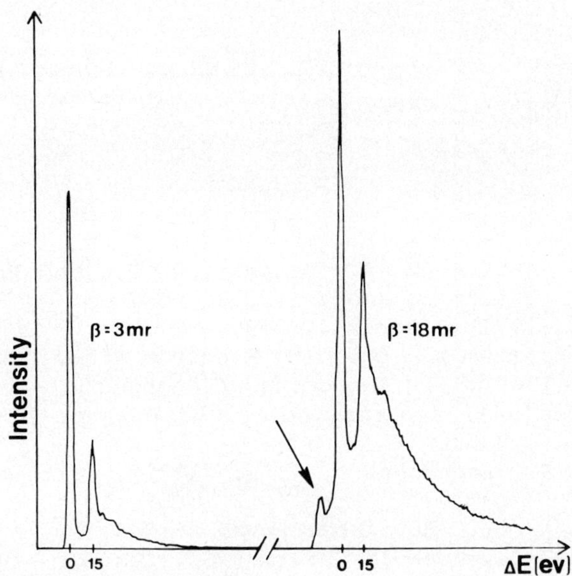

FIGURE 5. An illustration of spurious energy-loss structure (indicated by arrow) resulting in this case from post-specimen lens aberrations, but equally possible from diffracted beams if the diffraction is used as the spectrometer object.

pattern had been imaged at the dispersion plane using a total camera length of ∿ 0.4 mm (dispersion assumed to be = 1 μm/ev).

SUMMARY

Energy-loss spectrometry, along with its high energy resolution and high sensitivity, has a number of limitations which prescribe fairly tightly the operating conditions necessary for optimum results.

An awareness of the limitations of ELS is crucial not only to obtaining the full sensitivity of the technique but perhaps more importantly to avoiding the production of spurious results.

ACKNOWLEDGMENTS

This work was supported by NHLBI grants HL21371 and HL00472 (RCDA).

REFERENCES

Colliex, C., Cosslet, V. E., Leapman, R. D., and Trebbia, P. (1976). Ultramicroscopy 1:301.

Crewe, A. V. (1977). Optik 47:299.

Egerton, R. F. (1980). Scanning Electron Microscopy 1:41

Enge, H. H. (1967). In "Focusing of Charged Particles" (A. Septier, ed.), Vol. 2, p. 203. Academic Press, New York.

Isaacson, M. S., and Johnson, D. E. (1975). Ultramicroscopy 1:33.

Inokuti, M. (1971). Rev. Mod. Phys. 43:297.

Jeanguillaume, C., Trebbia, D., and Colliex, C. (1978). Ultramicroscopy 3:237.

Johnson, D. E. (1979). In "Introduction to Analytical Electron Microscopy" (D. C. Joy, J. J. Hren, and J. I. Goldstein, eds.), p. 245. Plenum Press, New York.

Johnson, D. E. (1980). Scanning Electron Microscopy 1:33.

Joy, D. C., and Maher, D. M. (1980). Ultramicroscopy 5:333.

DISCUSSION

SPEAKER: Dale E. Johnson.

MAHER: Didn't you do an analysis about a year ago on mass thickness effects in elemental mapping?

JOHNSON: Yes, we showed, for example, that a signal obtained as the ratio of contrasts before and at an energy-loss edge should be independent of mass thickness variations, and this is what is needed in any ELS mapping work. A number of approaches are possible, of course. Trebbia and Colliex have discussed the use of three detectors to accurately and on-line determine and eliminate the effects of background changes due to mass thickness variations. This same thing could also be accomplished, for example, by use of a three-step scan of the energy-loss spectrum for each image point.

SOMLYO: I think we ought to thank the physicists who spoke in this section. Not only did they bring us a gift but also they provided some instructions and warnings. And I don't know about the other biologists, but I am very much impressed, both by the potential power of EELS, and also by its potential danger. It's going to be probably more difficult to use, although potentially more promising than electron probe x-ray analysis was.

Part IV

ELEMENTAL CONCENTRATION DETERMINATION IN SINGLE ERYTHROCYTES

R. Gary Kirk

Department of Physiology
School of Medicine
Yale University
New Haven, Connecticut

Ping Lee

Department of Physiology
West Virginia University
Morgantown, West Virginia

INTRODUCTION

An elemental distribution study in blood and marrow frequently involves the analysis of hundreds of single cells. We have found that a large number of isolated cells can most easily be analyzed by smearing cells on thick carbon supports (Kirk et al., 1979). However, this method of x-ray microanalysis of single cells determines the amounts of various elements present in each of the cells and not their concentration. The concentration can be computed provided the individual cellular waters are known. If the cellular water is directly proportional to total major cations then this conversion is a relatively simple matter. In red blood cells and most other cells the major cations are K and Na. We have used three approaches to examine this relationship between (K + Na) and cellular water. (1) Red cells were separated into subpopulations according to their buoyant density by means of bovine serum albumin density gradient centrifugation. Cellular water and the (K + Na) contents were then determined in each fraction by conventional analytical methods. (2) Bone marrow cells were grouped into four different size categories by visual inspection of

ISBN 0-12-362880-6

individual cells during analysis. Their K and Na contents were determined and the relation between these contents and their sizes was examined. (3) Red cells from different species were analyzed by flame photometry and their (K + Na) contents were compared to their cellular volume. All three approaches support the relationship between (K + Na) and cellular water and, therefore, concentrations can be computed by dividing the amount of a given element by the cellular water estimated from the total (K + Na). Since the red cells contain about 65% water, the (K + Na) distribution can also be used as a volume distribution measurement.

METHODS

Preparation of Cells

Fresh blood samples from human volunteers and experimental animals were drawn into heparinized syringes. These were centrifuged and the plasma and buffy coating containing white cells were removed. Bone marrow cells were collected by flushing the marrow cavity of the femur with cold NaCl (0.155 M) containing heparin (1 U/ml). Both blood and bone marrow cells were washed four times with ice-cold $MgCl_2$ (0.12 M) buffered with 10% glycyl-glycine buffer (3.4g Mg carbonate, 5.1g glycyl-glycine in 100 ml H_2O pH 7.4) by repeated centrifugation (500 G, 5 min.) and resuspension of cells in the $MgCl_2$ solution. Washed cells were suspended (10% vol/vol) in sucrose solution (0.285 M) with 10% glycyl-glycine buffer (pH 7.4) and smeared onto polished pryolytic graphite blocks (2x2x10mm) cut from a one-inch diameter disc (Fullam, Schenectady, NY).

The smearing procedure has been described by Kirk et al. (1979). It consisted of: (1) dipping a 5 cm long, cotton-tipped applicator in the red cell suspension (10% Hct); (2) inserting wooden applicator dowel in the collet of a Moto-tool (Dremel, Racine, Wisconsin); (3) rotating slowly at approximately 300 rpm; and (4) passing a preheated carbon block (70°C) rapidly over the surface of the rotating cotton-tipped applicator. The block was then returned to a hot plate (70°C). Each block was subsequently examined with a metallurgical microscope equipped with a hot stage (70°C). Several warm blocks were anchored to warm SEM stubs with graphite in isopropanol (Dag-154 Acheson Colloids Co., Port Huron, Michigan) and stored in a desiccator with silica gel until analysis.

Determination of Elemental Content and Concentration

Methods for measuring elemental content and concentration have been previously described by Kirk et al. (1978, 1979). Calibrations were made by comparing the mean internal sodium and potassium concentrations from flame photometry and x-ray microanalysis. It has previously been found that there is a linear relationship between x-ray intensity and cellular elemental content in red blood cells (Kirk et al., 1979).

Electron Probe Microanalysis

Cells were analyzed one at a time with an electron probe (ETEC Autoscan, Hayward, California) using a raster containing single cells. Characteristic K_α x-ray lines were measured by two wavelength spectrometers using a lithium fluoride crystal for Fe analysis, a pentaerythritol crystal for K, and a rubidium acid phthalate crystal for Na. An electron beam current of 200 nA and a 15 kV accelerating voltage were used in this study. It was found helpful in the visualization of the cells to expose samples to 200 nA beam current at low magnification for 30 seconds to remove the sucrose layer coating the cell.

RESULTS

In order to establish the relationship between cellular water and (K + Na) content, it was necessary to obtain cells with a large range of these parameters. To achieve this we utilized the bovine serum albumin (BSA) density gradient centrifugation procedure to separate cells into subpopulations. It has been shown previously (Lee et al., 1980) that the more buoyant cells in BSA gradient tend to have larger amounts of water and cations. This BSA density gradient separation involves the preparation of a column of linear gradient with two isotonic BSA solutions of two different limiting specific gradients (1.105 and 1.070 g/cc), as previously described by Lief and Vinograd (1964). Human red cells washed in ice-cold 0.155 M NaCl were placed in gradients and centrifuged. The density gradients were then fractionated into 12 parts and each fraction was washed in $MgCl_2$ and analyzed for its cellular water (by wet and dry weight measurement) and (K + Na) content (by flame photometry).

Figure 1 shows the results of such measurements. The cellular water per unit dry weight is plotted against (K + Na) content per unit dry weight in each of the fractionated subpopulations. It was seen that the (K + Na) contents vary from 200 mEq/mg dry weight to about 330 mEq/mg dry weight and the cell water varies directly with the changes of (K + Na) content within this range. These results are consistent with those reported by Funder and Wieth (1960). Therefore (K + Na) content is a good index of cellular water and can be used for the computation of elemental concentration in human red blood cells.

In order to compare cells of drastically different sizes, we examined red cells of various species which range in volume from 33μm^3 in sheep to that of 105μm^3 in newborn dogs. When (K + Na) contents and mean volumes in these cells were compared, they exhibited a direct proportionality.

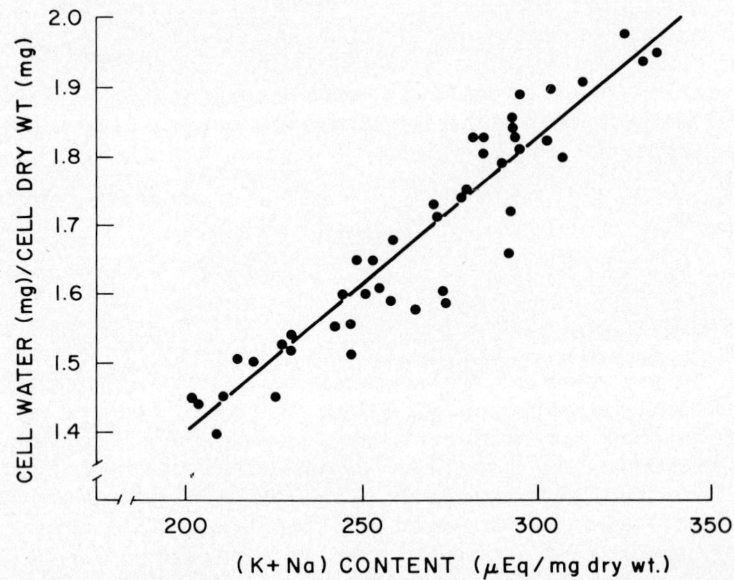

FIGURE 1. Cellular water and (K + Na) content in bovine serum albumin density gradient separated human red blood cells. Cellular water was measured by wet and dry weights and (K + Na) content was determined by flame photometry. Each point represents the mean value of cells in different density fractions from four experiments.

Figure 2 shows the relationship between cellular volume (determined from hematocrit and cell number) and (K + Na) content (determined by flame photometry) in man, rabbit, lamb, adult and newborn dogs. The single cell distribution of (K + Na) content in lamb and human measured by electron probe is shown in Figure 3. It is seen here that the smaller lamb cells ($33\mu m^3$) do in fact show a (K + Na) distribution well separated from that of human cells ($95\mu m^3$). Since cellular water constitutes 60% to 70% of the red cell volume and the hemoglobin concentrations in various animal species are fairly constant, one should be able to use (K + Na) contents to estimate both cellular water and volume.

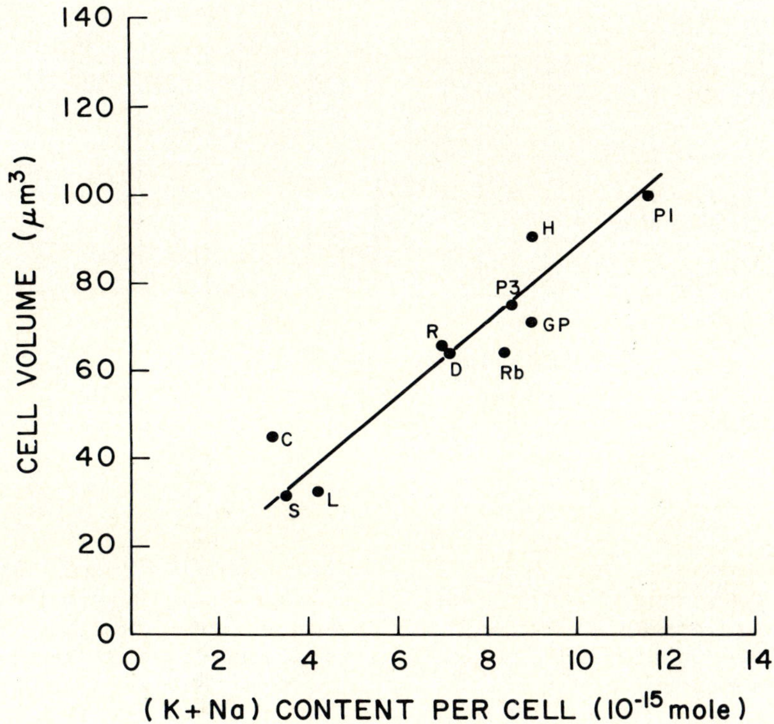

FIGURE 2. Relationship between cellular volume and (K + Na) content in red cells of various animals. Cellular volume was determined from hemocrit and the number of red cells in the sample. The (K + Na) content per cell was determined from flame photometric measurements and cell number. (R, rat; GP, guinea pig; H, human; C, cat; Rb, rabbit; S, sheep; L, lamb; D, dog; P3, three-week-old puppy; Pl, one-day-old puppy.)

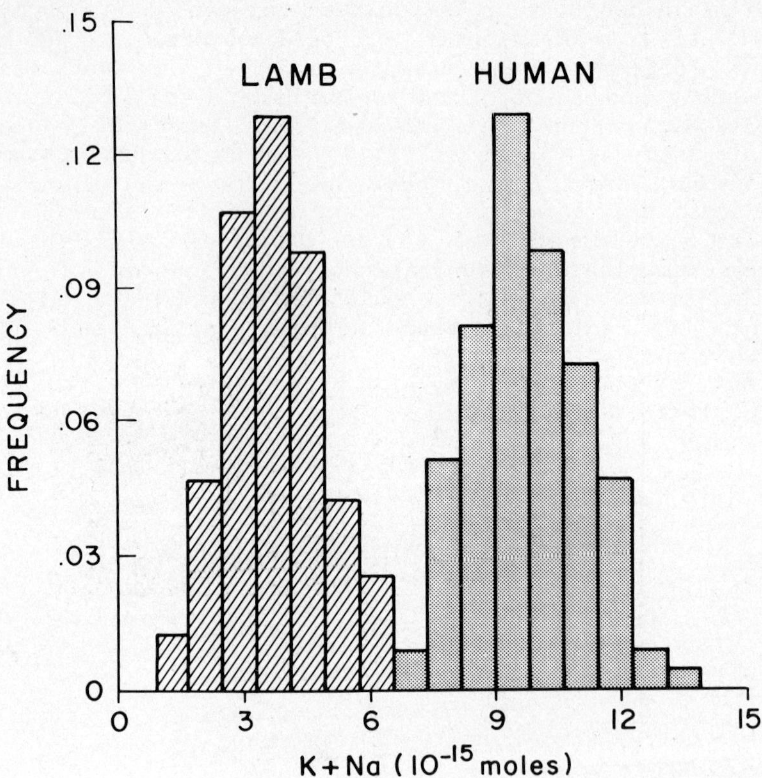

FIGURE 3. Distribution of (K + Na) content in single red cells of (a) lamb and (b) human, measured by x-ray microanalysis (15 KV, 200 nA).

Therefore, the (K + Na) is not only useful in arriving at elemental concentrations but also helpful in determining the cellular volume distributions. As an example, for the use of (K + Na) content in volume distribution studies, we report our studies of red blood cells in a genetically low-potassium type lamb that was three weeks old. Lambs are born with high-potassium red cells irrespective of their genetic type (i.e., high-potassium or low-potassium type). Six to eight weeks after birth, the genetically low-potassium lambs have their red cells converted to low-potassium cells. This is thought to be due to a replacement of the high-potassium fetal cells by the low-potassium adult cells produced after birth. As shown in Figure 4, the potassium concentration as estimated by the K/(K + Na) is bimodal, indicating that the replacement supposition is correct. The total (K + Na)

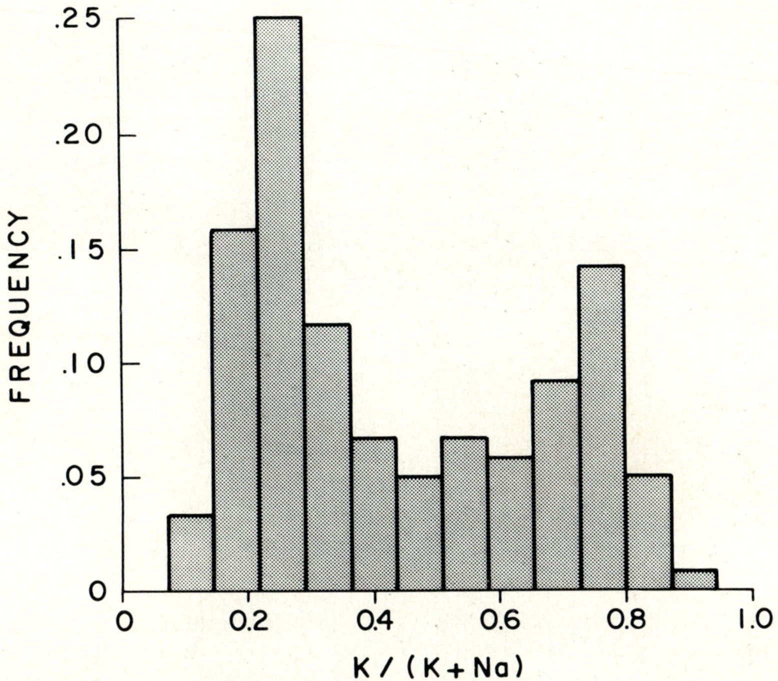

FIGURE 4. Distribution of K concentration estimated by K/(K + Na) in single cells of two-week-old lamb measured by x-ray microanalysis (15 KV, 200 nA).

distribution is single modal (Figure 3) and resembles the distribution of cellular volume obtained by electronic sizing with a Coulter counter. This indicates that the volumes of HK fetal lamb cells and LK adult cells are similar and that the loss of K in LK adult cells is replaced by equivalent amounts of Na ions, so that the overall (K + Na) content and volume do not change.

Because of our interest in studying the modification of cation composition during erythropoiesis, we have examined the cation contents of the rabbit bone marrow cells. These cells present a wide range of sizes, shapes, and hemoglobin contents, but, in general, the stem cells and the erythroblastic and myeloblastic cells are large and the red cells and reticulocytes (the immature red cells) are small. It was not our intention at this stage to measure their volume accurately. However, we have attempted to obtain a crude measure of their size by visually inspecting the cells during analysis in the electron probe. In so doing, we have

TABLE 1. X-ray microanalysis of individual rabbit marrow cells. Cellular diameters were visually measured during analysis and divided into four different size groups.

Cellular Diam. (μm)	K 10^{-15} mole	Na 10^{-15} mole	K+Na 10^{-15} mole	Fe 10^{-15} mole	Number of Cells Examined
I. 5-8	8.3	1.7	10.0	1.4	158
II. 13-16	25	1.6	26.6	0.2	37
III. 16-18	35	2.0	37.0	0.3	55
IV. 18-20	45	2.4	47.4	0.3	44

arbitrarily categorized the cells in rabbit bone marrow samples into four different groups: (I) 5-8 μm, (II) 13-16 μm (III) 16-18 μm, and (IV) 18-20 μm (Table 1). We have not attempted to ascertain the histological types of the large cells. However, it is quite certain that the small cells (5-8 μm) consist of a majority of the red blood cells, as shown by examination under the light microscope and also by the fact that these cells contain most of the Fe (i.e., hemoglobin content). In general, the large cells contain more K and Na, whereas the small cells (5-8 μm) have the least amount of K and Na. The result is consistent with the expectation that larger cells should have a larger amount (K + Na). From these studies, using the three different approaches to examine the relationship between (K + Na) content and cellular water, we have been able to show that (K + Na) is a suitable index for the cellular water and thus provide a basis for the calculation of elemental concentrations from electron probe x-ray data of smears of isolated cells in cells which contain hemoglobin and in cells which have no hemoglobin.

REFERENCES

Funder, J., and Wieth, O. (1960). Scad. J. Clin. Lab. Invest. 18:167-80.

Kirk, R. G., Brenner, C., Barba, W., Tosteson, D. (1978). Amer. J. Physiol. 235:C245-50.

Kirk, R. G., Lee, P., Duplinsky, T. G., and Tosteson, D. C. (1979). In "Microbeam Analysis in Biology" (C. P. Lechene and R. R. Warner, eds.), pp. 299-318. Academic Press, New York.

Lee, P., Kirk, R. G., and Hoffman, J. F. (1980). Submitted.

Lief, R.C., and Vinograd, J. (1964). Proc. Nat. Acad. Sci. 51:520-28.

DISCUSSION

SPEAKER: R. Gary Kirk.

WARNER: How did you say you were preparing your samples? Were you spraying them?

KIRK: We do not spray the cells because of the problems we have had with variations of the surface properties of the supports. The supports which we purchase now are more highly polished than those we used several years ago. The highly polished supports are not suitable for the spray technique. We decided to develop a technique less sensitive to the surface properties of the support. We tried (1) precooling graphite before spraying with liquid nitrogen; (2) freezing and freeze-drying smeared and sprayed cells; and (3) smearing cells on preheated supports. In the precooling-spray technique, we found that most of the cells were lost during the freeze-drying step. We have found that smearing cells on preheated supports to be the most reliable and convenient method. Freezing the cells after smearing with isopentane also works well but is not as convenient as the preheated support technique.

WARNER: How do you solve the contamination problem with plasma?

KIRK: The plasma is removed from our cells by washing with a sucrose solution prior to smearing. Some magnesium is added to this sucrose solution to maintain membrane integrity.

LECHENE: Did you try to analyze frozen hydrated cells?

KIRK: The plasma would contaminate them.

LECHENE: Not if you cut frozen hydrated sections.

KIRK: Brian Andrews and I have been planning to analyze sections of marrow, since in sectioned marrow the cell types could be more easily identified.

TORMEY: How do you smear the cells?

KIRK: Basically, the smearing procedure consists of dipping a cotton-tipped applicator in a cell suspension, inserting the applicator in the collet of a Dremel Moto-tool, and passing a preheated carbon block rapidly over the surface of the rotating cotton-tipped applicator.

TORMEY: For standards do you use microdroplets?

KIRK: No. We found a linear relationship between x-ray intensity and cellular content in loaded cells which had been prepared by the smear technique. We presently flame one of the samples in the experiment and use that as an anchor point.

CRYOSECTIONING OF BIOLOGICAL TISSUE FOR X-RAY MICROANALYSIS OF DIFFUSIBLE ELEMENTS

A. Saubermann

Department of Anaesthesia
Harvard Medical School at
Beth Israel Hospital
Boston, Massachusetts

Cryosectioning is an important preparative step for x-ray analysis of diffusible elements in biological tissue, yet cryosectioning is one of the most difficult and inconsistent preparative procedures. Furthermore, there are many questions as to whether or not the energy-requiring step disrupts elemental distribution. Therefore, an understanding of the cryosectioning process may help in obtaining consistent and predictable results when biological tissues are cryosectioned.

METAL MACHINING THEORY

Conventional metal cutting theory (Krar, Oswald, and St. Armand, 1969) can be considered as a theoretical foundation for cryosectioning. This model explains the surface structures of cryosections observed in the hydrated state (Saubermann, Echlin, Peters, and Beeuwkes, 1981) and direct measurements of the work of cryosectioning (Saubermann, Riley, and Beeuwkes, 1977). The major difference between metal cutting and cryosectioning is how the chip is handled. When cryosectioning, every effort is made to preserve the section (or chip), as opposed to metal machining, where every effort is made to eliminate the chip so as not to interfere with additional metal cutting. The metal machining theory is shown schematically in Figure 1.

ISBN 0-12-362880-6

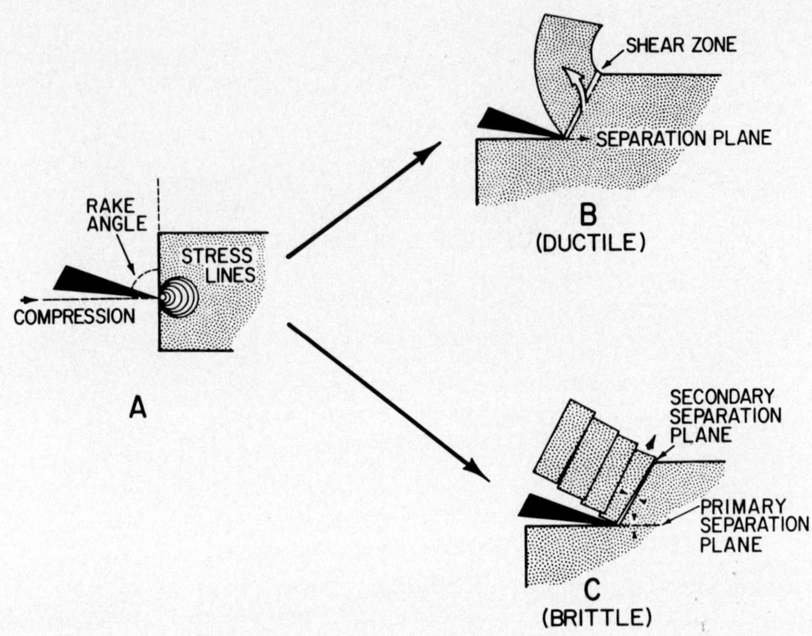

FIGURE 1. Schematic diagram of metal machining model of cryosectioning. Knife causes stress and compression in block along direction shown (A). Rake angle is defined as the angle between a line perpendicular to the cut surface and top surface of knife. The compressive forces are relieved by fracture or rupture along the fracture plane which is the plane of least resistance. A continuous chip is formed if the material is ductile (B) through plastic deformation or fracture in the plane of least resistance. Plastic deformation and plastic flow in the shear zone occur under ductile conditions. A discontinuous chip is formed if the block is brittle (C) with fracture along the primary separation plane as well as along a secondary separation plane in the shear zone.

During metal machining, a knife edge (tool edge) is forced into the block (work piece) causing compression and stress at this edge. Eventually, the force is relieved by rupture along crystal planes or planes of least resistance. Additional relief of these compressive forces occurs in the shear zone through plastic deformation. If, however, the material is brittle, the compressive force causes rupture along the shear angle prior to the movement of the chip over the knife. The type of chip depends upon whether there is plastic deformation with flow or rupture with slippage along the shear plane. Thus discontinuous or continuous chips are formed during cutting depending upon the ductility or brittleness of the material (Figure 1). Discontinuous chips are produced when very brittle material is cut; continuous chips are cut when the material is more ductile. It is the latter case in which chip formation is efficient, predictable, and requires the least work (Saubermann, Riley, and Beeuwkes, 1977). When discontinuous chips are formed, there is a slippage of the ruptured fragments over one another, causing an overlap of the chip pieces (Figure 1). However, when continuous chips are formed there is a continuous ribbon of material which usually curls on the upper surface of the tool or knife unless a chip breaker or flattener is used. The chip structure depends upon the ductility of the material in the shear zone during applications of stress (Krar, Oswald, and St. Armand, 1969). Since heterogeneous material may have many areas within the shear zone which are composed of widely differing ductilities, the "plane" of least resistance may not be a monodirectional plane. Instead one would expect that separation would occur between the matrix and a hard particle rather than through the hard particle. An analogy between a brittle ice crystal and a more ductile proteinous or lipid matrix surrounding the ice crystal can be drawn.

GENERAL CONSIDERATIONS

It has long been known that cryosectioning depends upon temperature, and in an earlier study it was demonstrated that force patterns obtained during cryosectioning change with temperature (Saubermann, Riley, and Beeuwkes, 1977). Most probably temperature strongly influences frozen specimen brittleness, since the work of cryosectioning is largely dependent on this physical property (Saubermann, Riley, and Beeuwkes, 1977). Observations of frozen hydrated sections (see Figure 2), cut at -80°C, -50°C and -30°C, are consistent and supportive of this machining concept (Saubermann, Beeuwkes, and Peters, 1981). There are a number of factors which can be considered for practical application of this machining theory to cryosectioning. These factors can be divided into two broad categories: first, those factors which can be altered or changed to improve cryosectioning, and second, those factors which are not under the control of the investigator. These latter factors include the composition of the tissue and its physical structure, such

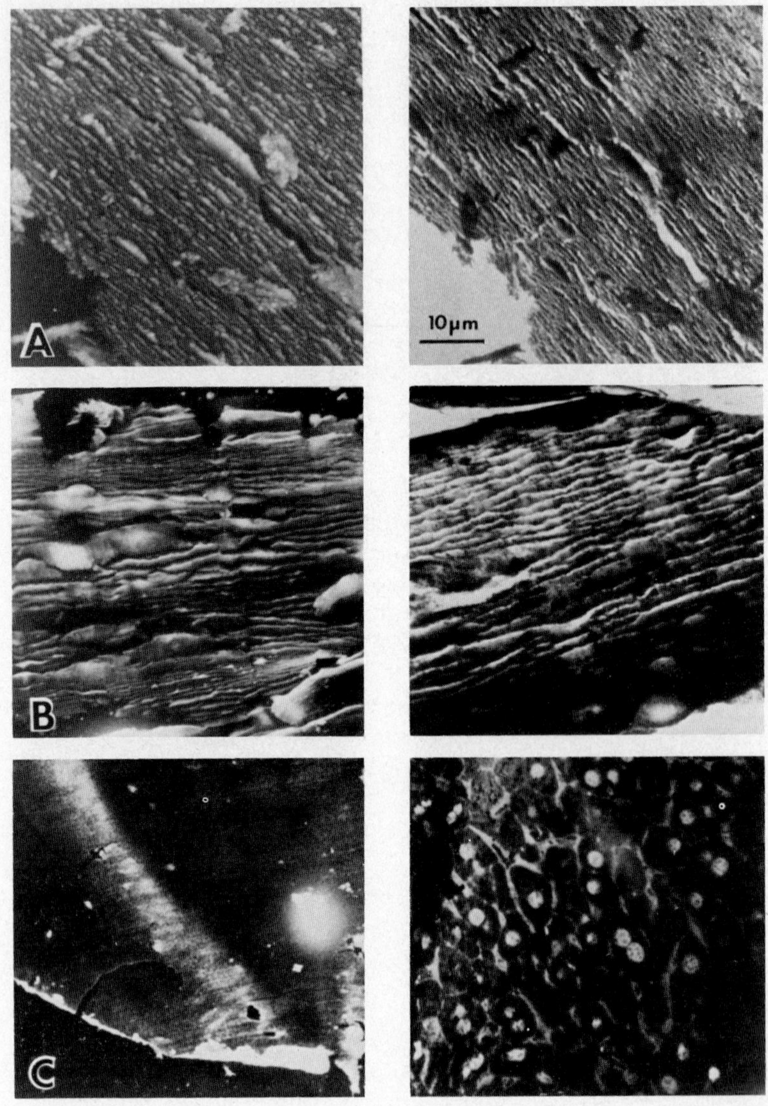

FIGURE 2. Secondary (left) and transmitted electron (right) images of 0.5 µm thick sections of mouse liver cut at -80 °C (A) with obscure morphological detail; sections cut at -50 ° C (B) show continuous chips beginning to form, but morphological detail is still unclear. In the continuous chip formed at -30° C (C) under low work conditions, morphological detail is easily seen. (From Saubermann et al., 1981).

as the presence of lumens or inclusions. Those factors which can be controlled include the freezing rate, the amount of water present (if prefreezing drying is acceptable), the temperature at which cryosectioning occurs, and the mechanical conditions under which the tissue is cryosectioned.

Tissue Properties

Unfortunately, frozen tissue is sometimes erroneously considered to be similar to a block of ice. However, it is more accurate to consider frozen tissue as a heterogeneous material having a variable degree of temperature dependent ductility and structure. Since ice is inherently brittle, and large ice crystal inclusions form large inclusions of brittle material, any reduction in the size or number of ice crystals formed during freezing would reduce the brittleness of the tissue block. The use of cryoprotectants and rapid freezing techniques for very small volumes of tissue can be used to promote better freezing (Dempsey and Bullivant, 1976; Franks, Asquith, Hammond, et al., 1977; and Somlyo, Somlyo, and Shuman, 1979). It is important to understand that if the tissue is partially dehydrated prior to freezing, either inadvertently or directly, smaller ice crystals will be formed and the tissue will behave in a more ductile fashion improving cryosectioning. These factors may explain the wide variety of cutting temperatures successfully used with different tissues.

Temperature

There is a common observation that there is an optimal temperature for cryosectioning. Appleton has described thermal conditions and mechanical conditions for optimal cryosectioning of a variety of tissue at which ribbons of sections can be formed (Appleton, 1974). The work of cryosectioning depends upon block temperature and upon the section structure (Saubermann, Riley, and Beeuwkes, 1977). For example, mouse liver cut at -80° C formed discontinuous chips and the work of cryosectioning was greater than when the same tissue was cryosectioned at -30° C (Saubermann, Riley, and Beeuwkes, 1977). At -30° C, cryosections of mouse liver formed continuous chips. Under these conditions the work of cryosectioning was one-third of that cryosectioning of the same tissue at -80° C. In spite of the fact that it appears that cutting temperature is one of the single most important factors in successful cryosectioning, raising cutting temperature until good sections are obtained has been considered potentially hazardous because of the possibility of creating (or enlarging) a melting zone, and the possibility of recrystallizing tissue ice. However, based upon the machining model, raising cutting temperature to improve tissue ductility would, ironically, promote low work conditions which are less likely to cause deleterious thermal effects to the frozen tissue block and the section.

Temperature of the block face determines the base temperature of the shear zone prior to introduction of additional heat from other sources. The temperature of the block face is primarily controlled by the atmospheric temperature surrounding the block (Christensen, 1971). Efforts to control tissue, block, or knife temperature separately gives illusory control while promoting instability from differential contraction or expansion. Conditions which promote differential temperatures between knife and block would tend to establish conditions which were not thermally stable and hence making reproducible section cutting difficult. The effective temperature of the knife edge and the block face are probably very close even in systems where the temperature of each of these positions is separately regulated. Control of the block face temperature would be more difficult in a nonequilibrium system than in a system where a true equilibrium exists. Cryostat-cooled systems are useful in establishing equilibrium cutting conditions (Appleton, 1974); however, this can also be achieved with a cooled constant flowing temperature controlled N_2 (Saubermann, Echlin, Peters, and Beeuwkes, 1981).

The section thickness appears to determine the temperature at which sections can be cryosectioned. This fact is predictable from metal machining and is related to the mass of the shear zone and the heat introduced into that region. Thin sections (< 100nm) require cutting at lower temperatures (-80° C or lower) (Appleton, 1974; Seveus and Kindel, 1974). Thicker sections require warmer temperatures (Saubermann, Echlin, Peters, and Beeuwkes, 1981). The explanation for this is that during cutting, sufficient energy is transferred to the shear zone to raise the temperature of the shear zone such that its thermally dependent ductility is improved to a point where continuous chip formation occurs. Since it requires the same amount of work to cut thick or thin sections (hence the same energy input), the effective cutting temperature will be determined by the amount of heat transfer to and from the block face surface and the shear zone. Although cutting temperature may be recorded at some point proximal to the tissue block face at, for example, -80°C, the effective cutting temperature at the shear plane, when it is only 100nm from the uncut block face surface may only be -30° C. Presumably the heat necessary to raise the temperature from -80° C to -30°C came from the work of cutting. Thus it is likely that the seemingly different temperature required to cut thick and thin sections is only apparent and does not reflect similar temperatures in the shear zone.

Cutting Tool

The chip (or section) must flow over the knife edge. Consequently, the characteristics of the knife and its angle are important. Since frozen biological material has a variable brittleness, the chip may fracture or bend as it is forced over the knife. There are two common ways of reducing bending forces. The first way is to minimize the knife angle and maximize the rake angle. The second is to counteract the bending forces which occur as the chip moves up over the knife surface through use of an antiroll plate. Use of an antiroll plate helps to flatten sections cut as continuous chips. The antiroll plate counteracts the plastic deformation of the continuous chip which tends to cause curling on the knife surface.

Knife angle affects the efficiency of material removal during machining. Since conditions which promote continuous chip formation are desirable, efforts to maximize rake angle are important. This angle can be maximized by reducing knife angle and minimizing clearance angle. Clearance angles in the range of 4° to 10° appear acceptable. By using metal knives with small angles the total cutting angle can be kept small (Saubermann, Echlin, Peters, and Beeuwkes, 1981). This improves the efficiency of sectioning and reduces resistance, hence lowering work. Since glass or diamond knives tend to have relatively large (40 to 50°) angles which increase the cutting angle and because of their poor thermal properties, such knives are less desirable than metal knives. Stainless steel blades have been used successfully and have theoretical advantages for cryosectioning.

Cutting Speed

The optimal cutting temperature for a given section thickness appears to depend not only on microtome temperature, but upon cutting speed. However, if cutting speed is increased, the rate of thermal input to the shear zero increases. Since heat transfer to and from the shear zone depends upon block temperature and composition as well as that of the knife, the effective cutting temperature at the shear zone will be dependent upon cutting speed. If heat input exceeds heat removal from shear zone, the temperature of this critical zone will increase, altering its ductility and hence cutting properties. Efforts to provide reproducible cryosectioning must take into account alterations in cutting speed. Since three variables--block temperature, cutting speed, and section thickness--can effect the temperature at the shearing zone, it is sensible to keep two of these variables constant and control the third in order to control cutting properties at the shear zone. Since section thickness is usually predetermined, cutting speed can be kept constant and the block temperature changed for optimum ductility and sectioning properties. This general approach has been advocated by Appleton (1974).

EFFECTS OF CRYOSECTIONING

The possibility of morphological disruption and elemental displacement during cryosectioning is of major concern in application of cryosectioned tissue for x-ray microanalysis. The issue is confused because there are limited tools for making measurements in such small compartments. A definition of the extent of elemental and morphological disruption is not yet complete. However, under certain circumstances some information and guidelines are available. We can consider the definition of the problem from two aspects. First, does cryosectioning introduce morphological artifacts in addition to that of initial freezing? Secondly, does cryosectioning cause elemental displacement of diffusible elements?

Morphological Artifacts

The major and most obvious morphological artifact is ice crystal damage. These artifacts can be observed in most frozen biological tissue, or freeze-substituted tissue (see Figure 3, next page). The relationship between freezing rate and ice crystal size is well known (Dempsey and Bullivant, 1976), as is the relationship between subsequent ice crystal growth after initial freezing to temperature and to time in certain test solutions. However, biological systems with

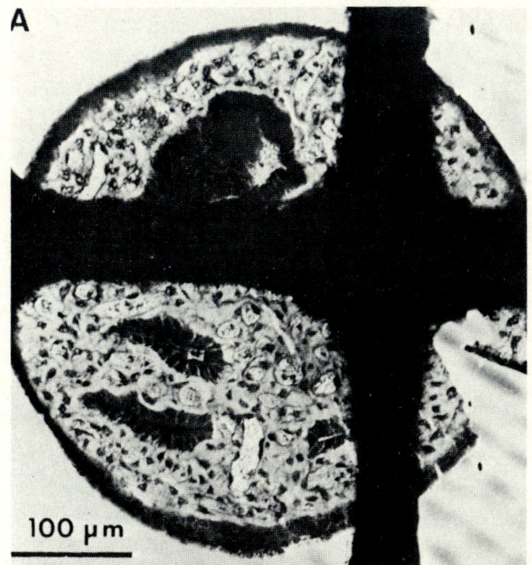

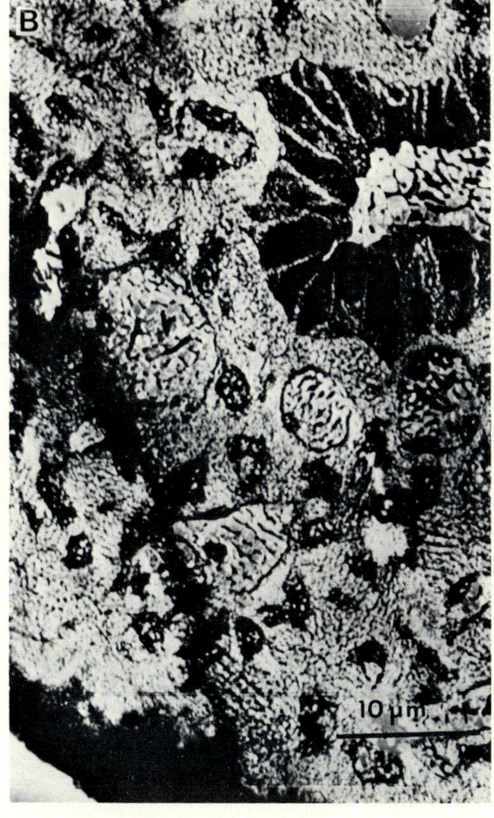

FIGURE 3. Scanning transmission micrographs of frozen-dried sections of rat renal papilla. A. low power of complete cross section very near papilla tip, B. high magnification. The reticulated appearance of the tissue results from ice crystals formed during the initial freezing step. (Reproduced from Bulger et al., 1981).

complex structure and composition complicate theoretical predictions based on ice crystal growth behavior in pure ice of biphasic systems (MacKenzie, 1972). In biological tissue, which not only has a complex chemical composition in the cell "sap," but also specific cytoskeletal and architectural features, practical guidelines precede theory.

Because we need some technique for observing ice crystal damage, initial freezing artifacts are generally observed after additional treatment, either freeze-drying or freeze-substitution or surface replication techniques. It is comforting that there is some general agreement between investigators as to ice crystal damage size between these various techniques. This suggests that subsequent treatment does not necessarily greatly affect those artifacts. In a recent study, we compared ice crystal size 10 micrometers from the freezing edge of a frozen drop of 20% albumin solution cut first at -80° C, then at -30° C (Saubermann, Beeuwkes, and Peters, 1981). In that study, a droplet of 20% serum albumin was rapidly frozen in Freon 12 at its melting point. Sections (0.5 micrometers thick) were cut at -80°C, transferred to the scanning electron microscope cold stage frozen hydrated, and then dried in vacuum by warming the cold stage to -50°C. The sections were then photographed and ice crystal damage artifacts were measured within a compartment 10 micrometers from the surface. The same block was then warmed to -30° C and a section cut, transferred, freeze-dried, and photographed and measured between these two groups. No difference in ice crystal size was observed between these two groups (-80° C: 0.40 ± 0.02 μm, -30° C: 0.42 ± 0.02 μm). This experiment took approximately two hours to complete, and it is possible that longer exposures to -30° C temperature would have increased ice crystal damage artifact size. However, it should be kept in mind that a small increase in ice crystal damage artifact radius would have meant a large increase in volume, and volume was not measured in these experiments.

Consequently, this diameter defines the limits of spatial resolution in this system. The size of these ice crystal damage artifacts was similar to the size of ice crystal damage artifacts observed in biological tissue containing 80% water (Bulger, Beeuwkes, and Saubermann, 1981). We do not know, of course, the structure or size of the actual ice crystals within that ice crystal damage artifact. Nor do we know the effects of temperature change and thermal input, albeit smaller with -30° C sections than with the -80° C sections upon elemental distribution within that ice crystal artifact. However, the minimal morphological compartment which could be resolved in this test system having no boundaries using the above criteria was limited to 0.4 micrometers. In systems having cell boundaries, there are clear differences between ice crystal damage artifacts in different compartments largely based upon the amount of water present in that compartment (Bulger, Beeuwkes, and Saubermann, 1981). A. P. Somlyo, A. V. Somlyo, and H. Shuman (1979) have clearly demonstrated that in ultrathin sections consistent compartmental differences can be observed between adjacent analytical compartments approximately 30 nm in diameter. They froze extremely small pieces of muscle tissue and were able to obtain specimens which were well frozen with minimal ice crystal damage artifacts present. However, others have sectioned tissue exhibiting large ice crystal damage artifacts and have demonstrated reproducible patterns in adjacent tissue compartments using rasters which exceeded ice crystal damage artifact size (Cameron, Smith, and Pool, 1979).

The possibility of measurable elemental redistribution occurring during cryosectioning is probably small. At worst it would appear to be a second order affect. Thus, a more important consideration is not whether major recrystallization and elemental redistribution occurs during thermal stresses to cut sections, but rather whether major redistribution occurs during the initial freezing damage. How long tissue can be kept at -30° C, for example, before ice crystal damage artifacts begin to exceed significantly the initial freezing damage dimensions is not clear and probably depends heavily upon water content and composition of each compartment. Practically speaking, cryosectioning very thin sections, as pointed out earlier, at low temperatures (i.e., -80 C) is probably similar to -30°C or -40° C sectioning of thicker sections. Consequently, data obtained from such sections would indicate that it is entirely possible to cryosection at similar effective temperatures without introducing obvious or unacceptable ice crystal damage artifacts exceeding that of the initial freezing (Somlyo, Somlyo, and Shuman, 1979). While recrystallization of ice crystals within the ice crystal damage artifacts seen in tissue may occur, such recrystallization does not appear to greatly effect the dimensions of those artifacts to the point where there is a significant deterioration in effective analytical compartment resolution. Therefore, the "tissue" defined limits of spatial resolution appear to remain essentially unchanged by cryosectioning at the optimum effective temperature for a particular section thickness.

Elemental Displacement

Displacement of diffusable elements is a more complicated question. What may appear as displacement during cryosectioning may in fact be instrumental effects (beam tailing, electron scatter, etc.) (Tormey and Platz, 1979). Such effects must be clearly understood and defined within a microprobe instrument before sectioning artifacts are considered to have been introduced (Tormey and Platz, 1971).

Two systems of biological tissue ought to be considered separately. Namely, boundary limited and unlimited systems. Both exist in nature. Boundary limited systems refer to membrane limited systems. Unbounded would be systems such as tissue matrix materials, ground substance, and extracellular spaces. Membrane limited systems are presumably less likely to permit diffusion of elements from thermal effects simply because of the mechanical barrier of the membrane. While a critical test of this concept has not yet been made, there is much circumstantial evidence indicating that membrane limited compartments retain their elemental profiles under a wide variety of cutting conditions and compartment sizes (Bulger, Beeuwkes, and Saubermann, 1981; Saubermann, Echlin, Peters, and Beeuwkes, 1981; Saubermann, Beeuwkes, and Peters, 1981).

A more critical situation is unbounded systems. We recently considered the effects of cutting temperatures on a standing diffusion gradient of 4 diffusable ions (Na, Cl, K, I) (Saubermann, Beeuwkes, and Peters, 1981). A standing diffusion gradient was obtained by allowing a drop of 15% sodium chloride and potassium iodide to diffuse for one minute on a 30% gelatin matrix. Serial 0.5 micron thick sections of the same block were cut at -80° C and then -30° C and analyzed at fixed intervals along the length of the diffusion gradient. Comparison of the curves thus obtained to show any significant difference between them (Figure 4), providing direct evidence that cryosectioning under low work conditions are unlikely to disturb the integrity of diffusible elements in unbounded systems (Saubermann, Beeuwkes, and Peters, 1981).

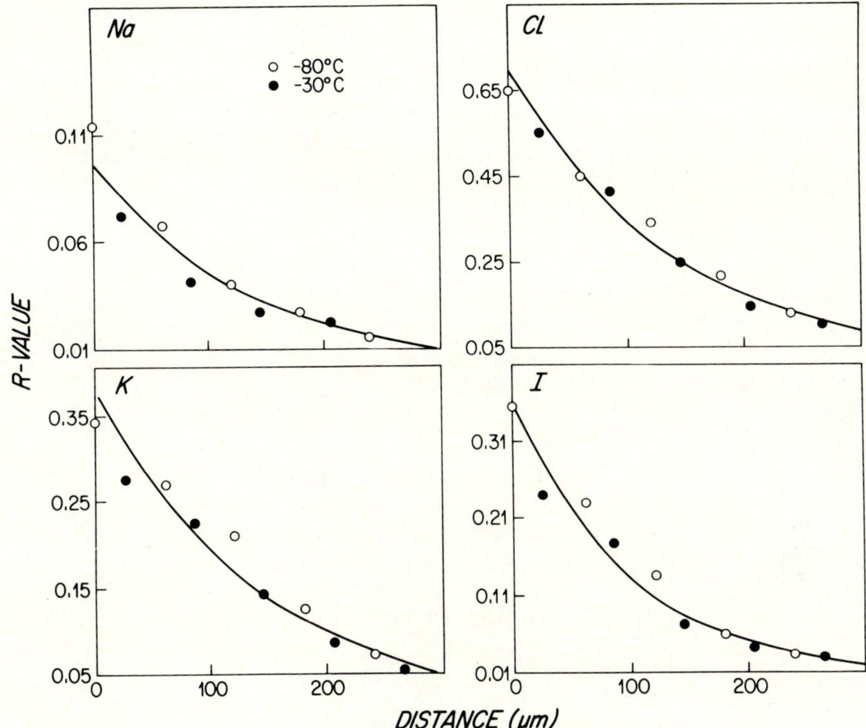

FIGURE 4. Observations and calculated best-fit experimental curves obtained by x-ray microanalysis of artificial electrolyte gradients. Gradients (Na, Cl, K, I) were established in gelatin blocks and were subjected to analysis in serial sections cut first at -80° C (open circles) and then at -30° C (closed circles). Statistical analysis showed no significant difference between the electrolyte gradients. (Reproduced from Saubermann et al., 1981).

SUMMARY

Cryosectioning is an important preparative step for many techniques for x-ray microanalysis of biological tissue. It has been a step which is poorly understood but thought to be a step which has great potential for disruption of both morphology and elemental distribution. Evidence suggests that cryosectioning tissue appears analogous to machining brittle metals. Based upon this model rather than the melt-freeze model proposed a number of years ago, there are theoretical advantages to cutting at "warmer" temperatures which reduce work input by improving tissue ductility. Such sections are theoretically less likely to have undergone thermal injury because there is less potential heat input. Direct observation of sections cut at different temperatures support this machining model. Furthermore, measurement of cutting work adds additional evidence for this model. When this model is applied to cryosectioning biological tissue, predictable and reproducible sections can be obtained. The fundamental principle of this model is that tissue brittleness is thermally dependent, hence controllable by determining the optimal temperature at which to cryosection. Factors such as section thickness, cutting speed, and heat removal will greatly influence the effective cutting temperature and hence the optimal temperature. Direct observation of the tissue section structure supports the hypothesis that there is dependence of tissue ductility upon cutting temperature. Available evidence strongly suggests that cryosectioning is a safe and acceptable preparatory step for diffusable element localization and that any additional morphological or elemental displacement damage from sectioning, if it occurs, is insignificant compared to damage from initial freezing.

REFERENCES

Appleton, T. C. (1974). J. Microsc. 100:49-74.

Bulger, R. E., Beeuwkes, R., III, and Saubermann, A. J. (1981). J. Cell Biol. 88:274-80.

Christensen, A. K. (1971). J. Cell Biol. 51:772-804.

Cameron, I. L., Smith, N. K. R., and Pool, T. B. (1979). J. Cell Biol. 80:444-50.

Dempsey, G. P., and Bullivant, S. (1976). J. Microsc. 106:261-71.

Franks, F., Asquith, M. H., and Hammond, C. C., et al. (1977). J. Microsc. 110:223-38.

Hodson, S., and Marshall, J. (1969). In: "Technology of Machining Tools," pp. 87-97. McGraw-Hill, Toronto.

Krar, S. F., Oswald, J. W., and St. Armand, J. E. (1969). In: "Technology of Machining Tools," pp. 87-97. McGraw-Hill, Toronto.

MacKenzie, A. P. (1972). In: "Scanning Electron Microscopy/1972," pp. 274-80. ITT Res. Inst., Chicago.

Saubermann, A. J., Echlin, P., Peters, P. D., and Beeuwkes, R., III. (1981). J. Cell Biol. 88:257-67.

Saubermann, A. J., Beeuwkes, R., III, and Peters, P. D. (1981). J. Cell Biol. 88:268-73.

Saubermann, A. J., Riley, W. D., and Beeuwkes, R. III. (1977). J. Microsc. 111:39-49.

Sevéus, L., and Kindel, L. (1974). Proc. 8th International Congress of Electron Microsc. Vol. 2, pp. 52-3. Canberra.

Somlyo, A. P., Somlyo, A. V., and Shuman, H. (1979). J. Cell Biol. 81:316-35.

Tormey, J. M., and Platz, R. M. (1979). In: "Scanning Electron Microscopy/1979," Vol. 2, pp. 627-34. Chicago.

DISCUSSION

SPEAKER: Albert J. Saubermann.

HALL: Could you mention the nature of the specimen?

SAUBERMANN: Our test specimen was a sodium chloride, potassium iodide 15% solution placed on the surface of a plate of 30% gelatin. Next a small sample was cored out, frozen, and turned so that the gradient was to one side and sections were cut first at -80° , and transferred into the microscope. The gradient was then measured. The block was then warmed in the microtome to -30°, another section cut and analysed as before. Details can be found in Saubermann, Beeuwkes, Peters (1981).

RICK: Isn't it a rather uncritical test because normally you would not expect that 1 or 2 micrometer of smearing would show up in such a test?

SAUBERMANN: No, we would expect any significant smearing to be magnified over the length of the gradients in such an unbounded system. In a bounded system small differences might not be detected.

LECHENE: You have determined the direction of the cutting?

SAUBERMANN: Yes. Furthermore, we turned the block by 90 increments and cut sections at each position without detecting any change in the gradient.

SAUBERMANN: I don't think anyone would seriously consider biological tissue to be a block of ice, since the composition is so completely different than ice, both chemically and structurally.

LECHENE: I certainly would! I would think that water--we will discuss it later--but I would certainly consider it a block of ice more than a bone or a piece of metal.

SAUBERMANN: I think if you believe that, then there are certain things to be considered with your method for mechanically sawing a piece of tissue with a rotary saw which are totally inconsistent with the concept of tissue being a block of ice.

LECHENE: The slides that you have shown indicate that you can cut at -30°, but where does it show that you don't redistribute ions?

SAUBERMANN: I am not trying to illustrate that point with those particular slides. I'm just showing some examples of different tissue and also making a point that the ice crystal size differs with the amount of water in each compartment.

LECHENE: I want to take argument because if what you are saying is that in order to do electron probe analysis one can cut at -30°, I don't believe your results and I think I don't believe them for two reasons. I don't believe your premises, because I don't think that one can apply metal theory on something which is 80% water, particularly in tissue where there is a lot of water, and at -30° I still believe in first physics principles where the edge of the knife will apply something like a ton of pressure on a phase diagram, so that is one point. Secondly, if you tell me that you don't melt, I am not convinced by the images that I have seen that there are sharp boundaries which are maintained between luminal compartment, i.e., on cell one interior so that you have not smeared whatever is in principally what your compartment over a cellular compartment.

SAUBERMANN: Well, your <u>beliefs</u> are obviously your own business. It is my intention to present our data and conclusions based on that data. However, if you have serious concerns about our data, then surely you must have similar beliefs about your data. This would be especially true where you take a kidney, freeze it, and prepare a surface for analysis with a diamond saw at 3000 rpm. This would affect your tissue according to the same principles I just described with metal machining, because it's almost exactly the same procedure--the difference being that in your preparation the amount of energy transfer to the tissue is enormous. I don't know if you've ever done any sawing, but if you touch the saw afterwards it is hot; also if you take a piece of pipe and cut it with a hacksaw, the ends of the pipe are hot. This is exactly what you're doing with your tissue. Melting by the mechanism of melting point depression from pressure is improbable, since the compressive forces from work (force times distance) are relieved through shearing or plastic deformation along plane of least resistance. However, if you want to consider the possibility that you're melting with pressure, then you have to consider that you're also melting protein, lipids, and all the other material present. This is contrary to freeze fracture data which provides evidence indicating that plastic deformation does occur, that plastic deformation is thermally dependent, and that it occurs in the lipoid and proteinacious structures present in a tissue while ice crystals remain prominent. Metal machining theory describes and explains these findings and it is the same process that I'm suggesting applies to cryosectioning. By using the metal machining concept, I am simply applying essential information on machining that has been around for a long time, namely, that separation occurs at the plane of least resistance, and separation occurs by plastic deformation or fracturing. The fact that you have curling in sections is another indication of plastic deformation.

LECHENE: When you compare cutting at -30° and cutting with a saw at -190° I think that you're making phase diagram which says that approximately at

-40 ° you can push as much as you want on ice and you will not melt it. Now there is a last point, if I am wrong, in determining this is so, it doesn't demonstrate that you're right in cutting it at -30°C.

RICK: I think that renal papilla is not a very good suggestion because essentially there are the intracellular, at least in hydropenia, the intracellular concentrations with respect to the composition most typically intracellular; they have high sodium, high chloride. The second reason that they are not a good tissue is that they have high osmolarity so you cannot compare freezing pattern and cutting pattern and all those things with a normal tissue that has normal osmolarity.

SOMLYO, Andrew: If I understand correctly from theory, then in cutting a 1000Å section, it should melt through, which of course would bother us no end, because most of our experiments are done on 1000Å sections. But there is a way to get a little handle on what the temperature is not where the thermocouple is, but where the cutting occurs and to find out experimentally whether such melting takes place. I found the experiment that was done pleasing but since Avril did the experiment, I would prefer that she present it, because I think it might answer the question experimentally where perhaps the theory is somewhat more flawed.

SOMLYO, Avril: With your calculations, would you predict 1000Å section to melt at a particular temperature? I realize you have to define the ambient temperature.

SAUBERMANN: These calculations were done at -30° C. If we cut a 1000Å section with this system, we have to go to a much colder temperature. If we try to cut at say -20° C, I don't think it melts and I think you are perfectly safe. What happens is that the section crinkles and is plowed up on the block face.

SOMLYO, Avril: I did some sectioning with toluene at -95° C and monitored the temperature of both the edge of the knife and the specimen. One could cut from 800-1000Å sections at -96° C. You really had to have the knife temperature at exactly the melting point of the toluene (-95°C) in order to start to see some melting on the sections and their transparencies change. We cut quite rapidly. The ambient temperature of the chamber, I realize that you have to define these parameters, was -130° C. We are convinced that under our conditions there was no evidence of melting.

SOMLYO, Andrew: In other words, the reasons I brought it up partly was that there is an experimental method if you have organic solvent, for example, with different melting points that you freeze and section, that allows you to directly and experimentally demonstrate that there is no melting under these experimental conditions.

HUTCHINSON: I am surprised that you do not completely dehydrate your half micron sections in a matter of seconds since the ambient atmosphere is nitrogen gas and can be considered a vacuum as far as the water vapor is concerned.

SAUBERMANN: Obviously you don't, but the reason we don't is that after you pick up the section they are placed on a cooled holder at -155° C.

HUTCHINSON: How quickly do you do that?

SAUBERMANN: Just 30 seconds or so. The other thing is that there is frost on top of the microtome so that the atmosphere is partially saturated.

HUTCHINSON: I don't believe it is partially saturated because you have cold dry nitrogen flowing. I mean you have a Sorvall unit, as I understand. There is cold dry nitrogen flowing all the time through this unit, and I would consider that to be very much tantamount to being a vacuum.

SAUBERMANN: Well, 95% of our sections are fully hydrated when we transfer them to the microscope.

HALL: Not fully.

SAUBERMANN: Fully.

HUTCHINSON: No.

SAUBERMANN: Fully.

MODELLING THE ULTRA-RAPID FREEZING OF CELLS AND TISSUES

Alan P. MacKenzie

Center for Bioengineering
University of Washington
Seattle, Washington

INTRODUCTION

A very obvious interest has attached to the growth of ice in cells and tissues during the course of this conference. The presence of microcrystalline ice has been noted by some speakers and inferred by others where it could not be detected directly. Most speakers have assumed that such a microcrystalline ice phase must be physically stable below a sufficiently low temperature during the time required to complete a microanalytical or other measurement. Little or no direct evidence has been accumulated to tell us the safe limits above which temperatures we may expect to see the grain-growth, or recrystallization of ice, the diffusion of small ions, and the possible precipitation of conventional eutectic mixtures where the solutes in question were preserved in amorphous states after very rapid freezing. This being the case, it may be helpful to consider the results obtained when certain gels were frozen at different rates and examined by differential thermal analysis during slow warming. We will see that the method proves to be most informative.

Numerous studies on aqueous gels appear to have been undertaken with a conviction that the gel may serve usefully to model the living cell in certain important respects. Gels possess a physical rigidity. They can be prepared to contain known quantities of water. Their molecular frameworks tend to bind rather well-defined numbers of water molecules.

ISBN 0-12-362880-6

Salts and other solutes can be introduced and shown to diffuse within the gel at measurable rates. The rapid freezing of aqueous gels appears to have been a matter of particular interest (Moran, 1926; Luyet and Gehenio, 1940; Luyet, 1957; Luyet and Rapatz, 1958; Meryman, 1958; Dowell, Moline, and Rinfret, 1962; Luyet, 1965; MacKenzie and Luyet, 1967; Persidsky and Luyet, 1975). Here, as in other work on aqueous gels, we may look to the biological implications.

We will examine the freezing behavior of a number of chemically different gels in the present study. All the gels were chosen for their partial resemblance to biological structures. We will report and compare the behavior of one such aqueous gel in the presence and absence of NaCl. We will describe the freezing behavior of aqueous solutions distinguished only from certain of the gels by the absence of the cross-links responsible for the maintenance of the gel structure. It will be seen that the cross-linking of the macromolecular constituents contributes to determine the freezing behavior and that the gel may be specially suited to serve as a model to study the freezing of whole cells. It will also be seen that very rapidly frozen gels are physically stable during slow warming to remarkably high temperatures.

MATERIALS AND METHODS

Materials

Cross-linked polyacrylamide gels were obtained in bead form from Bio-Rad Laboratories, Inc., Richmond, California, as Bio-Gel P-2, P-4, P-6, and P-10. The dry beads were soaked in distilled water for several hours at 20° C and washed repeatedly prior to use. Bio-Gels P-2, -4, -6, and -10 took up 1.5, 2.4, 3.7, and 4.5 g water per g dry gel to furnish gels containing 60, 70, 79, and 82% water, w/w, respectively.

Cross-linked dextran gels were obtained in bead form from Pharmacia Fine Chemicals, Inc., Piscataway, New Jersey, as Sephadex G-10, G-15, G-25, and G-50. The dry beads were soaked in distilled water for several hours at 20° C and washed repeatedly prior to use. Sephadex G-10, -15, -25, and -50 took up 1.0, 1.5, 2.5, and 5.0 g water per g dry gel to furnish gels containing 50, 60, 71.4, and 83.3% water, w/w, respectively.

A chemically modified cross-linked dextran gel prepared, according to the manufacturer, by the hydroxypropylation of Sephadex G-25, was likewise obtained from Pharmacia Fine Chemicals, Inc., Piscataway, New Jersey, as Sephadex LH-20. As with the other gels, Sephadex LH-20 was soaked and washed in distilled water prior to use. Sephadex LH-20 took up 2.0 g water per g dry gel to yield a gel containing 66.7% water, w/w.

Gelatin was obtained from the Central Scientific Company, Inc., Chicago, Illinois, as Cenco Gelatin catalog number C-2125. Gelatin gels of known w/w concentration were prepared by the careful addition of the dry gelatin to the distilled water. Mixtures were maintained at 70° C until clear and uniform in tightly capped flasks after which they were stored in the refrigerator.

Soluble dextran was obtained from Pharmacia Fine Chemicals, Inc., Piscataway, New Jersey, as Dextran T-110 having a weight average molecular weight of 110,000. Dextran solutions of known w/w concentration were prepared in sealed flasks in a boiling water bath. Dissolution occurred very rapidly at 100°C.

Methods

Differential thermal analysis (d.t.a.) was carried out with an apparatus patterned after the instrument described by Rasmussen and MacKenzie (1968, 1971). Copper/Constantan thermocouples made from 2 mil (0.05 mm) diameter wires were inserted in ca. 5 to 10 μl volumes of distilled water and experimental sample (drained gel beads, gelatin gel, and aqueous dextran solution) introduced into separate (but closely matched) 1.5 mm o.d. glass capillary tubes. These tubes were inserted into close-fitting holes located symmetrically in a massive (7.5 cm diameter) aluminum block. Suitably insulated, the block could be cooled (e.g., to -175°C) to receive the rapidly frozen specimens. The same block could also be cooled and warmed at controlled rates in the range: 0.1 to 10 deg C per minute. The difference between the temperature of the distilled water reference and the experimental sample was obtained as the difference of the thermocouple outputs. A second thermocouple placed in the distilled water/pure ice reference indicated the temperature of the reference (and allowed the precise calculation of the temperature of the sample). Temperature difference (ΔT) was recorded as a function of reference temperature (T) with an x - y chart recorder. Thermal events were read as perpendicular displacements of the recorded trace. The thermograms reproduced in the present study denote exothermic reactions (e.g., the formation of ice) by an upward motion of the differential signal. Endothermic reactions (e.g., the melting of a eutectic mixture, or of pure ice in the sample) are represented by a downward motion of the differential signal. The experimental set-up is shown schematically in Figure 1.

"Rapid" freezing (at rates of ca. 10^3 deg C per minute) was accomplished by the abrupt immersion of the loaded sample tube in liquid nitrogen. To achieve a "slow" freezing we transferred a freshly loaded sample tube to a warm apparatus. We cooled the block to -5 (supercooling the sample in the process). We raised and "seeded" the sample near the top with the touch of a forceps dipped in liquid nitrogen, lowered the sample, and kept the temperature at ca. -5 for another 10 minutes. Slow freezing was completed with further cooling at controlled rates of 1 to 2 deg C per minute to temperatures in the range: -125 to -175° C.

RESULTS

General

All the gels and aqueous solutions were subjected to the same sorts of thermal analysis to yield essentially similar sorts of thermograms. The precise shapes of individual thermograms proved, at the same time, to be sufficiently different (and sufficiently informative) that it was thought best to present each set in turn and to examine the thermal data sample by sample. We will present the findings in accordance with the chemical composition of the system.

Polyacrylamide Gels

Figure 2 reproduces four thermograms obtained when Bio-Gel P-2, -4, -6, and -10, rapidly frozen, were warmed at ca. 2 deg C per minute. The P-2 thermogram displays a rising (exothermic) baseline culminating in a small but sharp exothermic peak at -15.4° C. Such a peak can only represent the further conversion of water to ice on warming. The thermal trace returns to the baseline, runs level for several degrees C, and drops at -7.7 °C to signal the beginning of the

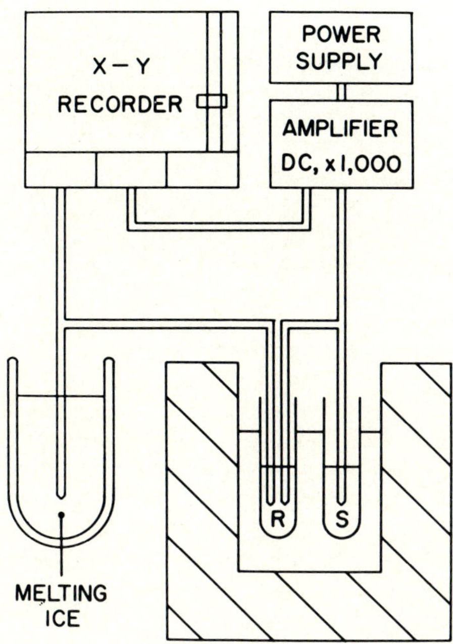

FIGURE 1. Apparatus for differential thermal analysis. R = distilled water/ice reference; S = experimental sample. Reference and sample tubes are located symmetrically in a massive metal block (plain area) surrounded by a ca. 10 cm thickness of foamed polystyrene insulation (shaded area). Cooling is effected by the aspiration of liquid nitrogen through a series of holes drilled vertically through the metal block. Warming is accomplished electrically with a Teflon-encased nichrome element. The differential thermal signal is amplified with the aid of an Analog Devices #261K operational amplifier incorporated in a Fairchild thermocouple amplifier circuit (details can be obtained from the author). The x - y recorder must possess an input sensitivity not less than 0.1 mV per inch to display sample/reference temperatures on the x axis. The author employs a Hewlett-Packard #7047A x - y recorder having a maximum input sensitivity of 0.05 mV per inch.

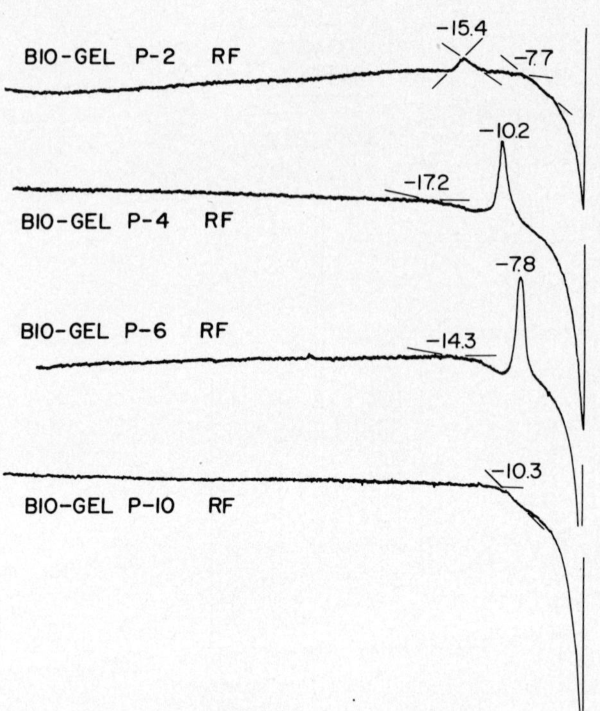

Figure 2. Differential thermograms obtained during the slow warming of aqueous polyacrylamide gels. (RF = rapidly frozen).

(seemingly thermodynamic) melting endotherm. The thermal analysis of the P-4 gel furnished a very stable, horizontal thermal trace indicative of a sample having a constant heat capacity up to -18.0° C after which we detect: (1) an endothermic trend, (2) a very sharp exotherm having a maximum at -10.4° C, (3) an immediate entry into the melting endotherm, and (4) the completion of the melting of all the ice within one or two tenths of a degree C of 0°C. We note that the further conversion of water to ice on warming required a higher temperature, was more extensive, and was followed much sooner by the start of the final melting in this, the second thermogram of the series. We note also that the area contained by the melting endotherm far exceeds the area contained by the preceding freezing exotherm (though we recognize that d.t.a. can only furnish semi-quantitative comparisons in this respect).

The P-6 thermogram reveals a very gradual (and most likely insignificant) exothermic trend terminated by a clearly significant endothermic transition (at -14.3° C) and the same exothermic/endothermic behavior we saw in the P-4 thermogram. Comparing the behavior of the P-4 and the P-6 thermograms we see that the temperatures of the several thermal transitions are higher in each case by several degrees C in the P-6 thermogram. In rather marked contrast, the P-10 thermogram displays a featureless thermal trace all the way to the start of a melting endotherm at -10.3° C. Nothing else was seen. We were, that is to say, unable to detect any additional conversion of water to ice during the slow warming of the P-10 gel (or of any other polyacrylamide gel having a still higher initial water content, though we had frozen all the gel samples the same way). Figure 2 describes a mode of warming behavior that would seem to be: (1) incompletely developed in the P-2 gel, (2) better developed in the P-4 and P-6 gels, and (3) absent in the P-10 gel. We will see in a moment that the pattern is largely duplicated in the corresponding series of cross-linked dextran gels.

Clues to the nature of the rapidly frozen state and the changes occurring during slow warming were obtained from series of experiments in which slow warming was interrupted prior to the total melting of ice. Samples warmed only to certain temperatures were cooled a second time (to -125°C or a still lower temperature) after which they were rewarmed to furnish new thermograms. These latter thermograms were compared with those obtained during the original warming and with others obtained from the same sorts of gels after "slow" freezing.

Figure 3 reproduces such a set of experiments with Bio-Gel P-4. The topmost curve depicts the results of a run duplicating the conditions employed to obtain the second thermogram seen in Figure 2. The second curve in Figure 3 demonstrates the thermal behavior when another P-4 sample, rapidly frozen, was rewarmed only to ca. -9° C. The third curve ("RF/RW") reproduces the thermogram obtained when the sample that was warmed only to -9 (and cooled again to -125) was warmed a second time. The bottom curve in Figure 3 depicts the thermal trace obtained during the warming of a slowly frozen sample. We see at once that the thermal cycling has caused the rapidly frozen gel to duplicate the behavior of the slowly frozen gel within the limits (or nearly within the limits) of the experimental errors inherent in d.t.a. The two lower thermal traces seen in Figure 3 reveal the same absence of thermal signal to ca. -10°C, the same abrupt endotherms at -9.8 and -9.2, the same minor exotherms and the same endothermic entries into the final

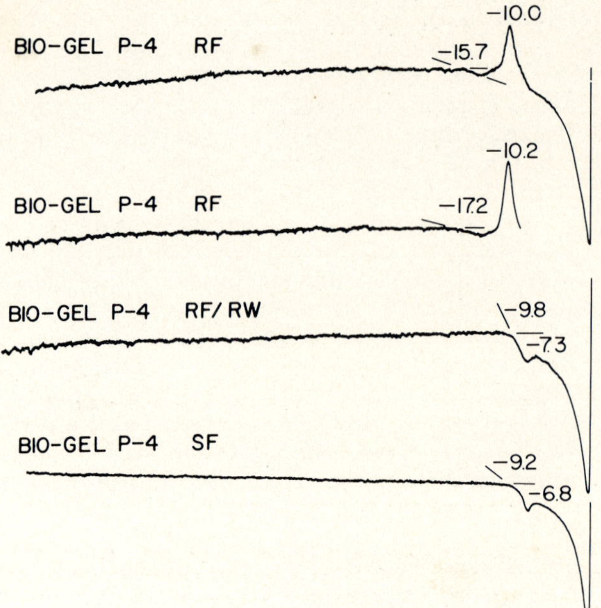

Figure 3. Differential thermograms obtained during the slow warming of Bio-Gel P-4. (RF = rapidly frozen; RF/RW = rapidly frozen and rewarmed; SF = slowly frozen).

melting process at -7.3 and -6.8° C, respectively, that characterize the warming of a totally frozen system. It would appear that we have shown that a single warming to a sufficiently high temperature suffices to cause a frozen P-4 gel to lose the metastability it acquired during faster freezing.

The results of the studies on the cross-linked polyacrylamide gels are summarized in Table 1. Anticipating a discussion of the findings in a later section we have employed the term "antemelting" to describe the first endothermic transition seen in most of the thermograms (Luyet and Rasmussen, 1968). Following the same nomenclature we have employed the term "antemelting temperature" (T_{am}), to designate the temperature at which the event is seen. The temperature at which the rate of further conversion of water to ice with slow warming attains a maximum has been called T_c. We see from Table 1 that T_{am} and T_c are each increased by several degrees C with each increase in original water content. We should also note the constancy of the T_c during three separate runs on the P-4 gel, rapidly frozen.

TABLE 1. Transition Temperatures in Frozen Aqueous Gels: 1. Cross-Linked Polyacrylamide (Bio-Gel P-)

gel	gH_2O/g dry gel	% w/w H_2O	treat-ment[a]	T_{am}[b]	T_c[b]
P-2	1.5	60	RF		-15.4
P-4	2.4	70	RF	-18.0	-10.4
P-6	3.7	79	RF	-14.3	-7.8
P-10	4.5	82	RF	-10.3	
P-4	2.4	70	RF	-15.7	-10.0
P-4	2.4	70	RF	-17.2	-10.2
P-4	2.4	70	RF/RW	-9.8	
P-4	2.4	70	SF	-9.2	

[a]We have followed the same designations that we employed in the Figures.

[b]T_{am} = antemelting temperature; T_c = temperature at the maximum of the exotherm that represents the further conversion of water to ice.

Dextran Gels

Figure 4 illustrates the results obtained when Sephadex G-10, -15, -25, and -50 were each warmed at ca. 2 deg C per minute after respective rapid freezings. We see at once that the G-10, -15, and -25 gels exhibit marked evidence of the conversion of water to ice with warming in the range: -40 to -10° C. The G-50 sample appears to exhibit a very small exotherm at ca. -6° C. The crystallization sharpens as we progress from G-10 to G-15 to G-25 (i.e., from lower to higher original water content). Antemeltings are seen during the warming only of the G-25 and the G-50 gels. The final melting process appears to begin in each case soon after the completion of the crystallization exotherm (and to be complete in a necessarily smaller temperature interval the higher the water content). None of the four thermograms furnishes any evidence of any thermal transition at a lower temperature than those already noted.

We chose Sephadex G-25 for further study and froze it rapidly, as before, to furnish a duplicate thermogram on warming through 0° C. The results are seen in the first

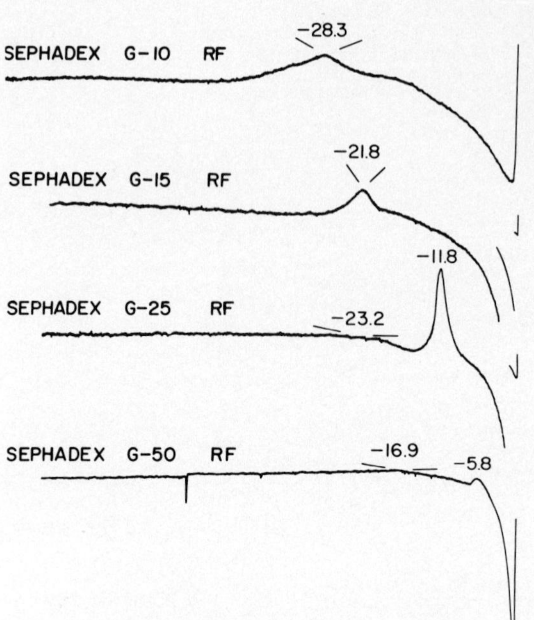

Figure 4. Differential thermograms obtained during the slow warming of aqueous cross-linked dextran gels. (RF = rapidly frozen).

thermogram in Figure 5. We froze another sample and warmed it only to the completion of the sharp exotherm (i.e., to ca. -11°C) to furnish the second thermogram in Figure 5. Cooling the same sample from -11 to ca. -100° C we warmed it a second time to yield the third thermogram. It will be seen that the third thermogram lacks the antemelting at ca. -25° C and the subsequent exotherm and that it reveals instead a sharp antemelting at -12.1 and a very small exotherm at -8.4° C after which the sample begins to melt. The fourth thermogram in Figure 5 depicts the warming of a slowly frozen sample of the same gel. Figure 5 shows how closely the Sephadex G-25 duplicates the behavior of the Bio-Gel P-4 (see again Figure 3). Rapid freezing results, in each case, in no more than a partial conversion of freezable water to ice. Slow warming permits an additional freezing. An interrupted warming and a subsequent cooling allows us to show that the freezable water has all been frozen and that the thermal cycling has caused the gels to assume respective physico-chemical states very similar to those achieved by a slow initial freezing. Table 2 lists the temperatures of the thermal transitions identified in Figures 4 and 5.

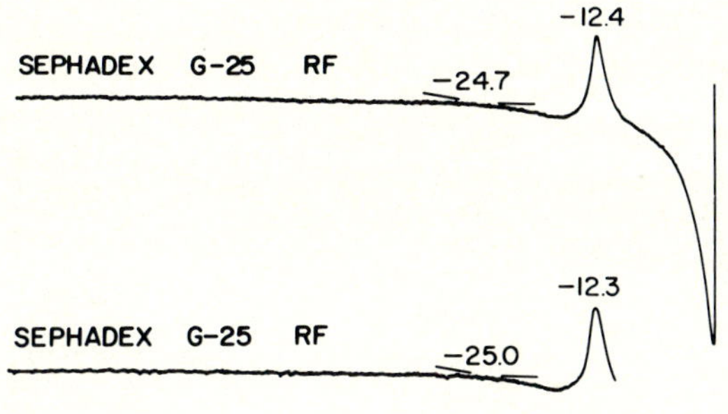

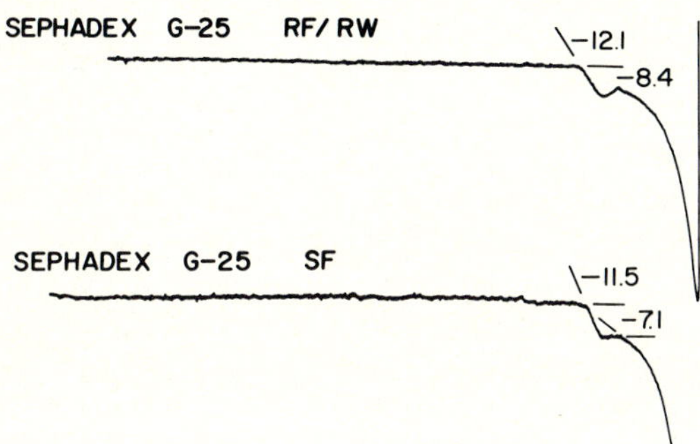

Figure 5. Differential thermograms obtained during the slow warming of Sephadex G-25. (RF = rapidly frozen; RF/RW = rapidly frozen and rewarmed; SF = slowly frozen).

Chemically Substituted, Cross-Linked Dextran

Sephadex LH-20 (hydroxypropyl-substituted, cross-linked dextran) was subjected to rapid freezing and to subsequent slow warming through 0° C to yield the first thermogram in Figure 6. One notes an antemelting at -29.7, a sharp exotherm with a peak at -20.9, and a smooth melting curve from ca. -19°C. A second thermogram reproduces the behavior of a duplicate sample warmed only to the beginning of the final melting (after which it was cooled to ca. -90° C). The third thermogram illustrates the rewarming behavior of the same sample; we see immediately that the antemelting temperature has hardly changed, that there is no detectable exotherm, and that the final melting curve still begins at ca. -19° C. A fourth thermogram illustrates the warming behavior of the same gel after a slow freezing. Figure 6 would seem to have a lot in common with Figures 3 and 5. Each of the three sorts of gel fails to freeze completely during rapid cooling. Each completes its freezing during slow rewarming. We would only note that the chemically substituted dextran gel behaves quite differently in one

TABLE 2. Transition Temperatures in Frozen Aqueous Gels: 2. Cross-Linked Dextran (Sephadex G-)

gel	gH_2O/g dry gel	% w/w H_2O	treat-ment[a]	T_{am}[b]	T_c[b]
G-10	1.0	50	RF		-28.3
G-15	1.5	60	RF		-21.8
G-25	2.5	71	RF	-23.2	-11.8
G-50	5.0	83	RF	-16.9	-5.8
G-25	2.5	71	RF	-24.7	-12.4
G-25	2.5	71	RF	-25.0	-12.3
G-25	2.5	71	RF/RW	-12.1	
G-25	2.5	71	SF	-11.5	

[a]We have followed the same designations that we employed in the Figures.

[b]T_{am} = antemelting temperature; T_c = temperature at the maximum of the exotherm that represents the further conversion of water to ice.

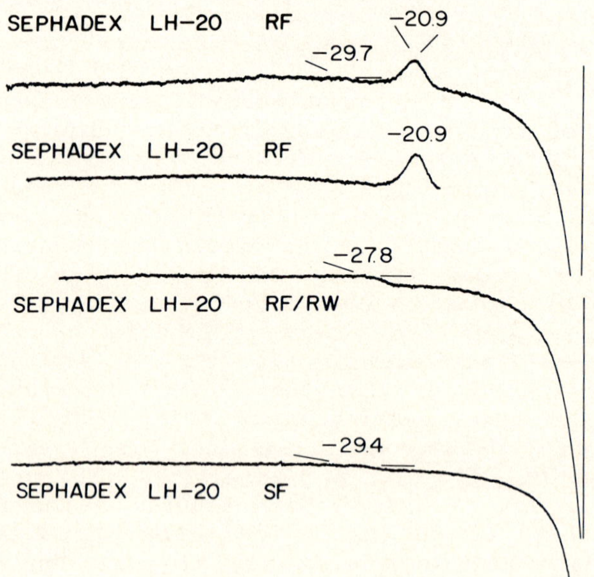

Figure 6. Differential thermograms obtained during the slow warming of chemically substituted, cross-linked dextran gel (Sephadex LH-20). (RF = rapidly frozen; RF/RW = rapidly frozen and rewarmed; SF = slowly frozen).

major respect. Thermal cycling and slow freezing raise the antemelting temperature in the polyacrylamide and the plain dextran gels by ca. 10 degrees C. The same thermal cycling and slow freezing are without effect on the antemelting temperature in the chemically substituted dextran gel.

Gelatin Gels

Mixtures of water and gelatin were made and warmed to yield gels with wt/wt gelatin concentration of 28.6 and 40.0% in an effort to duplicate, as far as possible, Sephadex G-25 and G-15, respectively. Slow warming after "rapid" freezing furnished the first and third thermograms in Figure 7. The 28.6% rapid freezing thermogram showed a gradual endothermic trend after -23.1° C with a pronounced acceleration at ca. -17.2° C. A small exotherm was seen at ca. -6.3° C (after which the final melting was very rapid). The 40% rapid freezing thermogram revealed a suggestion of an extended exotherm from ca. -45 to -25° C (or an otherwise unstable baseline signal) terminated by a presumed antemelting at ca. -21.2° C. This was followed by a small, well-defined exotherm at -9.8 and a rapid entry into final melting (which we saw terminate within one or two tenths of a degree of 0° C.)

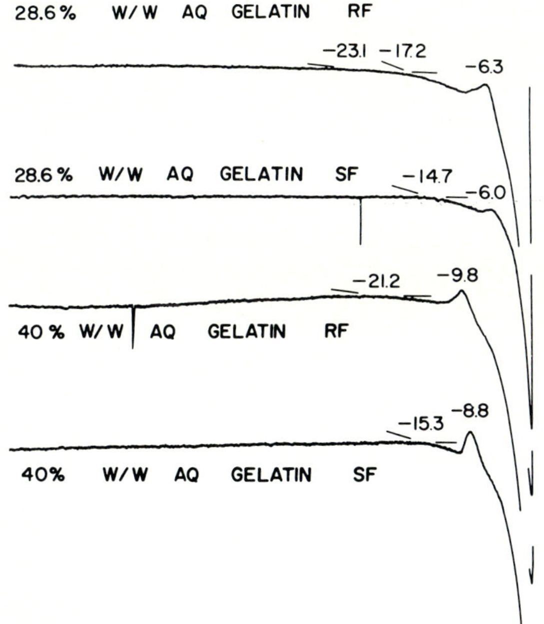

Figure 7. Differential thermograms obtained during the slow warming of aqueous gelatin gels. (RF = rapidly frozen; SF = slowly frozen).

Corresponding "slow" freezing thermograms obtained with gelatin gels of the same two wt/wt concentrations yielded the second and fourth thermograms seen in Figure 7. We observe in each case that the baselines are very straight indeed (suggestive of the total absence of thermal signals) to antemelting temperatures of -14.7 and -15.3° C, respectively. Antemelting is followed in each case by a small exotherm very similar to that seen after a respective rapid freezing. It would appear that the slower freezing straightens the baseline and raises the antemelting temperature, but that it does not affect the final exotherm which would seem, therefore, to characterize the material rather than the treatment.

Aqueous Dextran

A further set of comparative experiments was undertaken with plain aqueous dextran solutions. These solutions mimicked the chemical composition of the plain Sephadex gels. Only the chemical cross-links responsible for the maintenance of the gel structure were missing. Wt/wt dextran concentrations of 28.6 and 40% corresponded with Sephadex G-25 and G-15, respectively. Figure 8 reproduces the results

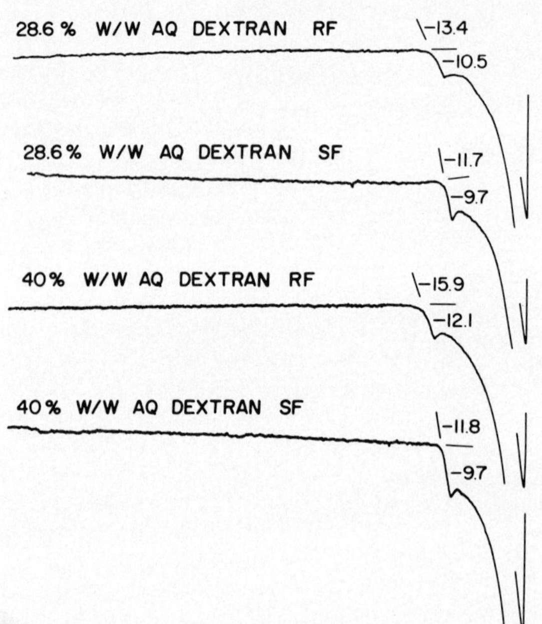

Figure 8. Differential thermograms obtained during the slow warming of aqueous dextran solutions. (RF = rapidly frozen; SF = slowly frozen).

obtained. The first and second thermograms illustrate the warming of the 28.6% solution. Neither thermogram shows any baseline instability. Antemelting and subsequent exothermic reaction are observed at -13.4 and -10.5, and -11.7 and -9.7°C, respectively. The slower freezing appears to raise both these temperatures. Findings from corresponding experiments with the 40% dextran solution are seen in the third and fourth thermograms. The baselines are once again very nearly straight. Antemelting and the subsequent exothermic reaction are seen at -15.9 and -12.1, and -11.8 and -9.7° C, respectively. Note the absence of any thermal events at lower temperatures. Clearly the two dextran solutions behave in essentially identical ways. Note especially the identity of the temperatures of the transitions in the second and fourth thermograms. We conclude that the slow freezing creates the same physicochemical state without regard to the initial concentration of the solution. Rapid freezing would appear, in contrast, to effect a greater reduction in the temperatures of the transitions in the 40% wt/wt dextran solution. We note the absence of any other differences in the thermograms attributable to the freezing rate. The findings are summarized in Table 3.

Aqueous Saline Dextran Gels

A final set of experiments was conducted to determine the effect the presence of a dissolved salt might have on the freezing/thawing behavior of one of the cross-linked dextran gels. Aqueous Sephadex G-25 was washed repeatedly with 5% wt/wt aqueous NaCl (on a Millipore type HA filter at ca. 20°C). The beads were drained and subjected to a "rapid" freezing (see, again, Materials and Methods). The sample gel was warmed to yield the second thermogram in Figure 9. This thermal trace exhibits a somewhat irregular baseline, a characteristic antemelting, a minute endotherm, and a large exotherm, after which the sample melts. All the thermal events appear to be very clearly defined. Second and third quantities of aqueous G-25 were washed repeatedly with 10 and 15% wt/wt aqueous NaCl, respectively, and subjected to the same careful drainage and rapid freezing. These sample materials furnished the third and fourth thermograms in Figure 9. It can be seen that these thermograms exhibit similarly well-defined features. The sample equilibrated with the 10% NaCl solution reveals an antemelting, an exotherm, a very small endotherm, and a subsequent final

TABLE 3. Transition Temperatures in Frozen Aqueous Gels: 3. Effect of Cross-Linking on Dextran

specimen	% w/w H_2O	treat-ment[a]	T_{am}[b]	T_c[b]
Sephadex G-15	60	RF		-21.8
Sephadex G-15	60	SF	-17.5	
Dextran T-110	60	RF	-15.9	
Dextran T-110	60	SF	-11.8	
Sephadex G-25	71	RF	-23.2	-11.8
Sephadex G-25	71	SF	-11.5	
Dextran T-110	71	RF	-13.4	
Dextran T-110	71	SF	-11.7	

[a]We have followed the same designations that we employed in the Figures.

[b]T_{am} = antemelting temperature; T_c = temperature at the maximum of the exotherm that represents the further conversion of water to ice.

melting. The gel exposed to the 15% NaCl exhibits an exotherm, a small endotherm, and a final melting.

The temperatures at which the various thermal events are observed in the presence of NaCl are best examined with reference to the behavior of the salt-free G-25. The first thermogram in Figure 5 has, to this end, been reproduced in Figure 9 to provide a sort of reference state. It will be seen that the antemelting temperature is lowered in the presence of 5% NaCl and that it is lowered again in the presence of 10% NaCl. It will be equally obvious that the exothermic reaction that denotes the further conversion of water to ice occurs at a lower temperature the higher the salt concentration. The essential constancy of the ca. 4 degree C gap between the temperatures of the successive exotherms in the four thermograms in Figure 9 bears testimony to the nature of the implied event. We are, it appears, seeing the same further conversion of water to ice in each case. Had we not seen the event in the first salt-free thermogram we might have been tempted to conclude a special relationship between the event and the presence of the salt.

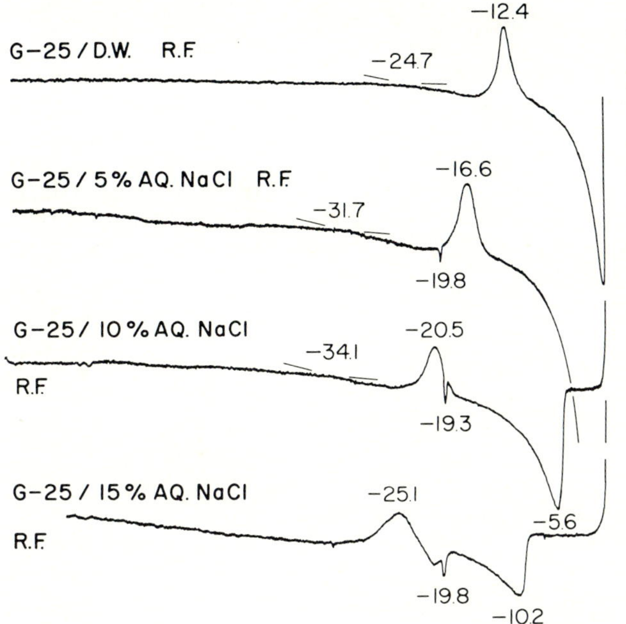

FIGURE 9. Differential thermograms obtained during the slow warming of Sephadex G-25 previously equilibrated with distilled water and with 5, 10, and 15% w/w aqueous NaCl, respectively. (RF = rapidly frozen).

The presence of a small endotherm at ca. -20 C in the second, third, and fourth thermograms in Figure 9 supports the notion that we are seeing the eutectic melting of a very small quantity of crystalline salt. The absence of any such peak in the first thermogram confirms the supposition. The size of the peak suggests that the bulk of the NaCl in the beads has not crystallized at any time during the rapid freezing or the subsequent slow warming. To test this contention we conducted one last thermal analysis with Sephadex G-25 we had washed with 5% NaCl but had not drained. Figure 10 compares the findings with the results we obtained when we drained the gel (the upper curve in Figure 10 has been taken from Figure 9). The eutectic melting seen in the undrained preparation is so much larger than that seen in the drained preparation that we concluded that the NaCl that yielded the eutectic meltings seen in Figure 9 (and in the first thermogram in Figure 10) must have crystallized at the surface of the drained beads and that the NaCl contained within the beads must indeed have remained dissolved at all times. The temperatures of the thermal transitions observed in Figures 9 and 10 are summarized in Table 4.

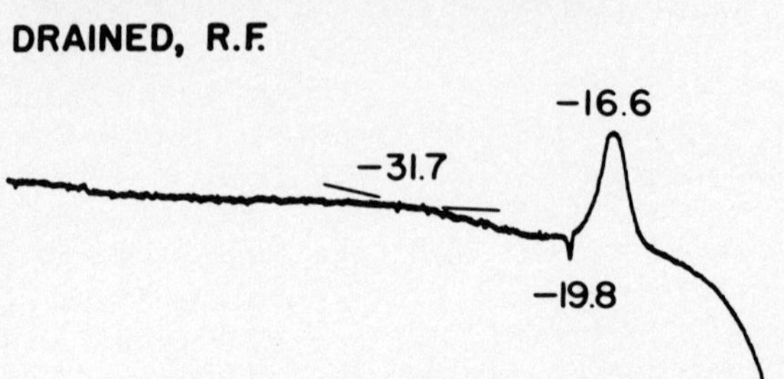

Figure 10. Differential thermograms obtained during the slow warming of Sephadex G-25 previously equilibrated with 5% w/w aqueous NaCl. Upper thermogram (taken from Figure 9): drained gel beads; lower thermogram: undrained gel beads. (RF = rapidly frozen)

TABLE 4. Transition Temperatures in Frozen Aqueous Gels: 4. Effect of NaCl on Sephadex G-25.

% aq. NaCl	treatment[a]	T_{am}[b]	T_c[b]	T_e[b]
0	RF	-24.7	-12.4	-
5	RF	-29.5	-16.6	-19.6
5	RF	-31.7	-16.6	-19.8
10	RF	-34.1	-20.5	-19.3
15	RF		-25.1	-19.8

[a] We have followed the same designations that we employed in the Figures.

[b] T_{am} = antemelting temperature; T_c = temperature at the maximum of the exotherm that represents the further conversion of water to ice. Also: T_e = the temperature of eutectic melting of frozen aqueous NaCl.

CONCLUDING DISCUSSION

Plain Aqueous Gels

Differential thermal analysis shows that cross-linked polyacrylamide, cross-linked dextran, and chemically substituted, cross-linked dextran gels are not completely frozen during cooling at rates of the order of 10^3 deg C per minute to -196°C. Judged by the same thermal criteria it appears that correspondingly concentrated gelatin gels and aqueous dextran solution are completely frozen during an equivalent cooling to the same temperature. Slow warming from cryogenic temperatures caused, as we saw, the "completion" of the freezing of the cross-linked polyacrylamide, cross-linked dextran, and chemically substituted, cross-linked dextran gels in a single thermal passage. Gel beads subjected to rapid freezing, slow warming to high sub-zero temperatures, slow cooling, and slow rewarming behaved in each case like slowly frozen samples during the last of these four steps. We conclude that a rapid freezing and a slow warming to a sufficiently high sub-zero temperature together create a physicochemical state identical in certain respects to that achieved by a nucleation of ice at -5° C and a subsequent slow cooling. We assume that a controlled cooling at ca. 1 deg C per minute furnishes a series of aqueous gels we can reasonably regard as "completely frozen."

We have seen that the further conversion of water to ice during warming was not observed in gelatin gels or in dextran solutions, though the water contents were the same as those of the Bio-Gel P-2 and -4 and the Sephadex G-25 and -25, and the freezing was in each case about equally rapid. It would appear that the covalently cross-linked Bio-Gel and Sephadex are not readily deformed by freezing and that their macromolecular structures harbor water that would, given time, find its way to the nearest ice crystal (with a corresponding accommodation on the part of the gel). The Bio-Gel P-10 and the Sephadex G-50 did not exhibit the further freezing on warming. These latter gels contained 4.5 and 5.0 g water per g dry gel solids, respectively, and were correspondingly less heavily cross-linked. One assumes that the less heavily cross-linked structures accommodated faster. Gelatin gels are cross-linked by hydrogen bonds that are broken and made again elsewhere with relatively little difficulty. Here again we have a basis for a faster physical accommodation. Dextran solutions can be expected to exhibit a macromolecular response faster than that of any gel.

Figure 7 reveals significant differences in the antemelting temperatures of the gelatin gels after rapid and slow freezing, respectively. Figure 8 shows that the dextran solutions behave in a corresponding way. We may reasonably conclude that freezing at ca. 10^3 deg C per minute to -196° C was fast enough to cause all the systems we examined to exhibit lowered antemelting temperatures, though the same freezing rate allowed the "complete" freezing of the gelatin gels and the dextran solutions. Antemelting would seem to be the more sensitive index of the physicochemical state of the frozen aqueous macromolecular system. We will return to the question of the nature of the antemelting.

We have seen from Figures 2 and 4 that antemelting and further freezing occur at higher temperatures the higher the total water contents of the respective gels. Had the rapid freezing achieved the same physical accommodation of the gel solids in each case we might have expected antemelting and further freezing always to occur at the same temperatures without regard to the composition of the system prior to freezing. Higher second order transition temperatures are associated with a greater dehydration of the solute phase in which they are observed (Luyet and Rasmussen, 1968; Rasmussen and Luyet, 1969; MacKenzie, 1977a). This would suggest a lesser resistance to deformation during rapid freezing the higher the initial water content. Put the other way round, the findings suggest a greater resistance to a molecular accommodation the greater the number of covalent cross-links per unit weight gel solids.

Summarizing the observations on aqueous gels and dextran solutions we may safely conclude that rapid cooling (at ca. 10^3 deg C per minute) to -196° C promotes a metastability maintained in every case during slow warming to remarkably high sub-zero C temperatures. Metastability is characterized by lowered antemelting temperatures and, in the case of the covalently cross-linked gels containing 1 to 3.7 g water per g gel solids, by the further conversion of freezable water to ice just prior to final melting.

Aqueous Saline Gels

It is immediately apparent from Figure 9 that the presence of the NaCl in the aqueous cross-linked dextran gel has not altered the way the rapidly frozen material behaves during slow warming. All the salt has done is lower the temperatures at which the samples exhibit antemelting, further freezing, and final equilibrium melting. These lowerings are in each case in approximate proportion to the concentration of the NaCl.

Simple eutectic behavior is widely recognized in aqueous NaCl during freezing and thawing, though the NaCl may supersaturate (MacKenzie, 1977b). We did not see more than a trace of simple eutectic behavior during the warming of the dextran gel beads we had equilibrated with aqueous NaCl (this despite the known permeation of the cross-linked dextran matrix by the sodium and the chloride ions). Figure 10 reminds us that we had no difficulty detecting regular eutectic behavior by d.t.a. when we failed to drain the gel beads prior to rapid freezing. Undrained samples contained ca. 25% volume/volume free aqueous NaCl. This was sufficient to give the eutectic melting endotherm seen in the lower thermogram in Figure 10. Had the NaCl within the gel bead undergone the same eutectic melting from the same crystalline state, the endotherms in the two thermograms at ca. -20° C would have been of a comparable size (roughly 1.4 times the heat would have been required to effect the eutectic melting of the undrained sample. The calculation allows for the exclusion of the NaCl from the dextran itself and from some of the more tightly bound water).

The findings in the presence of NaCl permit several tentative conclusions. It would appear, first, that the dissolved NaCl cannot nucleate within the gel, nor can its crystallization be induced by contact with salt that crystallized outside the beads. It is even possible, though it remains to be demonstrated, that, other things being equal, the dissolved NaCl represents the thermodynamically more stable state at all temperatures.

We conclude, secondly, that the observed behavior during slow warming is a consequence of the thermal properties of a "solute phase" that incorporates gel solids, sodium ions, chloride ions, and unfrozen water. Such an amorphous solute phase will concentrate during freezing. It will retain those water molecules that (1) hydrate the dextran and the NaCl, and (2) diffuse too slowly to reach the nearest ice in the time available. We estimate the weight/weight ratios, dextran solids/NaCl in the Sephadex G-25/5%, /10%, and /15% aqueous NaCl to be ca. 10:1, 5:1, and 3:1, respectively. We would expect to see more water remaining unfrozen the greater the weight/weight contribution by the NaCl. A picture emerges in which an aqueous saline dextran phase exhibits an antemelting transition at a lower temperature the more the phase consists of sodium ions, chloride ions, and water molecules. The sooner the antemelting process occurs on slow warming, the sooner the system acquires the freedom it needs to facilitate the further conversion of freezable water to ice.

We conclude, thirdly, that the mixed amorphous solute phase behaves in a more or less typical way during the final equilibrium melting process. While the solute phase cannot undergo a total dissolution, being constrained by the cross-linking of the dextran, the thermograms in Figure 9 show that the heat required to melt ice increases with increasing temperature, up to the final melting point, much as it does in simple aqueous solutions.

A question arises as to the mobilities of the sodium and the chloride ions and the unfrozen water below the so-called antemelting temperature. The thermograms we obtained to date with the model gels have not given us an answer. Other experiments, in which we measured the electrical resistance of various aqueous solutions during cooling to and warming from ca. -125° C suggest a residual mobility to a generally much lower temperature (MacKenzie, 1980). Correlating the results of the resistance experiments and the thermal analyses of aqueous saline solutions of globular proteins, we found a glass transition at ca. -80° C and a readily recognized antemelting at ca. -40° C (the temperatures varied by several degrees C with the weight/weight ratio: protein/NaCl). While we have not yet determined the electrical resistance of any of the gels as a function of temperature we suspect that their behavior will prove to be generally similar. Such a finding would imply a gradually increasing ionic mobility above a characteristic glass transition temperature, up to the antemelting above which temperature the mobilities of the respective ions would rise much faster with additional warming.

Practical Significance

We may well inquire the relevance of the studies on model aqueous gels to the problems encountered in cryogenic electron microscopy and microanalysis. Electron microscopists employ very rapid freezing, cold transfer, cryofracture, and cryoultramicrotomy. When they want to effect the sublimation of some or all of the ice they turn to freeze-etching and freeze-drying, respectively. Persons engaged in electron microscopy and microanalysis are particularly anxious to learn the limiting temperatures below which they must operate for best results. We think our model system studies can help in every case.

The models we employed possessed physical and chemical similarities to living cells and tissues. Water contents ranged from 1.0 to 5.0 g per g dry gel solids, or from 50 to 83% water (referred to the total wet weight). The human erythrocyte contains ca. 65% water by weight; other mammalian cells contain 70 to 85% water by weight, generally speaking. Plant cells may contain a lot more water. Many macromolecules are covalently linked, one to another, in the living cell; others are strongly associated by various sorts of non-covalent bonds.

The discovery that the thermal instabilities after rapid freezing were observed at such remarkably high temperatures--antemelting between -35 and -10 and further freezing between -30 and -6° C, according to the model--must be seen as most significant. Antemelting appears to represent the rather sudden acquisition of a new degree of translational motion in the gradually softening amorphous solute phase. Grain growth or migratory recrystallization of ice is observed to begin soon afterward with further warming (Luyet and Rasmussen, 1968; Rasmussen and Luyet, 1969; MacKenzie, 1977a). All the available evidence suggests that it is necessary only to keep a frozen specimen below its antemelting temperature to prevent a redistribution of ice (and a corresponding redistribution of the solute component). It would seem, by this reckoning, that it will suffice to maintain a rapidly frozen specimen below its antemelting temperature during, for example, frozen thin sectioning, to minimize "ice artifact." Such a conclusion would support Saubermann's argument (this conference) that frozen thin sectioning can be conducted safely up to, e.g., -30°C.

Other studies have established a close relationship between antemelting and the loss of structural detail during freeze-drying. Frozen aqueous solutions have been found to freeze-dry at and above characteristic threshold temperatures with a simultaneous "collapse" of the amorphous solute matrix (MacKenzie, 1975). The same solutions freeze-dry at lower temperatures to yield materials in which the distribution of the microcavities duplicates the microscopic distribution of the ice up to the time of its sublimation. The conversion from freeze-drying with good "retention" of matrix structure to freeze-drying with a complete "collapse" occurs rather abruptly (in an interval of 5 to 10 deg C, generally). "Collapse temperatures" have been correlated quite closely with antemelting temperatures, and it would appear that it will generally suffice to freeze-dry below an antemelting temperature to obtain an ultrastructurally satisfactory freeze-drying.

We have seen at the same time that the temperature at which antemelting was observed in Bio-Gel P-4, Sephadex G-25, gelatin gel, and aqueous dextran was lowered as much as 13 deg C (and as little as 2 deg) by "rapid" freezing. "Ultrarapidly" frozen gels could, conceivably, exhibit antemelting at still lower temperatures. We might reasonably suppose the antemelting temperature to be related linearly to the logarithm of the cooling rate. Were this to be the case we would need to establish lower maximum safe operating temperatures the faster we managed to freeze our sample materials. Clearly the subject merits further study.

The migration of diffusible ions in frozen aqueous systems would appear to raise other questions. Rates of migration will rise rapidly with warming above respective antemelting temperatures. But the same ions will, if we are to accept the results of the resistance studies (MacKenzie, 1980), have exhibited gradually increasing mobilities from much lower temperatures. The practical question will then relate to the actual rates. We will need to ask how long it will take at a given temperature to see a delocalization. We will need to determine the resistance as a function of temperature between the glass transition and the antemelting. In the absence of any such determination we will need to keep our samples below the respective glass transition temperature at all times. The relevant glass transitions would seem generally to occur in the range: -90° to -75°C.

REFERENCES

Dowell, L. G., Moline, S. W., and Rinfret, A. P. (1962). Biochim. Biophys. Acta. 59:158-167.

Luyet, B. J. (1957). Proc. Roy. Soc. B. 147:434-51.

Luyet, B. J. (1965). Ann. N.Y. Acad. Sci. 125:502-21.

Luyet, B. J., and Gehenio, P. M. (1940). In "Life and Death at Low Temperatures," Biodynamica pp. 1-341. Normandy, Missouri.

Luyet, B. J., and Rapatz, G. L. (1958). Biodynamica 8:1-68.

Luyet, B., and Rasmussen, D, (1968). Biodynamica. 10:167-91.

MacKenzie, A. P. (1975). In "Freeze-Drying and Advanced Food Technology" (S. A. Goldblith, L. Rey, and W. W. Rothmayr, eds.), pp. 277-307. Academic Press, London.

MacKenzie, A. P. (1977a). Phil. Trans. R. Soc. Lond. B. 278:167-89.

MacKenzie, A. P. (1977b). Cryobiology 14:705-06 (abstract).

MacKenzie, A. P. (1980). Cryobiology 17:615-16 (abstract).

MacKenzie, A. P., and Luyet, B. J. (1967). Biodynamica 10:95-122.

Meryman, H. T. (1958). Biodynamica 8:69-72.

Moran, T. (1926). Proc. Roy. Soc. A. 112:30-46.

Persidsky, M. D., and Luyet, B. J. (1975) Cryobiology 12:364-85.

Rasmussen, D., and Luyet, B. (1969). Biodynamica 10:319-31.

Saubermann, A. J. In this volume.

SUMMATION

T. A. Hall

Department of Zoology
University of Cambridge
Cambridge, U. K.

One naturally compares this meeting with the microprobe conference held in the same center in 1973. While the proceedings of the earlier meeting do include reports of biological studies, many participants clearly felt the need at that time to deal with major technical problems before they could pursue their biological goals. Two problems loomed large: the effects of radiation damage, and the problem of inadvertent alterations in the distributions of elements during specimen preparation.

In contrast, while many speakers in our 1980 meeting have concentrated on technical developments, we have heard many reports of substantial biological results as well. Some contributors have had to cope with the gratifying dilemma of describing both technical advances and interesting biology in a very limited time. Thus, while it seemed in 1973 that beam damage and preparative problems might be insuperable obstacles to the meaningful quantitative microanalysis of biological specimens, existing results have now proved that this is not true even for the highly diffusible electrolyte elements.

But it must be noted that the problems of preparation and beam damage are still far from final solution. Substantial beam damage (often up to 20-30% loss of total mass) still occurs during most analyses. Our understanding and control of beam damage are still primitive. For example, two speakers at this meeting have reported that beam damage seems to be drastically reduced in instruments which have a clean ultra-high vacuum (a finding first reported but generally considered unreproducible about five years ago). If the effect

ISBN 0-12-362880-6

is real, our conception of damage must change drastically and there may be a new route to the control of damage.

With respect to the problem of preserving the distributions of important elements during specimen preparation, cryopreparation has become established as the most generally suitable approach. The procedure starts with the quench-freezing of a small specimen. The most common alternatives then are to analyse the material after cryofracture or in the form of cryosections, either in the frozen-hydrated state or after dehydration by sublimation. This approach has had considerable success. For example, on the basis of one or another variant, biologically sensible measurements have been obtained for the concentrations of sodium, potassium, and chlorine in extracellular spaces and within cells and even within cellular organelles such as mitochondria. But we have to note that we still do not know how successful cryopreparation can be in preserving ionic gradients within cells or extracellular fluid channels, or how much redistribution occurs on a very fine scale (between cytosol and cytoskeleton) during freeze drying.

Alongside the advances in x-ray microanalysis, a major feature of this conference has been the enhanced stature of EELS (electron energy-loss spectroscopy) for biological microanalysis. It seems that even now elements as heavy as phosphorus and calcium (atomic number 15 and 20) can be studied by EELS with mass sensitivity and spatial resolution better than that achievable by x-ray microanalysis, and the advent of parallel recording in the near future should give a large further improvement. In readiness for biological application, EELS seems now to stand where x-ray microanalysis was in the 1960s and most of us will soon have to struggle to obtain, and come to terms with, EELS facilities on our microscope columns.

I wish to conclude this overview with a statement of thanks to Professors Hutchinson and Somlyo and their staffs for organizing this opportunity for us to come together to discuss our rapidly developing field.

Index

F

G

H

I

J

K

L

M

N